T0092158

Proton Transfer in Hydrogen-Bonded Systems

NATO ASI Series

Advanced Science Institutes Series

A series presenting the results of activities sponsored by the NATO Science Committee, which aims at the dissemination of advanced scientific and technological knowledge, with a view to strengthening links between scientific communities.

The series is published by an international board of publishers in conjunction with the NATO Scientific Affairs Division

A	Life Sciences	Plenum Publishing Corporation
B	Physics	New York and London
C	Mathematical and Physical Sciences	Kluwer Academic Publishers
D	Behavioral and Social Sciences	Dordrecht, Boston, and London
E	Applied Sciences	
F	Computer and Systems Sciences	Springer-Verlag
G	Ecological Sciences	Berlin, Heidelberg, New York, London,
H	Cell Biology	Paris, Tokyo, Hong Kong, and Barcelona
I	Global Environmental Change	

Recent Volumes in this Series

Volume 287—Coherence Phenomena in Atoms and Molecules
　　　　　in Laser Fields
　　　　　edited by André D. Bandrauk and Stephen C. Wallace

Volume 288—Intersubband Transitions in Quantum Wells
　　　　　edited by Emmanuel Rosencher, Børge Vinter, and Barry Levine

Volume 289—Nuclear Shapes and Nuclear Structure at Low Excitation Energies
　　　　　edited by Michel Vergnes, Jocelyne Sauvage, Paul-Henri Heenen,
　　　　　and Hong Tuan Duong

Volume 290—Phase Transitions in Liquid Crystals
　　　　　edited by S. Martellucci and A. N. Chester

Volume 291—Proton Transfer in Hydrogen-Bonded Systems
　　　　　edited by T. Bountis

Volume 292—Microscopic Simulations of Complex Hydrodynamic Phenomena
　　　　　edited by Michel Mareschal and Brad L. Holian

Volume 293—Methods in Computational Molecular Physics
　　　　　edited by Stephen Wilson and Geerd H. F. Diercksen

Volume 294—Single Charge Tunneling: Coulomb Blockade Phenomena
　　　　　in Nanostructures
　　　　　edited by Hermann Grabert and Michel H. Devoret

Series B: Physics

Proton Transfer in Hydrogen-Bonded Systems

Edited by

T. Bountis

University of Patras
Patras, Greece

Springer Science+Business Media, LLC

Proceedings of a NATO Advanced Research Workshop on
Proton Transfer in Hydrogen-Bonded Systems,
held May 21–25, 1991,
in Crete, Greece

NATO-PCO-DATA BASE

The electronic index to the NATO ASI Series provides full bibliographical references (with key-
words and/or abstracts) to more than 30,000 contributions from international scientists published
in all sections of the NATO ASI Series. Access to the NATO-PCO-DATA BASE is possible in two
ways:

—via online FILE 128 (NATO-PCO-DATA BASE) hosted by ESRIN, Via Galileo Galilei, I-00044
Frascati, Italy.

—via CD-ROM "NATO-PCO-DATA BASE" with user-friendly retrieval software in English, French,
and German (© WTV GmbH and DATAWARE Technologies, Inc. 1989)

The CD-ROM can be ordered through any member of the Board of Publishers or through NATO-
PCO, Overijse, Belgium.

Library of Congress Cataloging-in-Publication Data

Proton transfer in hydrogen-bonded systems / edited by T. Bountis.
      p.   cm. -- (NATO ASI series. Series B, Physics ; vol. 291)
    "Published in cooperation with NATO Scientific Affairs Division."
    Includes bibliographical references and index.
    ISBN 978-0-306-44216-2   ISBN 978-1-4615-3444-0 (eBook)
    DOI 10.1007/978-1-4615-3444-0
    1. Proton transfer reactions--Congresses. 2. Hydrogen bonding-
-Congresses.   I. Bountis, Tassos.   II. North Atlantic Treaty
Organization. Scientific Affairs Division.   III. Series: NATO ASI
series. Series B, Physics ; v. 291.
QD501.P8279   1992
541.3'93--dc20   92-14543
                                                         CIP

Additional material to this book can be downloaded from http://extra.springer.com.

ISBN 978-0-306-44216-2

© 1992 Springer Science+Business Media New York
Originally published by Plenum Press, New York in 1992

SPECIAL PROGRAM ON CHAOS, ORDER, AND PATTERNS

This book contains the proceedings of a NATO Advanced Research Workshop held within the program of activities of the NATO Special Program on Chaos, Order, and Patterns.

Volume 208—MEASURES OF COMPLEXITY AND CHAOS
edited by Neal B. Abraham, Alfonso M. Albano,
Anthony Passamante, and Paul E. Rapp

Volume 225—NONLINEAR EVOLUTION OF SPATIO-TEMPORAL STRUCTURES
IN DISSIPATIVE CONTINUOUS SYSTEMS
edited by F. H. Busse and L. Kramer

Volume 235—DISORDER AND FRACTURE
edited by J. C. Charmet, S. Roux, and E. Guyon

Volume 236—MICROSCOPIC SIMULATIONS OF COMPLEX FLOWS
edited by Michel Mareschal

Volume 240—GLOBAL CLIMATE AND ECOSYSTEM CHANGE
edited by Gordon J. MacDonald and Luigi Sertorio

Volume 243—DAVYDOV'S SOLITON REVISITED: Self-Trapping of Vibrational Energy
in Protein
edited by Peter L. Christiansen and Alwyn C. Scott

Volume 244—NONLINEAR WAVE PROCESSES IN EXCITABLE MEDIA
edited by Arun V. Holden, Mario Markus, and Hans G. Othmer

Volume 245—DIFFERENTIAL GEOMETRIC METHODS IN THEORETICAL PHYSICS:
Physics and Geometry
edited by Ling-Lie Chau and Werner Nahm

Volume 256—INFORMATION DYNAMICS
edited by Harald Atmanspacher and Herbert Scheingraber

Volume 260—SELF-ORGANIZATION, EMERGING PROPERTIES, AND LEARNING
edited by Agnessa Babloyantz

Volume 263—BIOLOGICALLY INSPIRED PHYSICS
edited by L. Peliti

Volume 264—MICROSCOPIC ASPECTS OF NONLINEARITY IN
CONDENSED MATTER
edited by A. R. Bishop, V. L. Pokrovsky, and V. Tognetti

Volume 268—THE GLOBAL GEOMETRY OF TURBULENCE: Impact of
Nonlinear Dynamics
edited by Javier Jiménez

Volume 270—COMPLEXITY, CHAOS, AND BIOLOGICAL EVOLUTION
edited by Erik Mosekilde and Lis Mosekilde

Volume 272—PREDICTABILITY, STABILITY, AND CHAOS IN
N-BODY DYNAMICAL SYSTEMS
edited by Archie E. Roy

SPECIAL PROGRAM ON CHAOS, ORDER, AND PATTERNS

Volume 276—GROWTH AND FORM: Nonlinear Aspects
edited by M. Ben Amar, P. Pelcé, and P. Tabeling

Volume 278—PAINLEVÉ TRANSCENDENTS: Their Asymptotics and
Physical Applications
edited by Decio Levi and Pavel Winternitz

Volume 280—CHAOS, ORDER, AND PATTERNS
edited by Roberto Artuso, Predrag Cvitanović, and Giulio Casati

Volume 284—ASYMPTOTICS BEYOND ALL ORDERS
edited by Harvey Segur, Saleh Tanveer, and Herbert Levine

Volume 291—PROTON TRANSFER IN HYDROGEN-BONDED SYSTEMS
edited by T. Bountis

Volume 292—MICROSCOPIC SIMULATIONS OF COMPLEX HYDRODYNAMIC
PHENOMENA
edited by Michel Mareschal and Brad L. Holian

Stephanos Pnevmatikos
1957-1990

PREFACE

Charge transport through the transfer of protons between molecules has long been recognized as a fundamental process, which plays an important role in many chemical reactions. In particular, proton transfer through Hydrogen (H-) bonds has been identified as the main mechanism, via which many biological functions are performed and many properties of such basic substances as proteins and ice can be understood.

In this volume, several of these important aspects of the H-bond are represented. As the division in different sections already indicates, present day research in proton teansfer in biochemistry, biology, and the physics of water and ice remains highly active and very exciting.

Nearly a decade ago, a novel approach to the study of collective proton motion in H-bonded systems was proposed, in which this phenomenon was explained by the propagation of certain coherent structures called solitons. In the years that followed, the approach of soliton dynamics was further extended and developed by many researchers around the world, into a legitimate and useful method for the analysis of proton transfer in H-bonded systems.

Dr. Stephanos Pnevmatikos, the original Director of this ARW, was one of the pioneers in the application of soliton ideas to the study of charge transport through H-bonds. Having used similar concepts himself in his research on energy transfer through molecular chains (and 2D lattices) he was convinced that solitons can play an important role in enhancing our understanding of protonic conductivity.

Stephanos was the one who conceived the idea of organizing a meeting in which physicists, biologists and chemists, theoreticians and experimentalists would come together to discuss the progress they had made in this field, compare results and outline the most promising directions for future research.

He invited, from around the world, a number of scientists who had done the best known and more widely appreciated work on this subject. Among them, besides the well-known and distinguished professors, there were also some younger researchers and graduate students, at the beginning of their career. Stephanos thought that in this way, science would be better served by the mutual interaction of the "old" and "new" generation of scientists at the Workshop.

As the final outcome of the conference shows - and this volume hopefully demonstrates - he was absolutely right. In the opinion of most participants, the interdisciplinary character of the Workshop, the vivid exchange of ideas between experimentalists and theoreticians and the testing of more established notions against new evidence, all contributed to the success of the meeting.

Unfortunately, Stephanos did not live to see his efforts come to fruition. His life was suddenly and tragically interrupted at a very early stage of his career, when his great ability as a researcher, organizer and teacher was starting to become world renowned. Although he died very young, he had already published several important papers and established a reputation for himself in the application of soliton dynamics to proton conductivity. Indeed, some of the articles in this volume are either based on work he did recently, or refer to some of his earlier research on this topic.

On behalf of all participants, I wish to dedicate this volume to his memory, as a small tribute to the exemplary man and scientist that he was, and to what he would have become, had he lived longer. As a close friend and collaborator of Stephanos', I felt that the organization of this Workshop and the publication of the present volume was the least I could do to show my deep appreciation for him.

In this task, I was not alone. The frequent advice and suggestions of the other members of the Organizing Committee, Professors G.Careri, J.F. Nagle and M.Peyrard proved very useful, before, during and after the conference. Professor E.N. Economou, the Director of the Research Center of Crete, where Stephanos worked the last 5 years of his life, also provided me with all the help I needed.

And last, but not least, I wish to thank Stephanos' secretary, Mrs. Mary Stavrakakis,whose exceptional skills were essential in making this meeting as well organized and smoothly conducted as it was.

Tassos Bountis
December,1991

CONTENTS

PROTON TRANSFER AND THE HYDROGEN BOND

Transfers of Protons as a Third Fundamental Property of H-Bonds
Importance of Cooperative Resononce Type in Cyclic Structures.... 1
Y.Marechal

Proton Transport in Condensed Matter................................. 17
J.F. Nagle

Extraction of the Principles of Proton Transfers by Ab Initio Methods... 29
S.Scheiner

Construction of the Model Strong Hydrogen Bond Potential Energy
Surface and Proton Transfer Dynamics............................... 49
N.D.Sokolov and M.V.Vener

PROTON TRANSFER AND SOLITON DYNAMICS

Thermal Generation and Mobility of Charge Carriers in
Collective Proton Transport in Hydrogen-Bonded Chains.............. 65
M.Peyrard, R.Boesch and I.Kourakis

Soliton Modelling for Proton Transfer in Hydrogen-Bonded Systems.... 79
St. Pnevmatikos, A.V. Savin and A.V. Zolotaryuk

Soliton Propagation Through Inhomogeneous
Hydrogen-Bonded Chains.. 95
Yu. Kivshar

The Concept of Soliton-Carrier Collective Variable For Proton
Transfer in Extended Hydrogen-Bonded Systems: Overview.......... 105
E.S. Kryachko and V.P.Sokhan

Dynamics of Nonlinear Coherent Structures in a
    Hydrogen-Bonded Chain Model: The Role of
    Dipole - Dipole Interactions... 121
I.Chochliouros and J.Pouget

Thermally Excited Lattice Solitons....................................... 131
N.Theodorakopoulos and N.C.Bakalis

Destruction of Solitons in Frenkel-Kontorova Models with
    Anharmonic Interactions... 139
A.Milchev

PROTON TRANSFER IN BIOCHEMISTRY AND BIOLOGY

Proton Polarizability of Hydrogen Bonded Systems due to Collective
    Proton Motion-With a Remark to the Proton Pathways in
    Bacteriorhodopsin.. 153
G.Zundel and B.Brzezinski

Percolation and Dissipative Quantum Tunneling of Protons in
    Hydrated Protein Powders.. 167
G.Careri

Light-Triggered Opening and Closing of An Hydrophobic Gate
    Controls Vectorial Proton Transfer in Bacteriorhodopsin............. 171
N.A.Dencher, G.Buldt, J.Heberle, H.-D. Holtje and M.Holtje

Proton Transfer in the Light-Harvesting Protein Bacteriorhodopsin.
    An Investigation with Optical pH-Indicators..................... 187
J.Heberle and N.A. Dencher

Diffusion of Protons in the Microscopic Space of the PhoE Channel....... 199
Y.Tsfadia and M.Gutman

Glass Transitions in Biological Systems.................................. 207
P.Pissis

Interaction Between Acetic Acid and Methylamine in Water.
    Molecular Dynamics and Ab Initio MO Studies........................... 217
J.Mavri and D.Hadzi

Crystal of Adenine and Dissociated and Nondissociated Chloroacetic
    Acid-A Promising Model for the Study of Proton Transfer
    Phenomena.. 229
J.Florian, V.Banmruk, J.Zachova and J.Maixner

Ab Initio HF SCF Calculation of the Effect of Protonation
On Vibrational Spectra of Adenine.................................... 233
J.Florian

PROTON TRANSFER IN WATER AND ICE PHYSICS

Studies of the Nature of the Ordering Transformation in Ice Ih............. 239
J.W.Glen and R.W. Whitworth

Defect Activity in Icy Solids from Isotopic Exchange Rates:
Implications for Conductance and Phase Transitions.................... 249
J.P.Devlin

Single-Molecule Dynamics and the "Optical Like" Collective
Modes in Liquid Water.. 261
D.Bertolini, A.Tani and D.Vitali

PROTON TRANSFER IN SOLUTIONS

Free Energy Computer Simulations for the Study of Proton
Transfer in Solutions.. 273
M.Mezei

A Kinetic Model for Proton Transfers in Solutions............................. 281
L.Arnaut

Raman Investigation of Proton Hydration and Structure In
Concentrated Acqueous Hydrochloric Acid Solutions................... 297
G.E.Walrafen, Y.C.Chu and H.R.Carlon.

Proton Charge Transfer in a Polar Solvent............................. 313
Yu.Dakhnovskii

GENERAL METHODS AND MODELS

A Density Functional Study of Hydrogen Bonds.................................. 325
M.Springborg

Theoretical Analysis of Isotopic Scrambling in Ion-Molecule Reactions
Involving Proton Transfers... 343
E.M.Evleth and E.Kassab

Two-Sublattice Model of Hydrogen Bonding at
Finite Temperatures... 351
O.Yanovitskii, N.Flytzanis and G.Vlastou-Tsinganos

Contributors.. 357

Index.. 363

TRANSFERS OF PROTONS AS A THIRD FUNDAMENTAL PROPERTY OF H-BONDS -IMPORTANCE OF COOPERATIVE RESONANCE TYPE IN CYCLIC STRUCTURES

Yves Maréchal

Département de Recherche Fondamentale sur la Matière Condensée
/SESM/PCM Centre d'Etudes Nucléaires de Grenoble
85X
F 38041 Grenoble Cedex France

INTRODUCTION

The idea to organize a meeting on transfers of protons through
H-bonds is due to S.Pnevmatikos whose tragic death most unfortunately
hindered us having his personal views on such a topics. It reveals a
strong capacity of anticipation and of thinking in the long term from
him. This topics "transfers of protons through H-bonds" corresponds to
a concept which is not new but which more and more appears as a general
crucial step in many reactions occuring in aqueous media and
particularly in biological ones. As may be seen when looking at the
different chapters of this book its importance slowly emerges and such
a mechanism is now strongly suspected to be vital in many reactions at
room temperature. We may think that within a short time it will be
invoked by a wide variety of scientists in such various fields as
physics, chemistry and biology.

Such transfers of protons may happen with no implication of
H-bonds. Their growing importance is, however, linked to the
importance, not always fully appreciated, of H-bonds. I shall
consequently begin this chapter by first briefly describing how two
basic properties of H-bonds of geometrical and thermodynamical nature
make these intermolecular bonds so central and vital at room
temperature and make water a species having exceptional physical,
chemical and biological properties. I shall then consider transfers of
protons through H-bonds which appear as a third basic property of these
H-bonds of a dynamical nature and shall discuss their importance. I
shall at the same time propose a classification for these proton
transfers through H-bonds, which will make some kind of an introduction
to the various topics which have been described and discussed during

this workshop. As the importance of this mechanism has only recently been recognized this classification should be taken as tentative and certainly not as definitive. I shall then describe a special class of transfers of protons through H-bonds which occurs in cyclic structures and is ascribed as cooperative resonance type. It is, in my opinion, a mechanism whose importance may soon appear as really fundamental.

H-BONDS: A WIDESPREAD INTERACTION OF CRUCIAL IMPORTANCE

H-bonds are electrostatic interactions linking two molecules. In view of simplification we shall write them $O-H \cdots O$ keeping in mind that both O atoms of the two linked molecules may as well be N atoms, or even (more scarcely) F or Cl atoms. They have two basic properties: they are directional interactions which make the three atoms O, H and O of this bond to lie on a same line at equilibrium. As a consequence they are at the origin of geometrically well defined molecular structures. Their second basic property is that their energies of formation are of the order of some kT at room temperature ($\simeq 300K$). As a consequence these molecular structures will be stable at this temperature but will require, to be transformed, an amount of energy of the order of some kT, which can be easily borrowed to the surroundings or, said differently, may be provided by thermal fluctuations. In other words H-bonds will be at the origin of molecular evolutive structures which will display special faculties of flexibility and adaptability. It consequently comes as no surprise that H-bonds are of central importance in biology when reactions are considered at molecular level.

A typical example of such a stable and evolutive structure is given by the well known double helix of DNA which keeps unchanged and transmits the genetical code of a living organism while duplicating it during cell replication. Another example of such stable but flexible structures is given by plasmic membranes[1] which isolate the inner parts of living cells from the outer plasma, or nuclear enveloppes which isolate nuclei inside the cell. The basic molecule is there a phospholipid which has a polar hydrophilic head (a glycerophosphate) able to establish H-bonds with water and is covalently linked to two paraffinic hydrophobic chains (unable to do so). In neutral water(p_H=7) these molecules assemble in bilayers having all their hydrophobic chains at the interior and their hydrophilic heads which define both outer and inner surfaces of the membrane facing water molecules. Similar structures are also encountered with soaps or detergents which also possess polar hydrophilic heads (carboxylates or phosphates) and a usually single paraffinic hydrophobic chain. At low concentration in water they form aggregates, called micelles, having all their polar heads on the outer part of the globule facing water molecules and hydrophobic chains in the inner part where they can accept molecules which are insoluble in pure water (Fig. 1). Another interesting organized H-bonded structure is that of proteins, made of polypeptides organized in helical coils.

This leads us to focus our attention on a particular species, water, which displays exceptional physical, chemical and biological properties which are all due to its exceptionally high density of H-bonds. Its basic molecule H_2O (or its usual isotopic analogs D_2O and HDO) has the peculiarity of having two covalent bonds while leaving two lone pairs of electrons on its O atom which enables it to accept two H-bonds from other H_2O molecules. It consequently has the possibility of forming assemblies of molecules having as many H-bonds as covalent bonds, which is an exceptional situation, nevertheless shared by another species: HF. In this latter case, these molecular assemblies

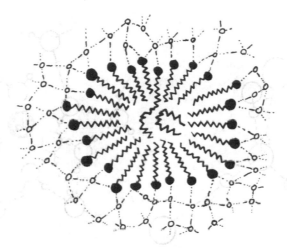

Fig. 1. Structure of a micelle in water.

are however restricted to one dimensional chains, whereas they are tridimensional in water (the central O atom in Fig. 2,a has a tetraedral basic symmetry). This tridimensional organization of molecules linked by so many H-bonds is responsible for the exceptional physical properties of water which, for instance, becomes denser when heated in the vicinity of its melting point. This is due to the possibility of diminishing the excluded volume between molecules by weakening H-bonds. It is the same mechanism which makes water denser than ice, which is a well known exceptional property of water which has many implications in everyday life. Water has many other exceptional physical properties[2,3].It becomes for instance more fluid when compressed whereas other liquids become more viscous. This is due to distorsion and weakening of H-bonds under pressure which makes H_2O molecules less linked to other ones and consequently more able to rotate. The same mechanism occurs in ice which melts when compressed at not too low a temperature, and is a most useful property for enjoying skating!

The unusually high density of H-bonds in water is also at the origin of its unique chemical and biological properties. It enables for instance water to build around an ion a flexible tridimensional screen (Fig.2,b) which efficiently hinders recombination of ions of opposite charges. Water has thus no equivalent to dissolve salts, acids or bases. It also appears as a practically infinite reservoir of H-bonds and this is at the origin of its exceptional biological properties: rupture and reformation of H-bonds, as in DNA or proteins (see S.W.Englander in a subsequent chapter), which are essential in many biological mechanisms, are of a much lower energetical cost when occuring in water than when occuring in another medium. They may consequently become feasible at room temperature in water.

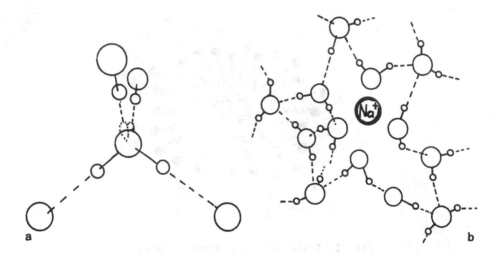

Fig. 2. Water. (a) An H_2O molecule in ice or water (where H-bonds may then be temporarily bent); (b) Organisation of H_2O molecules around a cation (After G.W.Neilson,in "Water and aqueous solutions", G.W.Neilson and J.E.Enderby eds, Adam Hilger, Bristol(1986).

The somewhat amazing point in our knowledge of the structure of water is that we satisfactorily understand how the extremely dense network of H-bonds is at the origin of anomalous properties but have much more difficulties understanding how this same network allows water to be a liquid[4]! It explains why research on water to establish a precise description of its constitution and dynamics at molecular level is so active, as is described in subsequent chapters by E.Whalley, J.P.Devlin, G.E.Walrafen, J.W.Glen and R.W.Whitworth, and also V.Petrenko and M.Mezei. The necessity to improve our knowledge of the structure of water is especially important in the problematics of transfer of protons through H-bonds (which is the central point of this workshop), because water is the main bulk system where such transfers

occur (next section) and also because, as will be seen in a subsequent
section, the presence of water may be necessary to establish favorable
conditions for these transfers to occur.

A THIRD PROPERTY OF H-BONDS

Beside these geometrical and thermodynamical properties which make
them so important at room temperature, particularly in aqueous media,
H-bonds have a third property, which make them appear even more
crucial: they enable transfers of protons from one molecule to another
one:

$$O-H\cdots O \rightleftharpoons O^- \cdots H-O^+$$

As it is written this reaction has, however, a very small
probability to occur, because of the appearance of net charges on the
two O atoms which is most of the time energetically too costly. One
notable exception is that of acid phosphates of the KDP family (See
subsequent chapter by H.Schmidt, and also by J.Dolinsek and R.Blinc)
which are well known as ferroelectric materials, and for which such net
charges may be at the origin of defects having energies of the order of
some kT only.

Even if the transfer of protons, as written precedingly, scarcely
occurs because of its high energetical cost, transfers of protons are a
frequently encountered mechanism, as there exists several ways to avoid
this formation of net charges. We shall describe one of them in details
in the subsequent section. Another one consists of having these charges
already created, before the transfer. This is what occurs, for instance,
in crystals of acid sulfate of sodium or potassium[5] where the H atom
of $(SO_4)_2H^{3-}$ anions may be transferred between the two SO_4 groups. This
system is certainly one of the most simple systems where such single
transfers may be studied and might become a model for them. A
theoretical approach of such model transfers will be depicted by
S.Scheiner, by A.Karpfen and by N.Sokolov et al. in subsequent chapters
while experimental approach, using isotope scrambling techniques will
be evocated by E.M.Evleth.

We may also obtain this precreation of charges by dissolving an acid
(or a base) in water, which leads to the creation of charged defects,
with only one type of charges belonging to the H-bond network. We may
then have a transfer which may be written as:
and is an elementary reaction in electrochemistry. Similar transfers
will be described in more details by G.Zundel in a subsequent chapter
(see also the chapter by L.G.Arnaut). They constitute mechanisms by
which conduction occurs in such proton conductors as UHP[6] or in
ionomers, that is membranes containing ions and which have pores inside
which water molecules may ensure this conduction[7]. The basic mechanism
for such transfers is quantum tunnelling of one proton through a
barrier. The effect of environment on this basic mechanism has been

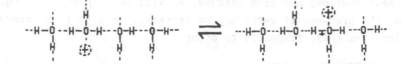

Fig. 3. Transfers of protons in acid water.

theoretically studied[8] (see also Y.I.Dakhnovskii in a subsequent
chapter). In most systems, such as water, the number of protons which
may successively be transferred is large, which implies an important
delocalization of charged defects.

Another field where these types of transfers of protons occur is
that found when studying such biological reactions as those encountered
in vision mechanism or photosynthesis. In both these reactions a
pigment is excited after capture of a photon in the visible region (see
subsequent chapters by N.A.Dencher et al., G.Soubignier, R.A.Dilley and
also Y.Tsfadia et al.). This excited state of the pigment relaxes
through somewhat complicated paths, one of them being the transfer of a
proton from one end of a protein (rhodopsin in the case of vision,
bacteriorhodopsin in the case of photosynthesis) to another end.The
accumulation of protons along one side of a membrane, which is an
efficient way to store energy after photosynthesis or constitutes the
starting signal for vision, is obtained in this way. The mechanism of
transfers of protons in these proteins has been the object of many
studies. The elementary step for such transfers is quantum tunnelling
of a proton through a barrier and transfer over a large distance is
assumed via a mechanism of propagation of a soliton (See subsequent
chapters by M.Peyrard, A.V.Zolotaryuk, Y.S.Kivshar, N.Theodorakopoulos
et al., E.Kryachko, J.F.Nagle, A.Askar, I.Chochliouros et al.,
M.Springborg, O.Yanovitskii et al., A.Milchev). This view of things has
been especially worked by S.Pnevmatikos whose contribution on this
subject[9] has been most remarkable and original. It constitutes an
interesting problem which is far from being completely solved (see
J.F.Nagle in a following chapter).

Viewing this problem differently the biochemical approach of these
transfers of protons leads to most interesting concepts such as that of
the water wire (see chapters by M.Akeson and D.W.Deamer; see also
J.F.Nagle) to bridge the gap between sites which are known to exchange
protons allowing for the establishment of a channel of H-bonded water
molecules through which protons may be transferred. A closely related
and most interesting concept, exposed by G.Carreri in a subsequent
chapter after conductivity measurements is that of the existence of a
threshold in water concentration for enzymatic activity to take place.
Below this threshold (or percolation transition point) this activity is
hindered by impossibility of protons to be transferred because of lack

of direct H-bonding linking sites between which such transfer should occur. Beyond this threshold activity occurs because of water establishing H-bonded bridges between active sites of the enzyme and thus allowing such transfers to occur between these sites.

The number of chapters devoted to this type of proton transfer being great I shall stop this enumeration. It indicates that these transfers of protons now begin to appear as a mechanism of a really vital importance and of a very great potentiality as a field of research. They really constitute a third fundamental property of H-bonds.

PROTON TRANSFERS IN CYCLIC STRUCTURES

Beside these delocalizing single transfers of protons there exists another type of proton transfers which occur in the ground state, with no requirement of having first to create defects or excited states. A typical example of such tranfers is found in carboxylic acid dimers which exist under two tautomeric forms:

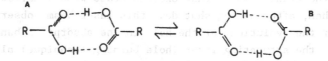

Fig. 4. The two tautomeric forms (A and B) of carboxylic acid dimers.

Passing from one tautomeric form to another one occurs via a cooperative transfer of the two protons of the H-bonds. No defect or excited state are created and consequently such transfers cannot delocalize them. They imperatively take place in cyclic structures where the absence of the apparition of net charges is obtained by a simultaneous or slightly delayed transfer of valence electrons.We shall examine what we know concerning these transfers in this chapter, closing it with remarks regarding its(potentially great)importance.

Evidence for these transfers has been obtained from IR spectroscopy of benzoic acid (R= a benzene nucleus in Fig. 4) by Hayashi and Umemura[10,11], who have shown, by studying their intense $\nu_{C=O}$ stretching band ($\vec{C}=\vec{O}$) that at a temperature of some K only one band, attributed to tautomer A (Fig. 4) existed. Raising temperature provokes the apparition of another band which has been attributed to tautomer B. In a crystal tautomers A and B may have different energies, and only that one which has the lowest energy is present at low temperature. This situation is different from that in the gas phase where both tautomers have the same energy and are in resonance. They are consequently equally populated at all temperatures. A comparison of intensities of $\nu_{C=0}$ bands due to tautomers A and B in benzoic acid crystals gave a value of some $35cm^{-1}$ for the difference of their energies. Further measurements of relaxation time T_1 of proton spins in these compounds[12] allowed an estimation of barrier height at not too low a temperature (the theory used is a phenomenological Arrhenius-type theory not valid at low temperatures) of some $400-500cm^{-1}$.

More recent optical measurements in the visible region have been performed on these systems. They consist of introducing in benzoic acid a small amount ($\simeq 10^{-5}$) of an impurity made of a molecule having an overall geometry very close to that of a benzoic acid dimer. It can then in the crystal take the place of a dimer while perturbing only weakly the other dimers. It should also display very sharp but intense absorption lines so that the very small modification of these lines ($<1cm^{-1}$) by transfers of protons in surrounding benzoic acid dimers may be detected. Thioindigo is an impurity which fulfills these requirements. Two-photons transition techniques, which consist of either directly populating the excited state of the impurity and then observing the corresponding fluorescence[13], or alternatively [14] saturating the transition to that excited state with a narrow band laser and then, after having shut down this laser beam, observing with a test laser the evolution of the hole in the absorption band at the frequency of the saturating laser (hole burning technique) allowed to measure the tunnelling frequency of protons in the surrounding benzoic acid dimers. It has been found equal to some $5\times10^9 s^{-1}$ at 2K, which is the first direct measurement of such a quantity. It has also been found that the symmetric structure of the thioindigo impurity lifted the energetical assymmetry of tautomers A and B of surrounding benzoic acid dimers and put them in resonance. Using other impurities which do not lift this assymmetry has shown that tunnelling hardly occurs when resonance conditions are not fulfilled.

From all these experiments a theoretical description of these resonance-type transfers of protons could be established. It is based on a double-well potential[15,16] for the symmetric stretching cooordinate q_1+q_2 of the protons (q_n is the coordinate representing the stretching motion O-$\vec{H}$ of proton n in the dimer) defining a sychronous transfer of both protons through their H-bonds, plus a term linear in q_1+q_2 defining the energetical assymmetry between tautomers A and B (which puts them off-resonance), plus a term also linear in this coordinate and depending on other degrees of freedom of the system and which represents modulation of this assymmetry by the surrounding of the considered dimer. The degrees of freedom of the surrounding are all

thermalized, that is sensitive to temperature, and play the role of a
heat bath.

Such a description allows for a fair account of IR, NMR and
multiphoton experimental results. They are however far from completely
describing these transfers of protons as they convey a difficulty which
is not yet overcome.It appears in the form of an inadequate ratio of
the energy of the maximum of the barrier to the energy of the $0 \to 1$
transition for the q stretching modes. It makes barrier heights so
high that tunnelling effects in these experiments should not be
observed. We have seen[12] this barrier height to be of the order of
500cm^{-1} from experiments. Ab-initio calculations on isolated formic
acid dimers, which are thought to be most similar to benzoic acid
dimers and are much more amenable to calculations, give values of the
order of 4000cm^{-1} or even more[17,18]. In the same way, calculations of
energy splittings due to tunnelling for q motions[19] in formic acid
dimers, give values of the order of 5×10^{-3}cm^{-1}, which is several orders
of magnitude too small when compared with experimental results.

An indication of the origin of this difficulty might be found when
looking at the structure of such dimers in the gas phase[20] (see also
Fig. 4). We may then recognize that transfers of protons implies moving
also other nuclei, even if the amplitude of these nuclear displacements
is much less than that of protons. It leads to the idea that such
transfers only occur for favorable relative positions of the two dimers
which can appear during intermonomer vibrations (relative vibrations of
the two monomers).This is called vibration-assisted tunnelling, which
is a mechanism which has been found necessary to explain huge isotope
effects on correlation times τ_C of the position of the proton[21].It
shows that transfers of protons are more complicated than a simple
one-dimensional tunnelling described by a double-well for a single
coordinate. They implicate several other nuclear displacements and
might better be seen as a tunnelling between valleys in a
multidimensional space spanned by coordinates of various vibrations.
What are these vibrations? The answer to this question is not yet
clear, even if several types of coordinates have been shown to be good
candidates to assist such transfers. The role of the antisymmetric
q_1-q_2 stretching coordinate has thus been emphasized[19]. It signifies
that the transfer of the two protons might not be exactly synchronous.
The role of various intermonomer modes have also been looked at. It has
been done for the totally symmetric $O \cdots O$ stretch[22] which may be
intuitively thought of as lowering the barrier height of the double
well in q_1+q_2, or that of the totally symmetric in-plane rocking[21]
which may facilitate this tunnelling by increasing the distance of the
two charged tunnelling protons. These theoretical analyses do not,
however,clearly identify which intermonomer vibrations really assist
tunnelling. It shows the necessity to have more precise information on
intermonomer modes of such cyclic H-bonded dimers, and in this context
recent FIR (far IR) measurements[23] which have put into evidence

anomalies of intermonomer vibrations in benzoic acid crystals when temperature is varied may give some further precious indication on the mechanism of transfer of protons in these systems. Let us note that, as these intermonomer vibrations are "thermal" vibrations, that is the energy separation $\hbar \times \Omega$ of their levels is less than kT at 300K, they have the effect of increasing the rate of transfer when temperature is increased.

Another point in the theory which is not so clear is that this transfer of protons, which represents a great amplitude nuclear motion, may be at the origin of a breakdown of the Born-Oppenheimer separation of motions of electrons and nuclei. It physically means that this transfer may be so rapid that electrons may not immediately follow proton motions but may require some delay to adapt themselves to new proton positions. This breakdown of the Born-Oppenheimer approximation has been invoked[24,25] to explain anomalous isotope effects on IR intensities in such cyclic dimers. It is a point which will certainly require further investigation.

It may be concluded that even for such simple model systems as H-bonded cyclic dimers the mechanism of transfer of protons which interchanges tautomer A and tautomer B (Fig. 4) is only partially understood. It implies moving not only protons but also many other nuclei, even if those other nuclei are far less moved than protons. It means that the image of a simple one dimensional tunnelling is largely insufficient. Vibration assisted tunnelling, which may be viewed as tunnelling between valleys through mountains in the space of nuclear coordinates, is clearly an improved image which suggests that at room temperature the rate of such transfer of protons is certainly some orders of magnitude faster than the 5×10^9 s^{-1} measured at 2K[13] by multiphoton spectroscopy. The identity of these assisting vibrations, however, still remains a point to clarify.

Importance of proton transfers in cyclic structures

As previously seen these types of transfers have been precisely and extensively studied and quantitative experimental results have been obtained on them. Their description has reached a satisfactory consistency, even if they still require deeper insight. Their importance is, however, not yet clearly established, which is not the case of previous single transfers of protons which we have seen to play a central role to delocalize and transfer charged defects in electrochemistry, or excited states in such bioprocesses as vision or photosynthesis. We may nevertheless think that these transfers in cyclic structures are of importance in such media as neutral water (p_H=7) at room temperature, where the concentration of charged defects or excited states is very low, but where some reactions require transfers of protons. As a consequence those required transfers are certainly cooperative resonance type transfers occuring within cyclic structures. These cyclic structures may well be established by water molecules which are able to close rings inside which such transfers may happen. It suggests that the role of water would then be very close to that of establishing "water wires", as suggested by Akeson and Deamer in subsequent chapters.

In order to illustrate this point let us take the example of an hydrolysis of an ester, amide or polypeptide (protein).This is a widespread type of reaction in aqueous media, especially in biomedia where it appears as an important step at molecular level in the metabolism of many organisms. Such a reaction is thought[26] to begin with the formation of a "tetraedral intermediate" which constitutes the rate determining step of the hydrolysis. In a basic medium (or in an acid medium as well, with slightly different steps) this formation writes, taking an ester as an example (an amide may be written as well):

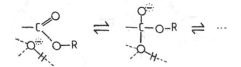

Fig. 5. The rate determining step (formation of a tetraedral intermediate) of the hydrolysis of an ester in a basic medium

As may be seen this reaction proceeds with no creation of net charges, but only transfer of charges. Such a reaction is not possible in a neutral medium because it would then require the apparition of net charges, which is a configuration of high energy, implying a very low rate:
The existence of a ring closed by water molecules, allowing for proton transfer inside this ring may however avoid the apparition of such net charges, and consequently make the reaction feasible in neutral water: This reaction requires somewhat severe steric conditions to occur and may become impossible for some esters and more often for amides or polypeptides. In the case of proteins,that is polypeptides with a well defined helicoidal structure, steric conditions make this reaction quite impossible and proteins are usually not hydrolyzed.They may become so (and this is necessary for digestion and many other metabolical reactions) under the action of such an enzyme as chymotrypsin which, thanks to its triad of amino-acids[27] Serine, Histidine, Aspartic acid (see also subsequent chapter by D.Hadzi and J.Mavri) is thought to act as a relay[28] for transferring protons[29] from one end of its active site to another one which may overcome, perhaps with the help of some water molecules, steric hindrances which inhibit formation of the tetraedral intermediate:

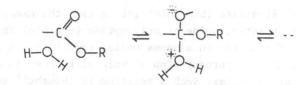

Fig. 6. Formation of a tetraedral intermediate during the
(hypothetical) hydrolysis of an ester in neutral water.

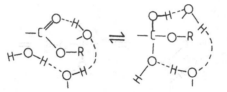

Fig. 7. Hydrolysis of an ester in neutral media involving proton
transfers enabled by several water molecules closing an
H-bonded ring. The bent dashed line represents such H_2O
molecules linked by H-bonds.

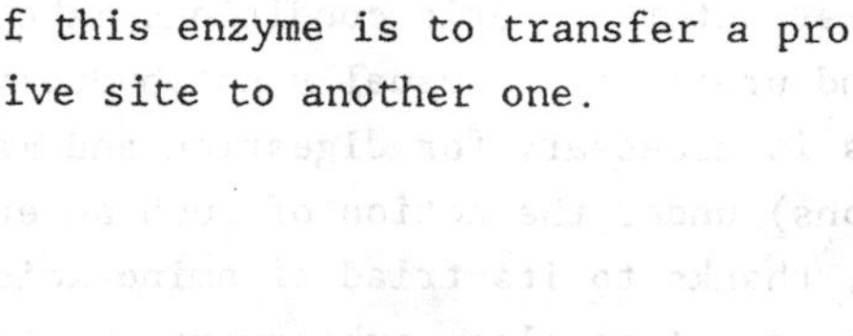

Fig. 8. Formation of the tetraedral intermediate with the help of an
enzyme, whose active site is represented by a zig-zag line.
The role of this enzyme is to transfer a proton from one end
of its active site to another one.

It illustrates the central importance that such transfers of protons in H-bonded cyclic structures may acquire when looking at reactions in aqueous media at molecular level. It requires, to become effective, having a minimum amount of water molecules which, by closing loops, allow for transfer of protons (Fig.8), thus enabling enzymatic activity.This is much in line with Carreri's observations (see subsequent chapter) that biological activity occurs beyond some percolation threshold for water concentration, so that water may bridge gaps between some sites, thus establishing a "water wire" which makes possible transfers of protons necessary for the process of photosynthesis to occur (see chapter by M.Akeson and D.W.Deamer) or closing H-bonded rings which also makes possible transfer of protons necessary for enzymatic activity. Because, as already seen, phonon assisted tunnelling has the effect of increasing the rate of such transfers when temperature is raised, biological reactions are much favoured when temperature is raised. This temperature cannot, however, be indefinetely raised, as the number of broken H-bonds also increases, hindering therefore such proton transfers and thus inhibiting many enzymatic reactions. We are therefore led to suspect that these transfers of protons may be at the origin of an optimum temperature (that of warm-blooded animals?) for biological processes to have an optimum efficiency.

CONCLUSION

The possibility for H-bonds to transfer protons is a property which more and more appears as crucial for many reactions and mechanisms occuring at room temperature, especially for reactions in aqueous media.Two basic properties of H-bonds, one geometrical (their linearities), the other one thermodynamical (their energies of formation are of some kT at room temperature) confer assemblies of molecules held together by such bonds stability, flexibility and exceptional properties of adaptibility at 300K. These two properties are also at the origin of exceptional physical properties of water, and, due to special faculties of adaptibility conferred by H-bonds, to the non less exceptional chemical properties of this species. Because water may be viewed in biology as an infinite reservoir of H-bonds, it is a unique and absolutely necessary medium in biology. The possibility of transferring protons between molecules, which appears as a third basic property of H-bonds, confer assemblies of molecules linked by such bonds exceptional reactivity in aqueous media. Several types of transfers have been distinguished in this paper. Besides the translational, single transfers, which require having first created charged defects or excited states, and which have been especially studied by S.Pnevmatikos, the cooperative resonance type which occur in ground state of molecules, but inside cyclic H-bonded structures, appear particularly important and is suspected to be crucial in many enzymatic reactions. It confers water a role even more central than usually thought, as it is able to bridge gaps between sites, or close

H-bonded loops, thus allowing for these transfers of protons to take place. We may then strongly suspect that transfers of protons through H-bonds, which still are in their early stage of experimentation and also of conceptualization have a great potentiality as central mechanisms for reactions in aqueous media, particularly for biological ones when considered at molecular level. It is certainly a new field of research which will soon reach its maturity.

REFERENCES

1. Y. Maréchal, La Recherche 20:480 (1989).
2. F. H. Stillinger, Science 209:451 (1980).
3. S. H .Chen and J. Teixeira, in: "Adv. Chem. Phys.", Vol. LXIV, I.Prigogine and S.Rice,eds., J. Wiley (1986).
4. Y. Maréchal, J. Chem. Phys. 95:5565 (1991).
5. F. Fillaux, A. Lautié, J. Tomkinson and G. J. Kearly, Chem. Phys. 162:19 (1989).
6. A. Novak, in: "Solid State Protonic Conductors III", J.B.Goodenough, J.Jensen and A.Potier, eds., Odense University press (1985).
7. M. Falk, in: "Structure and Properties of Ionomers", M.Pineri and A.Eisenberg, eds.,NATO ASI Series C, Vol. 198, D.Reidel, Dordrecht(1986).
8. D. C. Borgis, S. Lee and J. T. Hynes, Chem. Phys. Letters 76:88 (1989).
9. S. Pnevmatikos, Phys. Rev. Letters 60:1534 (1988).
10. S. Hayashi and J. Umemura, J. Chem. Phys. 60:2675 (1975).
11. S. Hayashi, M. Obotake, R. Nakamura and K. Mashida, J. Chem. Phys. 94:4446 (1991).
12. S. Nagaoka, T. Terao, F. Imashiro, A. Saika, N. Hirota and S. Hayashi, J. Chem. Phys. 79:4694 (1983).
13. R. M. Hochstrasser and H. P. Trommsdorff, Chem. Phys. 115:1 (1987).
14. A. Oppenländer, C. Rambaud, H.P.Trommsdorff and J.C.Vial, Phys. Rev. Letters 63:1432 (1989).
15. J. L. Skinner and H. P. Trommsdorff, J. Chem. Phys. 89:897 (1988).
16. R. Meyer and R. R. Ernst, J. Chem. Phys. 93:5518 (1990).
17. R. Nakamura and S. Hayashi, J. Mol. Struct. 145:331 (1986).
18. J. Lipinski and W. A. Sokalski, Chem. Phys. Letters 76:88 (1980).
19. N. Shida, P.F. Barbara and J. Almlöf, J. Chem. Phys. 94:3633 (1991).
20. O. Bastiansen and T. Motzfeldt, Acta Chem. Scand. 23:2848 (1969).
21. A. Stöckli, B. H. Meier, R. Kreis, R. Meyer and R. R. Ernst, J. Chem. Phys., 93:1502 (1990).
22. R. Nakamura, K. Mashida and S. Hayashi, J. Mol. Struct. 146:101 (1986).
23. H. R. Zelsmann and Z. Mielke, Chem. Phys. Letters 186:501 (1991).
24. Y. Maréchal, J. Chem. Phys. 87:6344 (1987).
25. Y. Maréchal, J. Mol. Struct. 189:55 (1988).
26. J. March, "Advanced Organic Chemistry-Reactions,Mechanism and Structure", Mc Graw Hill (1977).

27. D. Hadzi, J. Mol. Struct. 177:1 (1988).

28. C. Walsh, "Enzymatic Reaction Mechanism", Freeman (1979).

29. Y. Maréchal, J. Mol. Liquids 48, 253 (1991).

PROTON TRANSPORT IN CONDENSED MATTER

John F. Nagle

Departments of Physics and Biological Sciences
Carnegie Mellon University
Pittsburgh, PA 15213 USA

INTRODUCTION

This paper consists of some of my discussion comments on several issues regarding proton transport. The list in this paper is by no means exhaustive. Rather, it reflects a few of the issues that are of interest at the present time, as well as ones in which the author has done some work.

The first section reviews the principle of *charge partitioning* into two types of defects, which is the basic feature that makes proton transport richer than electron transport. Although this relates to some complex and controversial aspects of dielectric theory, the main theme of this section is that experiment and theory agree at the quantitative level, so there is a firm conceptual base for proton transport. The second section gives my view of some of the uncertainties regarding fundamental *charge transport mechanisms* in well characterized systems such as ice. This has been a difficult subject for a number of years, but one that should be resolvable if good data and good theories were to be combined. The third section discusses some related uncertainties that exist in our understanding of *proton transport in biological systems*, but it also emphasizes the even greater uncertainties due to not knowing the molecular structures very well.

CHARGE PARTITIONING

It is, by now, rather well accepted for solid materials such as ice or other hydrogen bonded crystals, that the fundamental protonic charge transport requires two sequential processes. One process involves hopping of ionic defects and the other involves turning of the molecules via Bjerrum defects to renew the original polarization state. These processes are sketched in Fig. 1 for a chain of water molecules, although it must be remembered that the defects need not be narrowly localized on one or two waters, but could be delocalized as shown by some recent calculations (Peyrard, St. Pnevmatikos and Flytzanis, 1987; Yanovitskii and Kryachko, 1988; Kryachko, 1988).

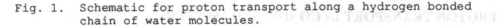

```
                    H    H    H    H    H
(a)   H+ --->     OH---OH---OH---OH---OH

                    H    H    H    H    H
(b)               HO---HO---HOH+--OH---OH      (+) ionic defect

                    H    H    H    H    H
(c)               HO---HO---HO---HO---HO    H+ --->

                    H    H    H    H    H
(d)               HO---HO---HO    OH---OH      (-) Bjerrum L defect
```

Fig. 1.   Schematic for proton transport along a hydrogen bonded
          chain of water molecules.

Here I focus on one very important quantitative result that does
not depend upon the details of defect delocalization, namely, the
partitioning of the full protonic charge, e, between the ionic
defect with charge e_I and the Bjerrum defect with charge e_B, so

$$e = e_I + e_B. \tag{1}$$

As can be seen from the first three rows of Fig. 1, while the
passage of an ionic defect results in release of a full protonic
charge at the end of the chain, the chain itself is polarized in the
opposite direction, so that the effective charge transfer is reduced
according to

$$Nde_I = Nde - 2\mu_{z,total}, \tag{2}$$

where d is the distance (projected onto the chain axis) separating
neighboring waters along the chain, N is the number of waters in the
chain, and $\mu_{z,total}$ is the *total* dipole moment of the chain.   Combining
Eqs. (1) and (2) yields

$$e_B = \lim_{N-->\infty} 2\mu_{z,total}/Nd. \tag{3}$$

The development in the preceding paragraph makes it clear that
the quantitative charge partitioning does not depend upon the
details of the defect structure.   Rather, it depends upon the *effects*
that are left behind after the defects have passed.   These effects
are quantitatively included in the dipole moment of the total chain.
It is, however, customary to describe the dipole moment in Eq. (3)
in terms of an **average** dipole moment,

$$\mu_z = \lim_{N-->\infty} \mu_{z,total}/N. \tag{4}$$

For ice this leads, using simple tetrahedral geometrical relations,
to a familiar equation (Onsager and Dupuis, 1962)

$$e_B = \mu 3^{1/2}/R_{H\text{-bond}} \tag{5}$$

where $R_{H\text{-bond}}$ is 2.76 Angstroms.   It is important to appreciate that
Eq. (5) actually defines a dipole moment μ for ice.

18

Using *ab initio* quantum chemical calculations Scheiner and Nagle (1983) calculated $\mu_{z,total}$ for N waters in a chain, where N was varied from 1 to 5. Eq. (3) was then used to estimate e_B. There were three principal technical considerations. The first is that, due to end effects, the limit in Eq. (4) for μ_z was clearly not reached even for N=5. However, the trend with increasing N was quite clear and led to a reasonable extrapolation to the N=∞ limit. The second consideration is that choosing different basis sets in the molecular orbital calculations leads to different estimates of μ_z. This is already apparent for N=1, i.e., for the dipole moment of a single water molecule which experiment tells us is 1.85 D. Therefore, for each basis set, all dipole moments were scaled to provide the correct dipole moment for a water monomer. The third technical consideration is one that applies when comparing the result of the calculation to ice which is not just a single linear hydrogen bonded chain of water molecules. However, exploration of the effects of adding additional water molecules that are not on the transport pathway indicated that the effects due to the actual three dimensional network are rather small. In this way, an estimate for e_B of 0.36±0.03 was obtained for charge partitioning in ice and/or linear hydrogen bonded chains of water molecules with hydrogen bonds of length 2.76 A.

The quantum calculations also illuminate a fallacy in following the naive approach that replaces each atom by an effective charge. The effective charge approach would demand that e_B be larger than e_I because the proton moves much further in the Bjerrum turning step than in the ionic hopping step. The reason this simple intuitive result is in error is that it does not take into account the fact that the electrons are not distributed symmetrically around the oxygen and the hydrogen atoms; therefore shifting the proton by defect transport also significantly changes the dipole moment of the electrons, including those associated with the oxygen.

A direct experimental estimate of e_B was obtained by Hubman (1979) by measuring the dielectric constant of ammonia doped ice and comparing to ordinary ice. Since Hubman's papers involve a good deal of theory, it is worth emphasizing how directly this measurement yields e_B. The primary point is that the ionic defect becomes the majority defect in ammonia doped ice, so it is the one that determines the dielectric constant, in contrast to ordinary ice where the majority defect is the Bjerrum defect. At frequencies below the Debye dispersion frequency but high enough that direct current can be ignored, the majority defects with charge e_M are in equilibrium with a driving force $e_M E$ due to the electric field E which is exactly balanced by a configurational polarization force p. The configurational polarization p is entropic and provides for more (fewer) paths for the defect to go in the opposite (same) direction as the E field. Since p balances the force from the electric field, it is proportional to $e_M E$. The actual polarization P is just pe_M, since the buildup of the configurational polarizational p requires passage of defects carrying charge e_M. Therefore, the dielectric constant $\varepsilon = P/E$ is proportional to e_M^2. Therefore, the ratio

$$\varepsilon_{\text{ammonia-doped}}/\varepsilon_{\text{pure ice}} = (e_I/e_B)^2. \tag{6}$$

Hubman measured this ratio to be 0.38. Combination with Eq. (1) yields $e_B = 0.38$ and $e_I = 0.62$. Closely similar values have also been obtained experimentally by Takei and Maeno (1987).

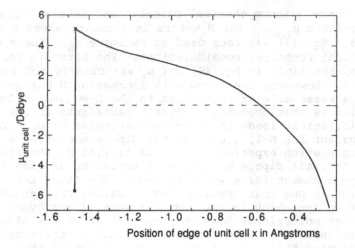

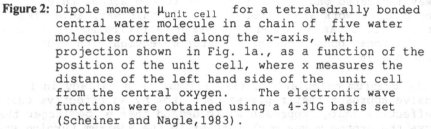

Position of edge of unit cell x in Angstroms

Figure 2: Dipole moment $\mu_{unit\ cell}$ for a tetrahedrally bonded
central water molecule in a chain of five water
molecules oriented along the x-axis, with
projection shown in Fig. 1a., as a function of the
position of the unit cell, where x measures the
distance of the left hand side of the unit cell
from the central oxygen. The electronic wave
functions were obtained using a 4-31G basis set
(Scheiner and Nagle,1983).

The fact that there is quite good agreement between theory and
experiment indicates that the quantitative degree of charge
partitioning is reasonably well known for ice. It also indicates
that quantum chemical calculations can be useful for questions where
experimental data are not available. However, there are some
further discussion comments worth mentioning. The fact that an
extrapolation to large N is required in estimating μ_z in Eq. (3)
indicates that there are substantial end effects in finite chains
involving charge transfer along the hydrogen bonded chain that would
be in excess of just reorienting the water dipoles. One might wish
to separate these end effects from an average dipolar effect, where
the dipoles would represent the bulk, or interior, molecules in the
chain (Karpfen and Schuster, 1984). Such a separation is fraught
with fundamental difficulties, however, because of the arbitrariness
of choosing a unit cell. This is emphasized in Fig. 2. This figure
shows the dipole moment for the central water molecule in a chain of
N=5 molecules as a function of the position x of the unit cell
chosen. As x decreases from 0 the calculated dipole moment
continuously changes due to the inclusion of electron density on the
left hand side of the unit cell and exclusion of electron density on
the right hand side. The discontinuous decrease in dipole moment as
x decreases through -1.4615 is due to exclusion of the proton on the
central water and inclusion of the proton from the water on the left
of the central water. The point of Fig. 2 is that there is no
obvious x value to use in calculating a true dipole moment for the
bulk. Indeed, the calculated dipole moment changes dramatically,
even when the choice of the unit cell contains all the nuclei of the
central water, including both positive and negative values for the
dipole moment as seen in Fig. 2. This problem is not restricted to
calculations for hydrogen bonded chains with open ends, but also

occurs in the same way for calculations of dipole moments when periodic boundary conditions are employed. This emphasizes the fact that there is no unique dipole moment for isolated molecular groups in hydrogen bonded crystals and that attempts to calculate such a bulk dipole moment are fundamentally incorrect.

Nevertheless, the preceding discussion leads to an interesting connection to another area that employs the concept of dipole moments, namely the dielectric constant. The theory of the dielectric constant in ice has been rather controversial (Nagle, 1979, 1982; Stillinger, 1982) and will not be thoroughly reviewed here. Suffice it to say that this author prefers the Onsager-Slater theory, which relates the dipole moment μ to ϵ through

$$\epsilon-\epsilon_\infty = 4\pi NG\mu^2/3VkT, \tag{7}$$

where ϵ_∞ is the high frequency dielectric constant, N/V is the concentration of water molecules and the G factor measures the sum of correlation functions and has been calculated to be close to 3 (Yanagawa and Nagle, 1979; Adams 1984).

A criticism of the Onsager-Slater formula in Eq. 7 is that dipolar interactions are not included in any way. In particular, the G factor calculation uses only the ice-rules and the factor of $3\epsilon_o/(2\epsilon_o+\epsilon_\infty)$ that appears in the Kirkwood-Frohlich theory is not present. This would appear to be inconsistent with the presence of a dipole moment in the water molecules comprising the ice crystal. However, Fig. 2 illustrates why the Onsager-Slater theory is consistent. One may choose the unit cell placement so as to make the dipole moment of each unit cell inside the crystal equal to zero! Of the several ways to do this, the preferred way (Nagle, 1979; Onsager and Dupuis, 1960) is to choose x=-1.4615 in Fig. 2 and to put just enough of the protonic charge on the neighboring water into the unit cell so that $\mu_{unit\ cell}=0$. This may be done for every unit cell in a chain provided that the ice-rules hold. Of course, when there is a high concentration of defects, one should consider corrections of the Kirkwood-Frohlich type, as has been discussed (Nagle, 1979). For the very small concentration of defects in real ice, this is a minor correction and so the assumption of dipolar-less unit cells necessary for the Onsager-Slater theory to be consistent is easily realizable by suitable choice of unit cell.

A possible source of semantic confusion with the preceding interpretation of the dielectric theory concerns the dipole moment μ that appears in Eq. (7). Clearly, this is not the dipole moment $\mu_{unit\ cell}$ which has just been chosen to be zero. Instead, it is the total dipole moment in Eq. (4) which includes the end effects. These end effects are just the excess surface charges that are typically drawn in discussions of dielectrics in elementary physics courses. These excess charges are simply related to the defect charges that must pass along the hydrogen bonded chain to bring about a polarization change. The dipole moment μ is then given quantitatively by Eq. (5). Although it is convenient, for the purpose of counting the net configurational polarization in the crystal, to associate this dipole moment with individual water molecules, it must be remembered that this is purely a counting device and that the precise meaning of the dipole moment is defined by Eq. (4).

It is worth emphasizing that the combined theory, involving both

the charge partitioning and the dielectric constant, is
overdetermined by the combination of Hubman's experiment and the
measurement of the dielectric constant. The fundamental quantities
to be determined are e_B, μ and ε and there are two relations, Eqs.
(5) and (7). Just using Hubman's experimental $e_B=0.38e$ yields
$\mu=2.94D$ and $\varepsilon=94.3$ (where $\varepsilon_\infty=3.1$) for T=265K. This calculated value
of ε is satisfying close to the measurements of ε of about 96 by
Johari and Jones (1978). Even the difference between the calculated
and the measured values are in the correct direction since the
theory does not take into account the interactions that eventually
bring about an ordered proton phase (Matsuo, Tajima and Suga, 1986;
Glen and Whitworth, 1991) at 70K and which would increase ε if they
were included in the equations.

CHARGE TRANSPORT MECHANISMS

The message in the previous section is that some basic concepts
exist regarding proton transport, that reasonable values have been
obtained for a number of quantities, and that there is some
consistency in experimental results. However, that section did not
deal with the underlying dynamic mechanisms for charged defect
transport. This is a subject of fundamental importance, but one
that is more difficult, both experimentally and theoretically.

The fundamental question concerns the dynamic mechanism for the
diffusion and mobility of both kinds of defects, Bjerrum and ionic.
There are several possible mechanisms, and it must be allowed that
different mechanisms may pertain to different defects, that
different mechanisms may dominate in different temperature regimes,
and that different mechanisms may occur for different hydrogen
bonded crystals and for different hydrogen bonded chains that might
occur in biological systems.

The conventional theory of dynamic mechanism assumes that the
defects are fairly well localized and that they move from site to
site in the lattice in a way that is consistent with thermal
activation. In this theory the motion of defects is very
incoherent; after each step any momentum is lost so no memory is
retained and the defect is equally likely (in the absence of a
field) to move randomly in any of the directions allowed by the
hydrogen bonding. The primary phenomenological quantity describing
this motion is the activation free energy, which includes an
activation entropy as well as an activation energy. In first
approximation the temperature dependence of the mobility should then
behave as $\exp(-E_{act}/kT)$, although one should also allow for the
possibility that E_{act} could depend upon T due to thermal expansion of
the lattice. By comparing the measured mobilities in ice with the
theory of transport in a protonically random medium, the basic
hopping rate near 265 K for ionic defects is calculated to be about
10^{12}/sec and about 10^{10}/sec for Bjerrum defects. (Nagle and
Tristram-Nagle, 1983).

Because ionic defect hopping involves such a short distance for
proton motion, quantum tunnelling theories have been considered.
For ice such theories must deal with two separate problems. The
obvious problem is to obtain a Hamiltonian for tunneling of an
excess proton between two neighboring water molecules. The second
problem is to take into account the fact that the defect motion is
dichotomously branching with each step because ice is a protonically

22

random three dimensional medium. Some time ago Onsager and colleagues (Chen et al., 1974) performed calculations on a very simple s-band tunnelling model with only one phenomenological parameter which was fitted to the mobility datum. The problem of the dichotomous branching was, however, rather carefully treated. This feature led to a finite mobility for ice, even though the same model for a hydrogen bonded chain would have an infinite mobility. However, the conclusion was that the dynamic mechanism for ionic defect transport is probably not pure quantum tunnelling on the grounds that the tunnelling energy required to obtain agreement with experiment is rather small compared to kT, so that tunnelling would be subject to phonon scattering. The temperature dependence of the conductivity for this basically non-dissipative quantum tunnelling mechanism is 1/T. In contrast Kunst and Warman (1980) have cited small polaron theory which involves multi-step dissipative tunnelling on a lattice, but with no consideration for the protonic disorder present in ice. This theory predicts that the mobility sharply decreases with increasing temperature in agreement with the inverse sixth-power found in Warman's experiments.

A theory of dissipative quantum tunnelling (Grabert and Weiss, 1984) has been cited by Careri and Consolini (1990) in connection with protonic conductivity on hydrated protein surfaces for temperatures between 230K and 270K. The theory, which deals exclusively with the first problem of tunnelling involving one transfer step and ignores any possible branching in a water network, predicts tunelling that strongly increases with temperature as $\exp(T^n)$ with n of order 2-4. The very strong difference between this temperature behavior and the behavior in ice emphasizes that different systems have different behavior and that water in contact with protein surfaces should not be thought of simply as immobilized ice-like water. It may also be noted that Careri and Consolini (1990) find that conventional thermal hopping is consistent with their data for temperatures higher than 270K.

Others in this conference will give an extensive accounting of soliton theory for the dynamic mechanism. Let me just say here that it is an appealing classical theory that appears to be more relevant for regular linear hydrogen bonded chains rather than for protonically disordered materials such as ice.

Experimentally, low frequency conductivity measurements yield the product of concentration c_m and mobility μ_m

$$\sigma = e_m c_m \mu_m \qquad (8)$$

where the subscript m refers to the *minority* defect. Since c_m is often very small (for ice c_{ionic} is of order 10^{-10}), most theories or simulations must inject a defect and then study the mobility. To compare to macroscopic measurements of σ, this then requires separating c from μ in Eq. 8. Furthermore, defects may become trapped in shallow traps (Woolbridge and Devlin, 1988), so the concentration of mobile defects is not easy to obtain. Finally, the identity of the minority carrier may change with temperature and doping, with crossover regions where neither type of defect is the majority defect. To this author's knowledge, the only hydrogen bonded system for which mobilities of both kinds of defects have been measured (with considerable variation from decade to decade) is ice, for example, Petrenko and Maeno (1987). However, ice is more difficult for theory because of the dichotomous branching in a protonically disordered system. I believe that progress might be made in determining proton transport mechanism if measurements of mobility were made in crystals such as lithium hydrazinium sulfate (Schmidt et al., 1971) or imidazole (Kawada et al., 1970) which have one-dimensional chains of hydrogen bonds. This would facilitate the comparison between theory and experiment, as well as broaden our data base for proton transport.

The alternative of studying protonic transport in protonically ordered systems with dimensionality higher than one involves a different complication. In such systems it is easy to see that defect pairs, once generated, will remained confined (Nagle, 1985), as on a rubber band. A simple example is the antiferroelectric F-model, represented in Fig. 3 with water molecules (though there is no known phase of water that emulates the F-model.) In the low temperature ordered phase, separation of an ion pair leaves a string of higher energy vertices, so the energy grows linearly with separation. While this makes an interesting analogy to confinement of quarks, it would not appear to be an auspicious direction for resolving the fundamental problem of dynamical mechanisms for proton transport.

```
     H            H            H
     O    HOH     O    HOH     O    HOH
     H            H            H

          H            H            H
   HOH    O    HO     HO+    HOH     O
          H            H            H

     H                        H
     O    HO-    HO    HOH     O    HOH
     H            H            H

          H            H            H
   HOH    O    HOH     O    HOH     O
          H            H            H
```

Fig. 3. The F-model with one ion pair connected by a string
         of two waters with higher energy (i.e., those with
         non-colinear HOH) in a sea of ground state waters
         (i.e., those with colinear HOH).

PROTON TRANSPORT IN BIOLOGY

The well established chemiosmotic paradigm of Mitchell tells us that differences in proton electrochemical potentials $\Delta\Psi$ are formed and maintained across biomembranes by proteins that tap external energy sources and that these proton potentials are then transduced into ATP or other forms of energy by the transport of protons back through the membrane from high potential to low potential. The formation of $\Delta\Psi$ could be accomplished by electron transfer, but for the well studied purple membrane (Stoeckenius and Bogomolni, 1982) and the mitochondrial membrane (Wikstrom et al., 1981) it is clear that there is proton transport. There is also experimental evidence (Dilley and Chiang, 1989) for proton transport laterally along chloroplast membranes. The importance of proton transport through membranes is attested to by the large number of researchers working on these topics. This provides a strong motivation for current research into the principles of proton transport and for this conference.

I have long maintained that, to understand biological proton transport, a data base for proton transport in simpler systems is needed. Already, calculations that use data from ice studies have led to important conclusions. From the mobility of charge defects in ice it appears that proton transport along hydrogen bonded chains could be very fast and efficient (Nagle and Morowitz, 1978; Nagle and Tristram-Nagle, 1983). Indeed, these calculations suggest that proton transport, while an essential step, may not be the rate limiting step in the biological processes of interest. This conclusion alleviates early concern that the entire picture of protons scooting through membranes was fundamentally impossible. Unfortunately, this conclusion also implies that it is difficult to learn about the details of proton tranport in biological systems from experimental data since those data are synchronized to the rate limiting steps that primarily reflect other events, such as protein conformational change. While I believe that this conclusion will stand, it is important to emphasize that it is really based only on data from ice. Although I would favor data from ice over data from any other single crystalline system, I would still prefer to have data from a variety of systems when discussing these theories. In particular, it would be possible that mobilities in ice are anomalously large due to a special transport mechanism that would not apply to biological systems. Conversely, it might be noted that a solitonic mechanism, which would not apply to ice because of its dichotomously branching random system, would tend to increase mobilities in regular linear hydrogen bonded chains. So if solitons were the dynamic proton mechanism in biological membranes, then the above conclusions would hold even more strongly.

The preceding paragraph emphasizes that our understanding of proton transport in biological systems shares the same uncertainty as our understanding for simpler hydrogen bonded crystals, namely the question of the detailed dynamic mechanism. However, there is an even more important uncertainty for biological systems, namely, that the structures or channels along which the protons are transported are poorly known. Even after many years the structure of the bacteriorhodopsin proton pump in purple membrane is known only to the 3.5-7.8 Angstrom level (Henderson et al., 1990). Also, functional studies of mutants have shown that a rather small number of amino acid side chains are essential to pumping (Khorana, 1988). Together, it can be concluded that any permanent channel must consist mostly of water molecules, and neutron diffraction studies (Papadopoulos et al., 1990) have identified about four water molecules that are closely bound in a location where they could participate in a proton pathway. This is consistent with the hydrogen bonded chain structure that I have favored, although my models envisioned a larger proportion of side chains and a smaller proportion of water molecules. In passing, it might be mentioned that a hydrogen bonded chain in a biological system would not be very regular. Even if one looked at a chain of water molecules, each water molecule will have a different environment of protein side chains which will perturb the hydrogen bonds between waters differently for different waters in the hydrogen bonded chain. These perturbations could seriously dampen any solitons, though they would be less disruptive for a conventional thermally activated hopping mechanism. It is therefore important that solitonic models be studied with impurities or randomness as is being done by Kivshar (1991).

In the preceding paragraph it was assumed that proton channels are permanent, consistent with the idea of a proton wire. Of course, a non-crystalline membrane system would be expected to have larger fluctuations than a crystal, but this alone is not inconsistent with the idea that there is a permanent proton wire. It does raise the possibility, however, that the channel is not connected from end to end part of the time or even all of the time. In this case, the proton could only be transported part way; it would then have to wait for a fluctuation to connect the portion in which it currently resides to the next portion. Finally, one must consider the possibility that there is *no* channel in the resting state of the protein. In this case, the channel would only be formed during parts of the proton pumping process. One possibility, consistent with time resolved x-ray diffraction results (Koch et al., 1991) which shows tertiary structural changes during proton pumping in bR, would be that conformational distortions and tilting of the alpha helices would open up a crevice into which water molecules would rush to form a transient aqueous channel during the parts of the proton pumping cycle when proton transport is needed (Dencher et al., 1991).

One example where transient hydrogen bonded chains would seem to be appropriate is for proton transport through lipid bilayers which are model membranes in which a permanent proton wire is most unlikely. *A priori* one would not suppose lipid bilayers to be especially permeable to protons, but Deamer and Nichols (1989) have shown in an extensive set of measurements that have now been confirmed by a number of groups that protons are notably more permeable than other small ions. As I have discussed elsewhere, the measured pH dependence of the conductivity eliminates the possibility that the protons are transported by water pores that act as bulk water channels. There are at least three different versions of transient hydrogen bonded chains that could account for these observations (Nagle, 1987), and the three versions are experimentally distinguishable by measuring conductivity with large electrical potential or large pH drops across the membrane. Unfortunately, as Deamer has reported at this meeting, none of the three versions accounts for the current data under these conditions, so there is still a mystery even for these simple model biological systems.

This section emphasizes that there are many uncertainties involved in proton transport in complex biological membranes. To help resolve these uncertainties, I believe there is a need to pursue both theoretical and experimental studies on the fundamentals of proton transport in simpler hydrogen bonded systems, such as crystals or even well characterized small synthetic proteins (Lear et al., 1988). There is a fair chance that such studies will be relevant to the complex biological membranes of biomedical interest, not to mention the intrinsic interest in such materials.

REFERENCES

1. D. J. Adams, J. Phys. C17, 4063 (1984).

2. G. Careri and G. Consolini, Ber. Bunsenges Phys. Chem. 95, 376 (1991).

3. M. S. Chen, L. Onsager, J. C. Bonner and J. F. Nagle, J. Chem. Phys. 60, 405 (1974).

4. D. W. Deamer and J. W. Nichols, J. Membrane Biology 107, 91 (1989).

5. N. A. Dencher, J. Heberle and G. Buldt, (1991) Contribution to this conference.

6. R. E. Dilley and G. G. Chiang, Annals of New York Academy of Sciences 574, 246 (1989).

7. J. W. Glen and R. W. Whitworth, (1991) Contribution to this conference.

8. H. Grabert and U. Weiss, Z. Phys. B 56,171 (1984).

9. R. Henderson, J. M. Baldwin, T. A. Ceska, F. Zemlin, E. Beckmann and K. H. Downing, J. Mol. Biol. 213, 899 (1990).

10. M. Hubman, Z. Physik B32, 127 and 141 (1979).

11. G. P. Johari and S. J. Jones, J. Glaciology 21, 259 (1978).

12. H. G. Khorana, J. Biol. Chem. 263, 7439 (1988).

13. A. Karpfen and P. Schuster, Canadian J. Chemistry 63, 809 (1984).

14. A. Kawada, A. R. McGhie and A. R. Labes, J. Chem. Phys. 52, 3121 (1970).

15. Yu. S. Kivshar, Phys. Rev. A43, 3117 (1991).

16. M.H.J. Koch, N.A. Dencher, D. Oesterhelt, H.-J. Plohn, G. Rapp and G. Buldt, EMBO Journal 10, 521 (1991).

17. E. S. Kryachko, Solid State Communications 65, 1609 (1988).

18. E. S. Kryachko, M. Eckert and G. Zundel, J. Mol. Structure (1991).

19. M. Kunst and J. M. Warman, Nature 288, 465 (1980).

20. J. D. Lear, Z. R. Wasserman and W. F. DeGrado, Science 240, 1177 (1988).

21. T. Matsuo, Y. Tajima and H. Suga, J. Phys. Chem. Solids 47, 165 (1986).

22. J. F. Nagle and H. J. Morowitz, Proc. Natl. Acad. Sci. (USA) 75, 298 (1978).

23. J. F. Nagle, Chem. Phys. 43, 317 (1979).

24. J. F. Nagle, Nature 298, 401 (1982).

25. J. F. Nagle, J. Phys. A 18, L181 (1985).

26. J. F. Nagle, J. Bioenergetics and Biomembranes 19, 413 (1987).

27. L. Onsager and M. Dupuis, Rendiconti S. I. F., X Corso, 294 (1960).

28. L. Onsager and M. Dupuis in Electrolytes (Pergamon Press, 1962) ed. B. Pesce, pp 27-46.

29. G. Papadopoulos, N. A. Dencher, G. Zaccai and G. Buldt, J. Mol. Biol. 214, 15 (1990).

30. V. Petrenko and N. Maeno, J. de Phys. (Colloque) 48, C1-115 (1987).

31. M. Peyrard, St. Pnevmatikos and N. Flytzanis, Phys. Rev. A36, 903 (1987).

32. S. Scheiner and J. F. Nagle, J. Phys. Chem. 87, 4267 (1983).

33. V. H. Schmidt, J. E. Drumheller and F. L. Howell, Phys. Rev. 4, 4582 (1971).

34. F. H. Stillinger in "Studies in Statistical Mechanics", Vol. 8, eds. J. L. Lebowitz and E. W. Montroll (North-Holland, NY, 1982). Chapter 6, pp. 341-431.

35. W. Stoeckenius and R. A. Bogomolni, Annual Review of Biochemistry 52, 587 (1982).

36. I. Takei and N. Maeno, J. de Phys. (Colloque) 48, C1-121 (1987).

37. M. Wikstrom, K. Krab and M. Saraste, Annual Review of Biochemistry 50, 623 (1981).

38. P. J. Woolbridge and J. Paul Devlin, J. Chem. Phys. 88, 3086 (1988).

39. A. Yanagawa and J. F. Nagle, Chem. Phys. 43, 329 (1979).

40. O. E. Yanovitskii and E. S. Kryachko, Phys. Stat. Sol. (b) 147, 69 (1988).

EXTRACTION OF THE PRINCIPLES OF PROTON TRANSFERS BY AB INITIO METHODS

Steve Scheiner

Department of Chemistry and Biochemistry
Southern Illinois University
Carbondale, IL 62901
USA

INTRODUCTION

The study of proton transfers in hydrogen-bonded systems goes back many years within the discipline of chemistry to early understanding of acids and bases. Renewed interest in this topic was generated by later advances in gas phase and very rapid measurement techniques.[1-4] The former permitted detailed investigation of the phenomenon unhindered by complicating effects of the environment while the latter opened up a new window into elementary steps in the overall process. Despite the accumulation of information about the proton transfer process resulting from these studies, other facets of the problem remained out of reach. Moreover, a need for additional insights has been coupled to the intriguing notion of solitons as a means of proton transport over long distances.[5-7]

Unlike experimental techniques which are largely limited to study of long-lived species, ab initio methods afford the opportunity to probe any desired arrangement of atoms and molecules regardless of whether or not it represents a minimum, or indeed any sort of stationary point, on the potential energy surface. Thus the highly transient but very important transition states can be identified and studied. The methods are also amenable to extraction of data relating to electronic rearrangements that accompany the proton transfer process. Such charge shifts can perturb proton transfers elsewhere within a larger system. If proper care is taken to choose appropriate levels of theory, ab initio methods can provide data competitive in accuracy with many experimental measurements. We present here some specific examples of insights into the proton transfer issue developed through the application of ab initio methods.

The first section discusses the energy barriers to proton transfer. The description focuses upon the dependence of this barrier upon the geometrical details of the hydrogen bond connecting the two species, in particular the H-bond length and angles. The sort of accuracy to be expected at various levels of theory is included. Since the size of the specific system that may be considered is inversely related to the level of theory that can be realistically applied, the effect of system enlargement upon the barrier height is considered explicitly. Also described is the relationship between the transfer barrier and the degree of asymmetry of the system, i.e. the difference between the proton donor and acceptor groups. The next section briefly outlines the shifts in the electron distribution that accompany the motion of the proton from one group to the other. The issue of proton migration over long distances by a series of hops of different protons along a sequence of hydrogen bonds is the topic of the following section which examines the likelihood that these hops will be simultaneous or sequential. The restriction of an externally fixed H-bond length or angle is removed in the final section where various systems are classified as containing either one or two distinct minima in their potential energy surface.

BARRIERS

Distance Dependence

In order to study the energetics of proton transfer, one first composes a system of the appropriate type. Figure 1 illustrates two water molecules plus a bridging proton which comprise the $(H_2O\cdot\cdot H\cdot\cdot OH_2)^+$ system. The two waters are held apart by some arbitrary distance R and the central proton allowed to move along the $O\cdot\cdot\cdot O$ axis. The energy of the entire system can then be calculated for every step of the proton along its path. For each H-bond length R, one then arrives at a proton transfer potential of the sort shown in Fig. 1. What is perhaps most important about each potential is the height of the energy barrier, $E^\dagger$, defined as the difference in energy between the top of the barrier and the minimum in the potential.[8,9]

The data illustrated in Fig. 1 show that the barrier is highly sensitive to the distance between the oxygen atoms. $E^\dagger$ is only 2-3 kcal/mol when R is 2.55 Å, increases to 7 when R is stretched by 0.2 Å, and increases further to 17 kcal/mol when R reaches 2.95 Å. This behavior is in fact not peculiar to the protonated water dimer or even to H-bonds between oxygen atoms but is rather a general feature of all H-bonds.[9] Figure 2 depicts the dependence of transfer barrier upon distance between donor and acceptor groups, as computed for a number of different systems. The OH→O designation is a shorthand for the aforementioned $(H_2O\cdot\cdot H\cdot\cdot OH_2)^+$ system, with similar notation applying to H_2S and H_3N. In all cases, the transfer barrier increases quite quickly as the H-bond is elongated. It is also interesting that the rate of increase is nearly uniform from one system to the next in that the various curves in Fig. 2 are almost parallel to one another. The notable exception is the transfer between two second-row (S) atoms where the barrier rises somewhat more gradually.

The effects of this energetic trend upon transfer kinetics are profound. For example, a stretch of the water-water distance from 2.6 Å to 2.8 Å raises the transfer barrier from 3 to 9 kcal/mol. If we assume an Arrhenius exponential form for the rate expression, then this 0.2 Å stretch slows down the process by over 5 orders of magnitude at 300° K. An additional 0.2 Å stretch to 3.0 Å produces a further slowdown by another 8 orders of magnitude. Of course, the small mass of the proton makes quantum mechanical tunneling a likely contributor to its transfer through the barrier in addition to the classical process. Even in such a case, the height

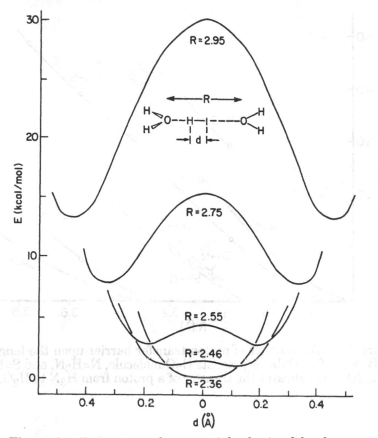

Figure 1. Proton transfer potentials obtained for the protonated water dimer at the SCF level using the 4-31G basis set. The two oxygen atoms are held apart by a distance R; d refers to distance of bridging proton from O⋯O midpoint.

of the barrier will have a very strong effect upon the rate of transfer.[10]

System Enlargement

For investigating the possibility of long range proton transfer along a chain of H-bonds, it is necessary to consider how the above results are affected if other members of the chain are present. In order to address this issue, the $(H_2O\cdots H\cdots OH_2)^+$ system was enlarged to $(H_2OH_2O\cdots H\cdots OH_2OH_2)^+$ by positioning one additional water molecule on each end of the dimer. The data in Table 1 reveal that this elongation has a very small effect on the dimer, increasing the barrier by no more than 1 kcal/mol in any case.[8]

Another type of modification of the system involves not addition of more molecules along the chain but rather enlargement of each group participating in the H-bond. For example, the hydroxyl group can occur not only on a water molecule but also on alcohols of various types. Comparison of HOH with the CH_3OH molecule allows one to elucidate the effect of replacing a hydrogen atom with a carbon as the atom immediately connected to the H-bonding O center. The results are presented in the last column of Table 1 where the methyl substitution can be seen to raise the transfer barrier by a small amount. This change was traced[11] not to any

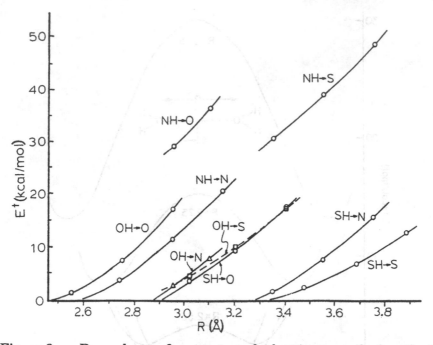

Figure 2.   Dependence of proton transfer barrier upon the length of the H-bond, R. O label represents H_2O molecule, $N=H_3N$, and $S=H_2S$. Thus, NH→O indicates the transfer of a proton from H_3N to H_2O.

fundamentally different chemical character of CH_3OH but rather to a natural contraction of the equilibrium OH bond length in methanol. It further stands to reason that additional elongation of the methyl group to ethyl, propyl, etc. will yield even smaller changes in the energetics of proton transfer. We hence conclude that computation of the transfer barrier in a small system such as $(H_2O \cdot\cdot H \cdot\cdot OH_2)^+$ is representative of results to be expected for hydroxyl groups in general or in longer chains of such molecules.

Accuracy

An essential issue which must be addressed concerns the accuracy of the quantum mechanical calculations. Unlike semiempirical or empirical methods, ab initio techniques do not involve arbitrary parameters that are

Table 1.   Barriers to proton transfer (kcal/mol) computed at SCF/4-31G level

R, Å	$(H_2O \cdot\cdot H \cdot\cdot OH_2)^+$	$(H_2OH_2O \cdot\cdot H \cdot\cdot OH_2OH_2)^+$	$(CH_3OH \cdot\cdot H \cdot\cdot HOCH_3)^+$
2.55	1.4	1.6	1.8
2.75	7.5	7.9	8.9
2.95	16.8	17.6	19.2

fit to experimental data. Nonetheless, the results can be highly sensitive to other choices that must be made. For example, there are a large number of ways of specifying the atomic orbitals that are used to describe the electronic structure of the system under study. While it is understood that more complete basis sets will lead to superior results, it is generally not feasible due to computational limitations to employ a basis set which approaches saturation, i.e. the Hartree-Fock limit. Since the quality of the molecular properties does not necessarily improve monotonically with each extension of the basis set, it is not uncommon for fairly small basis sets to provide results in better agreement with experiment than those of moderate size.

Even if it were possible to approach the Hartree-Fock limit, this approximation ignores the fact that their identical negative charges tend to keep the motions of electrons correlated with one another on an instantaneous level. Indeed, failure to include electron correlation can in many cases lead to completely spurious results.[12] This correlation can be incorporated completely only with a full configuration interaction calculation, again impractical for any system of size relevant to proton transfers. There are a number of means that have been developed over the years for efficient incorporation of partial electron correlation using a minimum of computer resources. Yet each suffers a number of drawbacks and limitations.

How is it possible to evaluate the accuracy of a given method? The first and most obvious is to compare the calculated data with a comparable experiment. Unfortunately, there are no systems which correspond to an isolated pair of molecules (relevant to the calculations) where the proton transfer barrier is known with any precision. Lacking this absolute point of comparison, the next best approach is to perform a calculation where the basis set is quite extensive and electron correlation is well accounted for. While this may in fact be possible for one or two geometrical configurations of a fairly small system, it is not generally feasible to apply this level to a larger number of geometries and becomes untenable for larger systems.

Our approach makes use of large-scale computations to serve as a yardstick by which to measure the reliability of less computer intensive methods which may be applied on a more routine basis. We consider the proton-bound water dimer as a prototype. An early study of this system[8] revealed that the transfer barrier obtained at the SCF level with a fairly small 4-31G basis set matched quite closely a barrier computed by an extensive CI calculation with a much larger basis set.[13] This calculation had suggested that increasing the size and flexibility of the basis set acts to increase the barrier while inclusion of correlation tends to lower it, and that the latter two effects can balance out one another under certain circumstances.

Indeed, this general conclusion reached for $(H_2O \cdot H \cdot OH_2)^+$ has since that time been verified in a wider range of systems. Of the various means of accounting for correlation, which include generalized valence bond and POL-CI,[14,15] the Moller-Plesset variety of perturbation theory has proven to be both accurate and cost effective. In a test on $(HO \cdot H \cdot OH)^-$, third-order MP theory reproduced to very high precision[16] the barrier obtained earlier by a 50,000 configuration CI using a fairly extensive basis set.[17] A more recent set of calculations taking the Moller-Plesset expansion up through fourth order, including triple excitations, concluded that this MP4 level matches quite closely coupled cluster results.[18] The data demonstrated as well that

MP2 offers a very good substitute for full MP4 at a fraction of the expense.

Due to computational limitations, most of the proton transfer potentials computed in this laboratory are obtained first at the SCF level. This prescription provides some estimate of the shape of the potential, number of minima, etc. Whenever possible, the height of the energy barrier is recomputed with electron correlation included. This correction is typically carried out at the MP2 level with as large a basis set as feasible. It has been found that while SCF barriers are frequently in error by several kcal/mol, they usually reproduce faithfully the trends in these properties occasioned by H-bond stretches and bends, changes in the nature of the proton donor or acceptor group, etc.

The general rules then would argue against accepting as accurate a proton transfer barrier computed at a low level of theory. On occasion, such a barrier may agree surprisingly well with a result calculated at a higher and more reliable level, due to the usual partial cancellation between the effects of basis set enlargement and correlation corrections. Nonetheless, such cancellation is not always the case and it would be prudent to compare the data with higher level calculations.

Since the problem concerns two groups, the donor and acceptor, competing for a proton, the experimental difference in proton affinity between the donor and acceptor groups must be reliably reproduced by whatever level of theory is adopted. Otherwise, the transfer potential will be distorted in one direction or the other, yielding an incorrect barrier and possibly even the wrong equilibrium position of the proton.

Asymmetric Systems

Most real systems in which one is concerned with a proton transfer are not completely symmetric wherein the proton donor and acceptor are identical and occur in the same exact environment. There are of course varying degrees of asymmetry. For example, the proton may be transferring between two water molecules but they might lie in slightly different environments or the internal geometry of one may be somewhat different than that of the other. A stronger asymmetry is introduced if the groups themselves are different as in the $(H_3N \cdots H \cdots OH_2)^+$ system mentioned above. Since the types and amounts of asymmetry are so varied, it would be a major step towards a systematic understanding if one could classify any given asymmetric system as a perturbation on a symmetric one.

There are in fact a number of approaches that have been taken to this problem, perhaps the most prevalent of which is the Marcus formulation, designed originally for electron transfer and then later adapted to proton transfers.[19-21] The Marcus equation writes the energy barrier in the asymmetric system as

$$E^\dagger = E^\dagger_0 + \Delta E/2 + (\Delta E)^2/16E^\dagger_0 \tag{1}$$

where $E^\dagger_0$ refers to the barrier in the symmetric system and ΔE to the difference in energy between the $(AH \cdots B)$ and $(A \cdots HB)$ states.[22] The premise upon which this equation is built is first that the motion of the proton away from one group, A or B, can be described by a parabola. It assumes secondly that the difference in basicity between groups A and B can be treated by lowering one parabola relative to the other on an energy scale. But a

34

further tenet of the theory, and a much less defensible one, is that while the bridging proton lies closer to A than to B, even if only marginally closer, the energetics of the entire transfer are completely unaffected by B's presence. At the moment that the proton passes the midpoint, A disappears and is replaced by B which completely controls the remainder of the transfer energetics.

Despite the obvious fallacy of the presence of only one group at a time, Marcus's equation has had good success in various applications over the years. The underlying reasons for its efficacy were probed by a rigorous quantum mechanical test.[22] A fully symmetric system like $(H_3N\cdots H\cdots NH_3)^+$ was taken as a suitable starting point. The energy barrier for proton transfer was computed for a given R(NN) distance and serves as the value of $E^{\dagger}_{0}$ as defined above. Asymmetry of varying degree was introduced by replacing one or more hydrogens on one of the ammonia molecules by alkyl groups methyl or ethyl. For each substitution, ΔE was evaluated as the difference in energy between the minimum with the proton on the unsubstituted ammonia versus the other minimum in the transfer potential, again at the same R(NN). The barrier $E^{\dagger}$ was predicted on the basis of Eq. (1) using $E^{\dagger}_{0}$, and ΔE. This value was compared to the barrier obtained from a quantum mechanical computation of the potential of the particular asymmetric system. The agreement was quite striking, not only at one H-bond length but at all three considered. In fact, very nearly perfect coincidence was seen as well when the oxygen analogues of water and its alkylated derivatives were studied.[22]

The superb performance of the Marcus equation is truly remarkable in light of the oversimplifications upon which it is based. Its accuracy for proton transfer reactions has recently been verified in a rigorous experimental measurement as well.[23] The source of this agreement is the subject of ongoing study in this laboratory. Regardless of the origin of its superior performance, the success of the Marcus theory provides a powerful tool in efforts to predict proton transfer energetics of complex systems. Consider for example an arbitrary H-bonded system composed of two amines substituted with various groups: $(ABCN\cdots H\cdots NXYZ)^+$. Even if this system is much too large for a serious quantum mechanical calculation, the transfer barrier can be estimated to some degree of confidence by using the Marcus equation. The first piece of information required is $E^{\dagger}_{0}$, which may be obtained quite accurately for the much smaller symmetric $(H_3N\cdots H\cdots NH_3)^+$. Various sources may be used to estimate ΔE. Gas phase proton affinities for ABCN and NXYZ would be most appropriate, or failing this, computed values may be substituted. If the two individual subunits are too large for ab initio calculations, semiempirical procedures can offer very good approximations. An alternative would be to use the measured pK difference between ABCN and NXYZ.

Angular Dependence

The systems considered up to this point have not contained any angular distortions in their H-bonds. That is, each pair of subunits have been free to adopt their most stable orientation relative to one another. In most cases, this leads to a linear arrangement in which the bridging proton lies directly along the X$\cdots$Y intermolecular axis. While this may be representative of dimers in the gas phase, when placed within the context of an

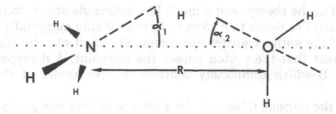

Figure 3.   Angular deformation measured by angles α_1 and α_2 which measure misorientation between N···O axis and symmetry axes of NH_3 and H_2O molecules, respectively.

intramolecular H-bond of a large molecule or of a crystal, the structural constraints imposed by the entire system would naturally lead to certain angular deformations of the H-bond.   These bends are analogous to the stretches that are normally observed in the various H-bond lengths noted extensively in the literature.[24,25] Since the intermolecular distances have been shown to exert a profound influence on the energetics of proton transfer, it is logical to presume that the angular attributes of the H-bond may affect the transfer potential as well.

For this reason, proton transfers were examined in systems as exemplified by that in Fig. 3 where an angular distortion has been imposed as follows.[26] The lefthand subunit, in this case NH_3, would, if optimized, be oriented such that its C_3 symmetry axis is collinear with the N··O H-bond axis. The group is turned so that these axes deviate from one another by an amount α_1; α_2 represents an analogous misorientation of the right-hand OH_2 molecule. Once a particular set of angles is chosen, the proton transfer potential is traced out by moving the proton in steps from left to right, allowing it to follow its least energy path.   Three different modes of distortion were examined. In the first, the left-hand molecule is left in its optimized alignment and the righthand group turned.   The second mode rotates both molecules by equal amounts and in the same direction while a disrotatory mode wherein $\alpha_1 = -\alpha_2$ constitutes the third.

The effects of these angular distortions upon the transfer barrier are presented in Fig. 4 where the notation as to the identify of each system is as in Fig. 2. Regardless of the nature of the proton donor or acceptor molecule, any type of angular deformation raises the energy barrier, making the transfer more difficult. The barrier increase is most gradual for the first mode involving misalignment of only one molecule and is largest for the disrotatory motion of both groups. These barrier increases are highly sensitive to the magnitude of the distortion. Taking the NH→N transfer between two amine groups as an example, the barrier in the undistorted conformation is 12 kcal/mol. A 20° disrotation raises this barrier by only 2 kcal/mol whereas a 40° misalignment yields a barrier of over 30. It is also worthy of emphasis that the dependence of barrier upon angular distortion is quite similar from one system to the next, indicating a general rule here. Finally, one can conclude from the data that angular considerations are just as important as the length of the H-bond in determining the height of the proton transfer barrier.

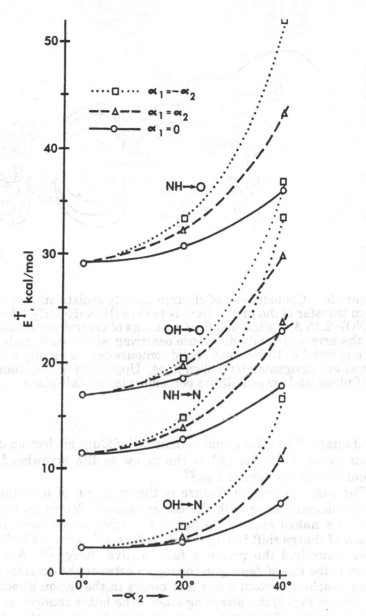

Figure 4.   Proton transfer barriers computed with 4-31G basis set for angularly distorted orientations of proton-bound dimers of water and ammonia. Identities of systems are notated as in Fig. 2. Intermolecular distance R=2.95 Å in all cases.

ELECTRONIC REARRANGEMENTS

The motion of a proton from one group to its partner across a hydrogen bond is expected to disrupt the original character of the H-bond and so occasion a series of shifts of electron density. Ab initio methods allow one to examine these charge redistributions in some detail. For example, Fig. 5 represents the changes of total electron density in various regions of space which result from transferring the proton from NH_3 to OH_2 in their proton-

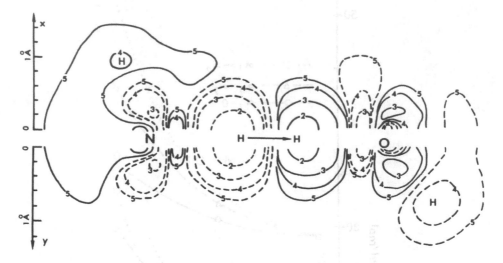

Figure 5.   Contour map of electron density redistributions resulting from transfer of the proton from N to O in $(H_3N \cdots H \cdots OH_2)^+$ for $R(NO)=2.75$ Å. Initial and final positions of central proton are connected by the arrow. Density difference occurring within each contour labeled by n is equal to $10^{-n/2}$ e/au^3. Solid contours denote density increase; losses are designated by broken lines. Upper half of map contains zx half-plane and lower half the perpendicular zy half-plane.

bound dimer. The solid contours indicate buildups of electron density as the proton moves from the tail of the arrow to the arrowhead; the broken contours represent density loss.[27]

The most prominent feature is the dragging of a certain amount of electron density along with the central proton, demonstrating it does not move as a naked proton. In fact, other calculations have quantified the amount of charge shift in $(H_2O \cdots H \cdots OH_2)^+$ as only some 64% of the total that would occur had the proton a full positive charge.[28] A second salient feature is the loss of density in the region between the acceptor atom (O) and the approaching proton; a loss also occurs in the region directly behind (to the right in Fig. 3) the accepting atom. The latter changes are matched by similar trends, albeit gains as opposed to losses, in the vicinity of the donating N atom. Note finally that electron density is gained in the region of the "π" orbitals of the O atom, perpendicular to the H-bond axis. Again, an opposite pattern is observed around the N. The trends described above have been analyzed in terms of the particular MOs which make up the total electron density.[27]

An important finding has been that as the proton moves from left to right, there is a concomitant shift of electron density in the opposite direction; i.e. from proton acceptor group to donor. This shift is consonant with experimental observation of "highly polarizable" hydrogen bonds[29,30] and also helps explain the high susceptibility of the transfer energetics to the presence of neighboring ions.[31-33]

It is immediately apparent from Fig. 5 that the electron shift patterns are not confined to the proton and its motion but that rearrangements occur throughout the system. The oversimplification that the only charge moving is that of the proton is clearly not acceptable, nor can one limit oneself to consideration only of electron density along the H-bond axis. The more complex but characteristic features of the sort of charge shift shown in Fig. 5 may in fact be useful in predicting the effects of substitution in various locations upon the proton transfer process.

MULTIPLE PROTON TRANSFERS

It has been suggested on numerous occasions that the transfer of a proton over a long distance can occur not only by the physical motion of one particular proton but more commonly as a series of single hops of a proton from one group of a H-bonded chain to the next, combined with the hopping of another proton to the succeeding group, and so on.[34-36] Evidence of such a mechanism comes from liquid and solid phases of water, crystalline phases of other materials, and biological phenomena.[37,38] One very important question that arises from this multiple proton transfer mechanism is the degree of synchronicity between one hop and the others. In other words, does the second proton wait for the first to complete its transfer before beginning its own or do the two move at the same time?

In an effort to address this question, a short chain of three water molecules was put together and an extra proton added to the first in the chain.[8] Combination of the hopping of this proton to water 2 and transfer of another proton from water 2 to 3 results in the net transfer of a proton from 1 to 3. By calculating the energetics of configurations in which the two protons are in various stages of transfer, it was possible to construct a contour plot as depicted by Fig. 6. The small x in the upper left corner corresponds to the $(H_2OH^+ \cdots HOH \cdots OH_2)$ starting point and the x in the lower right corner to the end product $(H_2O \cdots HOH \cdots {}^+HOH_2)$ after transfer of both protons. A straight-line diagonal path between these two points would represent a fully synchronous motion of both protons. In contrast, the stepwise motion of H_a followed by transfer of H_b would start at the first x, go straight down to the lower left corner, representing $(H_2O \cdots {}^+HOH_2 \cdots OH_2)$, before the second transfer to $(H_2O \cdots HOH \cdots {}^+HOH_2)$.

The results indicate that a fully simultaneous transfer across the diagonal of Fig. 6 is much too demanding energetically to normally occur: the barrier for this path is quite high. In contrast, a stepwise path which first goes straight down before curving over to the right is much less energetically costly. In fact, when the chain of three waters is lengthened by an additional molecule on either end to reduce distorting end effects, the barrier for this stepwise transfer is no higher than for a single transfer at the same R(OO) while the concerted double transfer barrier is approximately double this height. One may extend this finding to state that the multiple proton hops required in the mechanism described above are more likely to occur by one proton moving at a time rather than all protons moving at the same time. Godzik has recently fit these calculations to a multiproton potential function and verifies that the degree of coupling between the protons is quite small.[7]

This observation is not unique to chains of water molecules. For example, analogous calculations were carried out in which the central water is replaced by HCOOH.[39] Not only is the carboxyl group different than the hydroxyl of water but the first proton approaches an entirely different O atom than the oxygen from which the other proton departs. In this case, the R(OO) distance was held at 2.75 Å, as compared to 2.95 Å in

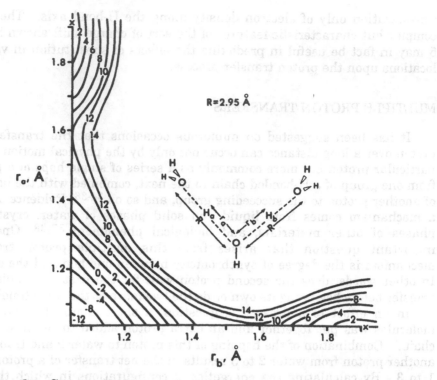

Figure 6.   Contour map representing the energetics (in kcal/mol) of transfer of two protons, H_a and H_b, between the three water molecules in $(O_3H_7)^+$. All hydrogen bonds are linear with the proton lying directly along the indicated O··O axis. Contours higher than 14 kcal/mol are not shown.

the water trimer. Despite these differences, the simultaneous transfer of both protons is much more costly energetically than is the stepwise transfer. To be more quantitative, the transition from $(H_2OH^+··HCOOH··OH_2)$ to $(H_2O··HC(OH)_2^+··OH_2)$ and then to $(H_2O··HCOOH··^+HOH_2)$ via two single transfers passes through two barriers of some 15-20 kcal/mol. A synchronous transfer of both of these protons occasions a transition state of the type $(H_2O··H··HCOOH··H··OH_2)^+$ which is nearly 40 kcal/mol higher than the starting point. There are numerous possible conformations of this system in which each of the two waters can be either syn or anti, or even lie directly along the C=O axis. In every case examined, the stepwise double proton transfer proceeds at less cost in energy than the simultaneous process.[39]

Other investigators have considered multiple proton transfers computationally in various other types of systems[40,41] and arrived at much the same conclusions. This situation should be clearly contrasted with cyclic systems wherein all molecules involved in the chain return to their original protonation state following completion of all proton transfers as, for example, in the cyclic arrangement of $(HF)_3$ or the formic or benzoic acid dimer.[42-45] In these cases, simultaneous multiple proton transfer is required to prevent large buildups of opposite charge on the various groups.

In summary, then, if the system under investigation includes a chain of H-bonded groups plus an excess proton which is being transported, stepwise transfers are to be expected. Synchronous transfers are more likely when the H-bonded chain forms a closed topological loop and all molecules are electrically neutral.

POTENTIAL ENERGY SURFACES

Most of the work described above has examined H-bonds in which the distance between the two subunits has been fixed at some particular value. Such a situation may be appropriate for a H-bond in some sort of rigid environment like a protein molecule or a crystalline matrix, but within the context of a simple dimer in the gas phase, there would be no constraint keeping the intermolecular separation from achieving its optimal distance.

With the intermolecular distance free to change, there are two possible situations, both of which are pictured as contour plots of the energy in Fig. 7. The horizontal axis corresponds to this distance R while the vertical axis refers to the motion of the bridging proton along the H-bond axis. Specifically, the parameter (r_b-r_a) represents the relative distance of the proton from the A and B groups; it is zero when the proton is midway between them. One possibility is that the potential energy surface may contain a single minimum, as on the left side of Fig. 7a. In such a case, the least energy path for proton transfer is traced out by the squiggly curve which takes the system from $AH^+ + B$ to $A + BH^+$, passing through the minimum $A\cdots H^+\cdots B$ along the way. In such a situation, the most stable conformation of the system has the bridging proton nearly equidistant between the A and B groups. An example is the proton-bound water dimer. The energetics of this path are presented as the "adiabatic" curve on the right which illustrates its single-well character.

Even if the full potential contains a single well, a slice through it at fixed R (dotted vertical line in Fig. 7) can trace out a double well potential. That is, the path cuts through a low energy region, then rises in energy until going back down again. The high energy point or "transition state" of this frozen path is designated by the **f** label both in the surface and on the right where the energetics of this path are illustrated.

The other possibility consists of a true double well potential, as pictured in Fig. 7b. The lowest energy, or adiabatic path, has the AH^+ and B groups first coming together to form a stable $AH^+\cdots B$ complex. Proton transfer leads to the other minimum $A\cdots H^+B$, passing through transition state **a**. The latter minimum can then dissociate to A and H^+B. Just as in the single-well potential above, a transfer that takes place under the restraint of a frozen R intermolecular separation may lead to a double-well potential. It is important to note, however, that the transition state **f** for this process is located at a different point on the potential energy surface than the transition state **a**. In either case 7a or 7b, a slice through the potential for fixed R less than the equilibrium separation will lead to a single-well potential.

The systems studied previously by ab initio methods may be categorized as belonging to either the single or double-well type of Fig. 7. A summary of the results is presented in Table 2 where the left side of the table lists symmetric systems, i.e. A and B are identical. In the case of single well

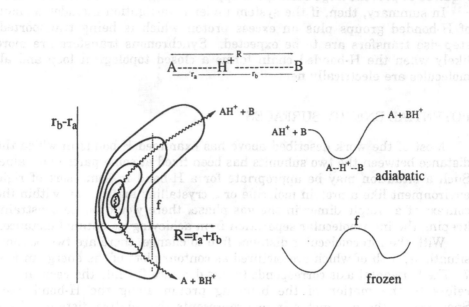

a) single well

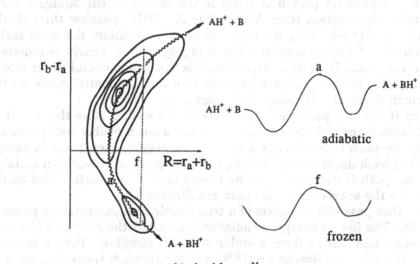

b) double well

Figure 7.   Schematic diagrams of proton transfer potentials as energy contour plots in terms of intermolecular separation R and H-bond asymmetry paramater r_b-r_a. Lowest energy path for adiabatic proton transfer is represented by squiggly line and the dotted line illustrates a constrained path in which R is frozen to some particular value. The locations of transition states are identified by **a** and **f** labels for the two types of transfers, respectively. On the right are shown the energetics along each path.

Table 2. Properties of proton transfer potentials, in Å and kcal/mol.

SINGLE-WELL

symmetric	R^a, Å	ref
$H_3O\cdots H^+\cdots OH_2$	2.35	[46]
$HO\cdots H^+\cdots OH^-$	2.42	[16]
$CH_3OH\cdot H^+\cdots HOCH_3$	2.37	[11]
$C\equiv N^-\cdots H^+\cdots N\equiv C^-$	2.75	[49]

asymmetric	R^a, Å	ΔPA^b	ref
$H_3N\cdot H^+\cdots OH_2$	2.71	35	[14]
$H_3N\cdot H^+\cdots SH_2$	3.35	40	[47]
$H_2CO\cdot H^+\cdots OH_2$	2.52	7	[48]
$HCOOH\cdot H^+\cdots OH_2$	2.55	13	[50]
$HO^-\cdot H^+\cdots OCOH$	2.77	50	[51]
$N\equiv C^-\cdots H^+\cdots N\equiv C^-$	3.01	18	[49]

DOUBLE-WELL

symmetric	R_{eq}^a	R_{TS}^c	$E^{\dagger,d}$	ref
$H_3N\cdots H^+\cdots NH_3$	2.68	2.59	1	[46]
$H_2C=N\cdots H^+\cdots N=CH_2$	3.03	2.49	2	[52]
$C\equiv N\cdots H^+\cdots N\equiv C$	2.75	2.52	0	[49]
$H_2S\cdots H^+\cdots SH_2$	3.48	3.37	0.6	[53]
$H_3C\cdots H^+\cdots CH_3$	4.03	2.88	15	[55]
$H_2C=CH\cdot H^+\cdots CH=CH_2$	3.73	2.88	13	[52]
$HC\equiv C\cdots H^+\cdots C\equiv CH$	3.35	2.81	8	[49]
$HN=CH\cdots H^+\cdots CH=NH$	3.52	2.86	12	[52]
$N\equiv C\cdots H^+\cdots C\equiv N$	3.20	2.79	5	[49]

asymmetric	R_{eq}^a	R_{TS}^c	$E^{\dagger,d}$	ref
$H_2O\cdots H^+\cdots SH_2$	3.02	2.87	2.6(1.9)	[47]
$H_2CHN\cdots H^+\cdots NH_3$	2.82	2.58	5.9(3.1)	[54]

aequilibrium distance between nonhydrogen atoms.
bdifference in proton affinity between two subunits participating in H-bond
cdistance in transition state along proton transfer pathway
dtransfer barrier in going from left to right; reverse barrier for asymmetric system in parentheses

potentials, the equilibrium separation is reported as R. In the case of asymmetric single well potentials, the nature of the listing makes it clear with which of the two groups the proton's equilibrium position places it in closer association. For example, the higher proton affinity of NH_3 vs. OH_2 makes $H_3NH^+ \cdots OH_2$ the preferred arrangement. In fact, a big difference in proton attracting power such as this is one factor in a double well potential acquiring single-well character. For this reason, the difference in proton affinity, ΔPA, between groups A and B is also reported in Table 2.

In the case of the double-well potentials, Table 2 lists the $A \cdots B$ separation in both the (lower energy) minimum and in the transition state $\mathbf{f}$ that separates them. Comparison reveals the trend that R is typically considerably smaller in the transition state. That is, proton transfer is usually coupled with a drawing together of the two groups, which pull away from one another again after the transfer has passed the midpoint. The amount of this H-bond contraction is related to the height of the energy barrier, $E^\dagger$. For example, the intercarbon distance in $H_3C^- \cdots H^+ \cdots CH_3^-$ diminishes by 1.15 Å when the proton moves to the midpoint, passing through a barrier of 15 kcal/mol. In contrast, the much lower barrier in $H_3N \cdots H^+ \cdots NH_3$ is associated with an R(NN) contraction of only 0.1 Å.

Two asymmetric systems with double wells are listed on the right side of Table 2. In each case, the A and B groups have quite similar proton affinities and the double-well potential is not highly skewed. The barrier for transfer from left to right is not very different than the reverse barrier (in parentheses).

A few patterns contained in Table 2 are worth discussion. We note first that as the electronegativity of the atom decreases, there is a tendency for the transfer barrier to rise. For example, transfers between oxygen atoms in $H_2O \cdots H^+ \cdots OH_2$ or in $HO^- \cdots H^+ \cdots OH^-$ take place within a single well potential with no barrier. A barrier, albeit a small one, appears when the oxygen is replaced by nitrogen in $H_3N \cdots H^+ \cdots NH_3$ or $H_2C=N^- \cdots H^+ \cdots N=CH_2$, and a much larger one when the transfer occurs between C atoms as in $H_3C^- \cdots H^+ \cdots CH_3^-$ or $HN=CH^- \cdots H^+ \cdots CH=NH$. Similarly, the $H_2S \cdots H^+ \cdots SH_2$ system contains two minima, as opposed to the single minimum in $H_2O \cdots H^+ \cdots OH_2$.

A more subtle influence on barrier is the multiplicity of bonding in which the atom participates. The triple-bonded nitrogen in $C \equiv N^- \cdots H^+ \cdots N \equiv C$ is involved in a single-well potential whereas two minima appear when the bonding is reduced to double or single. Similarly in the case of carbon, a barrier of 5 kcal/mol separates the two minima in $N \equiv C^- \cdots H^+ \cdots C \equiv N$; this barrier rises to 12 when the bonding is of the double type in $HN=CH^- \cdots H^+ \cdots CH=NH$ and up to 15 in $H_3C^- \cdots H^+ \cdots CH_3^-$ where the C is single-bonded only.

One can connect the preceding phenomena of bonding multiplicity and electronegativity through the vehicle of orbital hybridization. A carbon atom participating in a triple bond is thought of as sp-hybridized which makes it more electronegative than the sp^3 hybridization of single bonding due to the greater proportional participation of the s-orbital, lower in energy than p.

ACKNOWLEDGEMENTS

I am indebted to my many coworkers who have contributed long hours and inspirational ideas to this project. The work has been supported financially by the National Institutes of Health (GM29391).

REFERENCES

1. A.C. Legon and D.J. Millen, Gas-phase spectroscopy and the properties of hydrogen-bonded dimers: HCN···HF as the spectroscopic prototype, Chem. Rev. 86:635 (1986).

2. K.I. Peterson, G.T. Fraser, D.D. Nelson, Jr., and W. Klemperer, Intermolecular interactions involving first row hydrides: Spectroscopic studies of complexes of HF, H_2O, NH_3, and HCN, in: "Comparison of Ab Initio Quantum Chemistry with Experiment for Small Molecules", R.J. Bartlett, ed., Reidel, New York (1985), pp. 217-244.

3. P. Herbine and T.R. Dyke, Rotational spectra and structure of the ammonia-water complex, J. Chem. Phys. 83:3768 (1985).

4. C.-C. Han and J.I. Brauman, Electron transfer competes with proton transfer in gas-phase acid-base reactions, J. Am. Chem. Soc. 110:4048 (1988).

5. E.S. Kryachko, Collective proton transfer in the (A-H...)$_\infty$ system with double-Morse symmetric potential. I. Model of proton kink defect, Chem. Phys. 143:359 (1990).

6. Y. Kashimori, T. Kikuchi, and K. Nishimoto, The solitonic mechanism for proton transport in a hydrogen bonded chain, J. Chem. Phys. 77:1904 (1982).

7. A. Godzik, An estimation of energy parameters for the soliton movement in hydrogen-bonded chains, Chem. Phys. Lett. 171:217 (1990).

8. S. Scheiner, Proton transfers in hydrogen-bonded systems. Cationic oligomers of water, J. Am. Chem. Soc. 103:315 (1981).

9. S. Scheiner, Theoretical studies of proton transfers, Acc. Chem. Res. 18:174 (1985).

10. S. Scheiner and Z. Latajka, Kinetics of proton transfer in $(H_3CH\cdot\cdot CH_3)^-$, J. Phys. Chem. 91:724 (1987).

11. E.A. Hillenbrand and S. Scheiner, Effects of molecular charge and methyl substitution on proton transfer between oxygen atoms, J. Am. Chem. Soc. 106:6266 (1984).

12. A. Szabo and N.S. Ostlund, "Modern Quantum Chemistry", Macmillan, New York (1982).

13. W. Meyer, W. Jakubetz, and P. Schuster, Correlation effects on energy curves for proton transfer. The cation $[H_5O_2]^+$, Chem. Phys. Lett. 21:97 (1973).

14. S. Scheiner and L.B. Harding, Molecular orbital study of proton transfer in $(H_3NHOH_2)^+$, J. Phys. Chem. 87:1145 (1983).

15. S. Scheiner and L.B. Harding, Proton transfers in hydrogen-bonded systems. 2. Electron correlation effects in $(N_2H_7)^+$, J. Am. Chem. Soc. 103:2169 (1981).

16. M.M. Szczesniak and S. Scheiner, Moller-Plesset treatment of electron correlation effects in $(HOHOH)^-$, J. Chem. Phys. 77:4586 (1982).

17. B.O. Roos, W.P. Kraemer, and G.H.F. Diercksen, SCF-CI studies of the equilibrium structure and the proton transfer barrier $H_3O_2^-$, Theor. Chem. Acta. 42:77 (1976).

18. Z. Latajka and S. Scheiner, Correlated proton transfer potentials. (HO-H-OH)$^-$ and (H$_2$O-H-OH$_2$)$^+$, J. Mol. Struct. (Theochem) in press.

19. R.A. Marcus, Theoretical relations among rate constants, barriers, and Bronsted slopes of chemical reactions, J. Phys. Chem. 72:891 (1968).

20. R.P. Bell, "The Tunnel Effect in Chemistry", Chapman and Hall, London (1980).

21. W.J. Albery, The Application of the Marcus Relation to Reactions in Solution, Ann. Rev. Phys. Chem., 31:227 (1980).

22. S. Scheiner and P. Redfern, Quantum mechanical test of Marcus theory. Effects of alkylation upon proton transfer, J. Phys. Chem. 90:2969 (1986).

23. S.S. Kristjansdottir and J.R. Norton, Agreement of proton transfer cross reaction rates between transition metals with those predicted by Marcus theory, J. Am. Chem. Soc. 113:4366 (1991).

24. I. Olovsson and P.-G. Jönsson, X-ray and neutron diffraction studies of hydrogen bonded systems, in: "The Hydrogen Bond: Recent Developments in Theory and Experiments", P. Schuster, G. Zundel, and C. Sandorfy, eds., North-Holland, Amsterdam (1976) p. 393.

25. E.N. Baker and R.E. Hubbard, Hydrogen bonding in globular proteins, Prog. Biophys. molec. Biol. 44:97 (1984).

26. S. Scheiner, Comparison of proton transfers in heterodimers and homodimers of NH$_3$ and OH$_2$, J. Chem. Phys. 77:4039 (1982).

27. S. Scheiner, Proton transfers in hydrogen-bonded systems. VI. Electronic redistributions in (N$_2$H$_7$)$^+$ and (O$_2$H$_5$)$^+$, J. Chem. Phys. 75:5791 (1981).

28. S. Scheiner and J.F. Nagle, Ab initio molecular orbital estimates of charge partitioning between Bjerrum and ionic defects in ice, J. Phys. Chem. 87:4267 (1983).

29. B. Brzezinski, G. Zundel, and R. Krämer, An intramolecular chain of four hydrogen bonds with proton polarizability due to collective proton motion, Chem. Phys. Lett. 157:512 (1989).

30. B. Brzezinski, H. Maciejewska, G. Zundel, and R. Krämer, Collective proton motion and proton polarizability of the hydrogen-bonded system in disubstituted protonated Mannich bases, J. Phys. Chem. 94:528 (1990).

31. S. Scheiner, P. Redfern, and M.M. Szczesniak, Effects of external ions on the energetics of proton transfers across hydrogen bonds, J. Phys. Chem. 89:262 (1985).

32. I.J. Kurnig and S. Scheiner, Additivity of the effects of external ions and dipoles upon the energetics of proton transfer, Int. J. Quantum Chem. QBS 13:71 (1986).

33. S. Scheiner, R. Wang, and L. Wang, Perturbations of proton transfer potentials caused by polar molecules, Int. J. Quantum Chem. QBS 16:211 (1989).

34. J.F. Nagle, Theory of passive proton conductance in lipid bilayers, J. Bioenerg. Biomemb. 19:413 (1987).

35. J.F. Nagle, M. Mille, and H.J. Morowitz, Theory of hydrogen bonded chains in bioenergetics, J. Chem. Phys. 72:3959 (1980).

36. H.J. Morowitz, Proton semiconductors and energy transduction in biological systems, Am. J. Physiol. 235:R99 (1978).

37. L. Glasser, Proton conduction and injection in solids, Chem. Rev. 75:21 (1975).

38. E. Pines and D. Huppert, Kinetics of proton transfer in ice via the pH-jump method: Evaluation of the proton diffusion rate in polycrystalline doped ice, Chem. Phys. Lett. 116:295 (1985).

39. E.A. Hillenbrand and S. Scheiner, unpublished results.

40. J. Fritsch, G. Zundel, A. Hayd, and M. Maurer, Proton polarizability of hydrogen-bonded chains: An ab initio SCF calculation with a model related to the conducting system in bacteriorhodopsin, Chem. Phys. Lett. 107:65 (1984).

41. M. Eckert and G. Zundel, Energy surfaces and proton polarizability of hydrogen-bonded chains: An ab initio treatment with respect to the charge conduction in biological systems, J. Phys. Chem. 92:7016 (1988).

42. S.I. Nagaoka, N. Hirota, T. Matsushita, and K. Nishimoto, An ab initio calculation on proton transfer in the benzoic acid dimer, Chem. Phys. Lett. 92:498 (1982).

43. T. Yamabe, K. Yamashita, M. Kaminoyama, M. Koizumi, A. Tachibana, and K. Fukui, Dynamics of double proton exchange in the formamidine-water system, J. Phys. Chem. 88:1459 (1984).

44. M.T. Nguyen, A.F. Hegarty, and T.-K. Ha, Ab initio study of the hydration of carbon dioxide: Additional comments based on refined calculations, J. Mol. Struct. (Theochem) 150:319 (1987).

45. D. Heidrich, M. Rückert, and H.-J. Köhler, The course of proton exchange reactions: A theoretical study, Chem. Phys. Lett. 136:13 (1987).

46. S. Scheiner, M.M. Szczesniak, and L.D. Bigham, Ab initio study of proton transfers including effects of electron correlation, Int. J. Quantum Chem. 23:739 (1983).

47. S. Scheiner, Proton transfers between first and second-row atoms: $(H_2OHSH_2)^+$ and $(H_3NHSH_2)^+$, J. Chem. Phys. 80:1982 (1984).

48. S. Scheiner and E.A. Hillenbrand, Comparison between proton transfers involving carbonyl and hydroxyl oxygens, J. Phys. Chem. 89:3053 (1985).

49. S.M. Cybulski and S. Scheiner, Hydrogen bonding and proton transfers involving triply bonded atoms. HC≡CH and HC≡N, J. Am. Chem. Soc. 109:4199 (1987).

50. E.A. Hillenbrand and S. Scheiner, Analysis of the principles governing proton-transfer reactions. Carboxyl group, J. Am. Chem. Soc. 108:7178 (1986).

51. S.M. Cybulski and S. Scheiner, Hydrogen bonding and proton transfers involving the carboxylate group, J. Am. Chem. Soc. 111:23 (1989).

52. S. Scheiner and L. Wang, to be published.

53. S. Scheiner and L.D. Bigham, Comparison of proton transfers in $(S_2H_5)^+$ and $(O_2H_5)^+$, J. Chem. Phys. 82:3316 (1985).

54. E.A. Hillenbrand and S. Scheiner, Analysis of the principles governing proton-transfer reactions. Comparison of the imine and amine groups, J. Am. Chem. Soc. 107:7690 (1985).

55. Z. Latajka and S. Scheiner, Energetics of proton transfer between carbon atoms $(H_3CH--CH_3)^-$, Int. J. Quantum Chem. 29:285 (1986).

CONSTRUCTION OF THE MODEL STRONG HYDROGEN BOND

POTENTIAL ENERGY SURFACE AND PROTON TRANSFER DYNAMICS

N.D.Sokolov, M.V.Vener

Semenov Institute of Chemical Physics Ac. Sci. USSR
Moscow, USSR
Lumumba University of the People Friendship, Moscow, USSR

INTRODUCTION

Results of theoretical treatment of the proton transfer dynamics depend strongly on the potential energy surface (PES) used for the hydrogen bonded system description. The quantum chemical calculations[1-15] of PES are however still very approximate especially when a condensed phase is considered. The environmental effect on the PES shape can be very strong whereas the quantum chemical calculation deals usually with isolated systems. Practically some simple analytical function of nuclear coordinates, having a double well, is preferred for proton transfer treatment.

The problem arisen in this connection is how to choose the values of fitting parameters entering in this potential. In most investigations they are defined from the experiment to be interpreted. Unfortunately the potentials determined in such a way are as a rule unable to describe properly other experimental regularities and data (spectroscopic, structural) known for the same compounds. Moreover an agreement between the theory and experiment disappears if the fitting parameter values are taken e.g. from spectroscopic data[16].

In the second part of this communication slightly improved two–dimensional model PES, constructed earlier for the quasi–symmetric OHO fragment[17], will be considered in short. A most characteristic feature of this potential is it takes into account strong coupling of the proton movement with the O...O intermolecular vibration.

Using the PES an approximate theory of the proton transfer dynamics will be developed. In the third part the related wave equation will be solved in adiabatic approximation. The fourth part is devoted to the proton transfer as such. Its main aim is to study the mechanism of the proton transfer reaction promoted by excitation of the low–frequency mode (the O...O vibration) being coupled strongly with the proton movement. The vibrationally assisted proton tunnelling (as well as the H–atom) was studied experimentally by many authors[18-24]. Some theoretical models were suggested for interpretation of the proton transfer processes in which heavy atom vibrations are participating[23,25-31]. We will use the model in which the proton transfer occurs by quantum jumps between the vibrational levels belonging to both low

and high frequency modes. A similar model, including only the proton vibrational levels, was investigated by several authors[32-38]. A coupling with the
low-frequency vibrations was considered in[23,28-31]. In contrast with[28,29,31]
and similarly to[23,30] a dynamic asymmetry of the proton double-well potential
will be taken into account. We will show that the probability of the proton
tunnelling, occurring from every vibrational level of the "slow" subsystem,
increases with its quantum number at a definite condition. The interpretation
and discussion of the results obtained will be given in the fifth part.

POTENTIAL ENERGY SURFACE DESCRIPTION

Consider crystals with almost linear OHO fragments the proton moving in a
field with approximately central symmetry. It is perturbed slightly by the

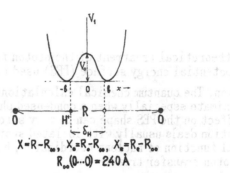

Fig. 1.   Designations used in the paper. R_0 – the O...O
interatomic distance corresponded to the minimum
of the PES, R_e – a true equilibrium value, 2b – a
distance between the points of minimum proton potential energy, δ_H – a distance between the points of
maximum proton density in the two wells.

environment action resulting in dynamic asymmetry.

The designation used are shown in Fig. 1 (for symmetric case).

The following empirical regularities known for such crystals were used
for construction of the PES : the proton stretching frequency (ν_H) dependence
on the equilibrium O...O separation $X_e = R_e - R_{00}$ ($R_{00} = 2.40$ Å); the
nonmonotonious dependence of the isotope frequency ratio $\gamma = \nu_H/\nu_D$ on ν_H ; the
Ubbelohde effect ($\Delta R = \Delta R_e^D - \Delta R_e^H \neq 0$); sharp decrease of the isotope
difference $\Delta \delta = \delta_D - \delta_H$ with X_e, δ_H (δ_D) being a distance between the points of
maximum proton (deuteron) density in the two potential wells.

The experimental point scattering in the respective figures[17] is rather large. It means that these properties depend not only on one variable (X_e) but also on some other characteristics of the related compounds: the chemical composition, crystal structure and especially bending of the OHO fragment. Nevertheless, we will suppose that X_e is its main characteristic. Accordingly, similarly to the Ref.[17], we will suggest that the OHO fragment potential energy is a function only of two coordinates: x and X. The obtained "empirical" potential energy function U(x,X) is in fact a PES for some typical compound with strong hydrogen bond specifying by a definite O...O equilibrium interatomic distance X_e. This PES automatically allows for the environment effect and evolves correctly with X_e: it has no potential barrier at $X_e = 0$, the barrier and two wells appear at $X_e > 0$ and the barrier increases with the value of X_e. Then in a symmetric case at high barrier a tunnel splitting becomes small and at some X_e practically disappears. At this X value the wells become to be isolated and the hydrogen bond becomes identical with the asymmetric OH...O hydrogen bonds of a mediated strength. Accordingly, the functions, expressing the stretching frequency ν_H and ν_D dependence on X_e for strong hydrogen bonds, are continuously joined with respective functions for the intermediate (asymmetric) hydrogen bonds OH...O.

The analysis shows that the above conditions can be satisfied only if the potential barrier height increases very steeply (exponentially) with X_e. In particular such a dependence is the only one permitted to reproduce adequately the experimental regularities for frequency ν_H and for isotopic frequency ratio $\gamma = \nu_H/\nu_D$. Due to such a steepness the PES shape is very sensible to the O...O distance variation.

The dynamic asymmetry is taken into account in the model PES described below. The latter differs slightly from the PES suggested in the Ref.[17] by the parameter values also. As a whole the improved PES reproduces the above regularities somewhat better than the previous one.

The suggested PES U(x,X) contains three terms:

$$U(x,X) = V_1(x,X) + V_2(X) + V_3(x,X) \tag{1}$$

where $V_1(x,X)$ is a proton potential depending on both x and X, $V_2(X)$ is the O...O interaction energy and $V_3(x,X)$ takes into account the dynamic asymmetry.

The proton potential V_1 consists of three parts. The central part represents the potential barrier of a width 2b and a height V_0, the two side parts describe the proton repulsion from the left and the right oxygen atoms:

$$V_1(x,X) = V_0(X)\cos^2(\pi x/2b(X)), \ |x| \leq b, \tag{2}$$

$$V_1(x,X) = (1/2)k(X)(|x|-b)^2, \ |x| \geq b. \tag{3}$$

The functions $V_0(X)$, b(X) and k(X) are chosen as:

Table 1. The characteristic values of PES parameters.

a/cm^{-1}	$\lambda/Å^{-1}$	b_0	$k_0/cm^{-1}Å^{-2}$	$\lambda_1/cm^{-1}Å^{-3}$	$k_\sigma/cm^{-1}Å^{-2}$	$c/cm^{-1}Å^{-2}$
243.0	19.5	2.5	$3.52\ 10^5$	$7.36\ 10^5$	$6.45\ 10^4$	310.0

$$V_0(X) = a\,(\exp(\lambda X) - 1) \tag{4}$$

$$b(X) = b_0 X \tag{5}$$

$$k(X) = k_0 + \lambda_1 X \tag{6}$$

The parameter a, λ, b_0, k_0 and λ_1 values are listed in the Table 1.

The function $V_2(X)$ can be approximated by the harmonic potential

$$V_2(X) = (1/2)k_\sigma(X-X_0)^2; \tag{7}$$

where X_0 ($\neq X_e$, see below) corresponds to the minimum of the PES and is defined by the condition $(\partial U/\partial X)_{|x|=b} = 0$. The k_σ value is given in the Table 1.

The dynamical asymmetry function $V_3(x,X)$, introducing a difference between depths of the two wells, is approximated as

$$V_3(x,X) = cxX \tag{8}$$

This expresion is supposed to be valid in the range $0 \leq X \leq 0.18$ Å. At $X > 0.18$ Å V_3 is assumed to be independent on X. The c value can be defined only very approximately. For sufficiently high barrier the tunnel splitting is small and dynamical asymmetry alone determines the two lowest energy level separation. This quantity, designated by 2A, is defined experimentally. Usually it varies from 30 to 80 cm^{-1} and higher[40-41]. We assume that its typical value for compounds with $X_e > 0.18$ Å is 2A = 50 cm^{-1}. Then using the equality $2A \simeq 2V_3(x_e,X_e)$ one obtains $c \simeq 3.1\ 10^2$ cm^{-1}Å^{-2}.

Note the PES described by the equations (2) – (8) provides with the barrier lowering when the reaction path involves a proton shift along x and a simultaneous O...O distance reduction. A similar effect was found in many quantum chemical computations also[2-14,29,31].

SOLUTION OF THE WAVE EQUATION

The two dimensional Schroedinger equation with potential (1) was solved in the adiabatic approximation. The full wave function was written in the form:

$$\Phi_{vm}(x,X) = \varphi_v(x,X)\psi_{vm}(X),\tag{9}$$

where $\varphi_v(x,X)$ and $\psi_{vm}(X)$ are the wave functions of the "fast" subsystem (proton or deuter on) and of the O...O "slow" subsystem respectively; v and m are the respective quantum numbers.

The Hamiltonian for the "fast" subsystem is

$$\hat{H}_1 = -(\hbar^2/2m^*)(\partial^2/\partial x^2) + V_1(x,X) + V_3(x,X)\tag{10}$$

where m^* is the proton mass, X being considered as a parameter. The respective wave equation

$$\hat{H}_1\varphi_v(x,X) = \mathscr{E}_v(X)\varphi_v(x,X)$$

was solved by numerical method described in[42]. The eigenvalues and eigenfunctions were calculated for v = 0 and v = 1 in the range $-1.2\ \text{Å} \le x \le 1.2\ \text{Å}$ and $0 \le X \le 0.18\ \text{Å}$. At $X = X^*$ ($X^* \simeq 0.25\ \text{Å}$) the function $\mathscr{E}_0(X)$ was smoothly joined with the proton adiabatic energy $\mathscr{E}_0^*(X)$ of the asymmetric OH...O hydrogen bond with a single well[39].

Knowing $\mathscr{E}_v(X)$ the true O...O equilibrium distance $X_e^{(v)}$ can be found from the equation

$$dU_v/dX = 0,\tag{11}$$

where according to (1), (7) and (10)

$$U_v(X) = \mathscr{E}_v(X) + (1/2)k_\sigma(X-X_0)^2.\tag{12}$$

When $d\mathscr{E}_v/dX > 0$ the $X_e^{(v)}$ value is easily seen to be less than X_0. For the ground state the index "v" at X_e will be omitted later on.

In Fig. 2 and 3 the functions $U_v(X)$ (v=0,1) are depicted at $X_e = 0.12\ \text{Å}$ and $X_e = 0.24\ \text{Å}$ respectively. Some anharmonicity of an "anti–Morse type" is observable especially for $X_e = 0.24\ \text{Å}$ for v=0 (Fig. 3) whereas at $X_e = 0.12\ \text{Å}$ a "Morse-type" anharmonicity is seen in Fig. 2. Equilibrium values X_e and $X_e^{(1)}$ are different ($X_e^{(1)} > X_e$) especially in the Fig. 2.

The wave equation for the "slow" subsystem (O...O vibrations) is

$$\hat{H}_2\psi_{v,m} = E_m\psi_{v,m}, \text{ where } \hat{H}_2 = (\hbar^2/2M)(d^2/dx^2) + U_v(X)\tag{13}$$

It was solved numerically at $X_e = 0.24$ Å and $X_e = 0.12$ Å for $v=0$ and $v=1$ with the potential (12) and the reduced mass M being 100 amu. The Numerov's[43] method with 600 gridpoints was used. The calculated low-frequency vibrational levels for $v=0$ and $v=1$ are shown in the Figs. 2 and 3. The solid lines ($v=0$) correspond to the left well (quantum numbers are designated by m), the broken ones ($v=1$) to the right well (quantum numbers n).

PROTON TRANSFER DYNAMICS

The quantum mechanical model of the proton population thermal relaxation in the double-minimum well in condensed phase has been suggested originally by Pschenichnov and Sokolov[32,33]. These authors supposed that the proton transfer is brought about by the step wise transitions between the vibrational levels in the double minimum potential under the random force action from environment. The master equation was solved to obtain the relaxation time. Later, several other authors have used a similar model[34-38,23,30]. Unfortunately, the used PES has never been able to describe the main empirical "static" regularities which are observed for strong hydrogen bonds.

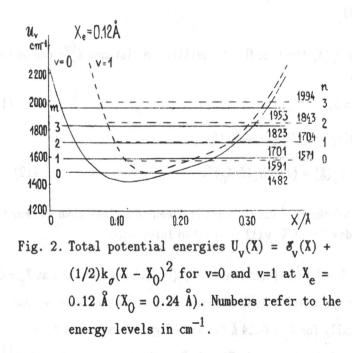

Fig. 2. Total potential energies $U_v(X) = \mathcal{E}_v(X) + (1/2)k_\sigma(X - X_0)^2$ for $v=0$ and $v=1$ at $X_e = 0.12$ Å ($X_0 = 0.24$ Å). Numbers refer to the energy levels in cm^{-1}.

One of the important results formulated in[32,33] was the resonance condition: the "active" vibrational frequency in the hydrogen bonded system has to differ not strongly from the phonon frequency of environment. For one dimensional proton movement in the double-well quasi-symmetric potential this condition is not fulfilled. It becomes clear that some low-frequency mode, coupled strongly with the proton vibration, has to be taken into account. In our model this is the O...O symmetric stretching vibration.

As was mentioned in the introduction the proton transfers assisted by the low-frequency excitation were studied experimentally and theoretically by many authors[18-31]. Numerous findings were interpreted theoretically but some important problems remain: by what mechanism the low-frequency level excitation does assist the proton tunnelling in condensed media? Under which conditions can this assistance be expected? We will try to answer to these questions using rather simplified model.

Proton transfer from one well to another is physically meaningful only when the proton is localized strongly in both the initial and the final states. This conditions is fulfilled for sufficiently large X_e (~0.24 Å). At shorter X_e this condition is not satisfied. For instance, at $X_e = 0.12$ Å $V_0 \simeq 2,3 \cdot 10^3$ cm^{-1},

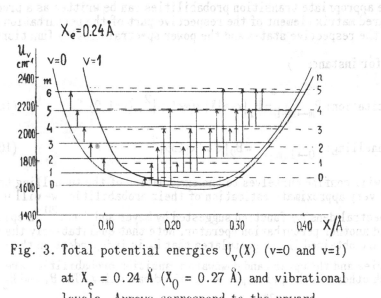

Fig. 3. Total potential energies $U_v(X)$ (v=0 and v=1) at $X_e = 0.24$ Å ($X_0 = 0.27$ Å) and vibrational levels. Arrows correspond to the upward transitions $|m\rangle \rightarrow |m+1\rangle$ in the left well (v=0) and to the tunnelling transitions $|0m\rangle \rightarrow |1n\rangle$ from the left to the right well.

the asymmetry is $2A \simeq 20$ cm^{-1} and the tunnelling splitting is $2J \simeq 85$ cm^{-1}; with this only about 60% of the proton density is localized in the left well in the ground state (v=0). Hence, the transition v=0 $\rightarrow$ v=1 is not a complete proton tunnelling but only a redistribution of the proton density over two wells. This effect can be in principle observed by some physical experiments (NMR, dielectric relaxation etc.). Both cases can be treated uniformly.

The perturbation, inducing the quantum jumps in a system "hydrogen bond plus heat bath", is supposed to be of the form:

$$V_4(x,X,y) = \sum_k f_1(t)\Delta X \Delta y_k + \sum_k f_2(t)\Delta x \Delta y_k \qquad (14)$$

where Δx, ΔX and Δy_k are the small deviations from equilibrium positions, y_k being coordinate of the k-th "active" oscillator of the heat bath, $f_1(t)$ and

$f_2(t)$ are random functions of time. Two kinds of transitions exist in the system considered:

1.   The "slow"–subsystem excitations:

$|0m> \rightleftharpoons |0(m+1)>$ in the left well,

$|1n> \rightleftharpoons |1(n+1)>$ in the right well.

2.   The proton tunnelling transitions, accompanied by the heavy atoms vibrational state changes

$|0m> \rightleftharpoons |1n>$

Here $|vm>$ (v=0,1) designates the function (9).

The appropriate transition probabilities can be written as a product of the squared matrix element of the respective part of the perturbation operator between the respective states and the power spectral density functions B_1 and

B_2 (see for instance[36])

$$\text{excitation: } P_{m \rightarrow m+1} = B_1 |<vm|X|v(m+1)>|^2, \; (v=1,2) \qquad (15)$$

$$\text{tunnelling: } P_{0 \rightarrow 1, m \rightarrow n} = B_2 |<0m|x|1n>|^2 \qquad (16)$$

We will confine ourselves to the examination of the tunnelling transitions. For very approximate estimation of their probabilities we will use the power spectral density function suggested by Meyer and Ernst[30] in spite of that they used another perturbation operator. Note that qualitatively the same results are obtained if one postulates that B_2 is independent on the transition frequencies and the upward and downward transition probabilities are connected with each other via the principle of detailed balancing. The B_1 and B_2 functions suggested in the Ref.[30] are consistent with this principle.

Using (9), the tunnelling matrix element in (16) can be written as:

$$<0m|x|1n> = \int dX \, \psi_m(X) \, \psi_n(X) \, L(X) \qquad (17)$$

where

$$L(X) = \int dx \, \varphi_0(x,X) x \, \varphi_1(x,X) \qquad (18)$$

Calculate the squared matrix elements $|<0m|x|1n>|^2$ and use the equation (3.38) and Fig. 5 in the Ref.[30] for the estimation of B_2. The transition probabilities obtained in such a way are summarized over all n for different m. The resulting quantities

$$C_m = \sum_n P_{0 \rightarrow 1, m \rightarrow n} \qquad (19)$$

are total probability   tunnelling transitions from different levels m of the 0...0 vibration.

The results are depicted in Fig. 4 for $X_e = 0.24$ Å for two temperatures: T =

300 K and T = 40 K. The "slow" subsystem excitation is seen to promote the proton tunnelling transfer. At high temperature the excitation of the levels 1 and 2 is particularly effective. When T = 40 K the upward transitions become to be less probable and tunnelling is provided chiefly with the downward transitions alone. A rise of the tunnelling transition probability from second level is seen in Fig. 4. The broken line in this figure will be discussed below.

With $X_e = 0.12$ Å the quantity C_m can be considered as a total probability of the proton density redistribution due to transitions $|0m> \rightarrow |1n>$. Fig. 5 shows that C_1 and C_2 are remarkably less than C_0. This means that no vibrational promotion of the proton density redistribution can be expected for very short hydrogen bonds ($X_e \cong 0.12$ Å).

INTERPRETATION AND DISCUSSION

To interpret the above results consider the tunnelling matrix element (17). It depends on the overlap of the three functions: $\psi_m(X)$, $\psi_n(X)$ and $L(X)$. As examination shows an overlap of the product $\rho_{mn}(X) = \psi_m(X) \psi_n(X)$ with $L(X)$ is most important, varying strongly with m, n and X_e. This is illustrated in Fig. 6. One can see that $L(X)$, independent on m and n, falls very steeply at $X > 0.12$ Å and almost disappears at $X_e = X^{**} \simeq 0.16$ Å. At $X_e = 0.24$ Å for m=n=0 the overlap of the functions $L(X)$ and $\rho_{00}(X)$ is small since $\rho_{00}(X)$ is rather strongly localized near the point $X_e > X^{**}$ (Fig. 6). Therefore, the squared matrix element

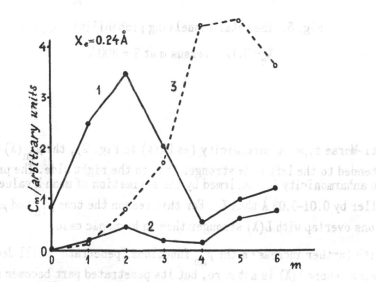

Fig. 4. The total tunnelling probabilities $(C_m = \sum_n P_{0 \rightarrow 1, m \rightarrow n})$ for $X_e = 0.24$ Å versus m for T = 300 K (1) and T = 40 K (2). The broken line (3) refers to the harmonic potential $U^h(X)$.

$|<0|L|0>|^2$ is also small ($0.14 \ 10^{-3} \ \text{Å}^2$). For m=1 and m=2 the functions $\rho_{1n}(X)$ and $\rho_{2n}(X)$ become to be more extended and theirs overlap with L(X) increase. This is shown in Fig. 6 for ρ_{22} ($|<02|L|12>|^2 = 3.7 \ 10^{-3} \ \text{Å}^2$).

If the function $U_v(X)$ were harmonic ($U_v^h(X)$), $\rho_{22}(X)$ would had the shape shown by broken line in the Fig. 6 (ρ_{22}^h). Its overlap with L(X) is rather small at low m ($|<2|L|2>|^2 = 0.9 \ 10^{-3} \ \text{Å}^2$) but increases monotonously with the increase of m to m=5 (see the broken line in Fig.4). If the $U_v(X)$ potential has

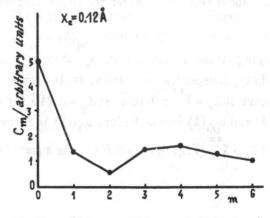

Fig. 5. The total tunnelling probabilities C_m for

$X_e = 0.12 \ \text{Å}$ versus m at T = 300 K.

an "anti-Morse type" anharmonicity (as $U_0(X)$ in Fig. 2), the $\rho_{2n}(X)$ functions are extended to the left side stronger than to the right side. The presence of such an anharmonicity is confirmed by the estimation of mean X value $<m|X|m>$: it is smaller by 0.01–0.03 Å than X_e. For this reason the true ρ_{1n} and ρ_{2n} functions overlap with L(X) stronger than in harmonic case.

With further increase m the ρ_{mn} functions "penetrate" still deeper into the region, where L(X) is not zero, but its penetrated part becomes smaller owing to the wave function normalizing. This explains why C_m falls at m > 2.

This factor also effects on the harmonic ρ_{mn}^h function though only at higher m (m > 5, Fig. 4).

At shorter O...O distances ($X \simeq 0.12$ Å) the L(X) function (18) has the same shape as that for large X_e. If X_e is less than X^{**} the behavior of the matrix element $\langle 0m|L|1n \rangle$ differs appreciably from that when $X_e > X^{**}$. Indeed, as Fig. 7 shows, the overlap of $L(X)$ with $\rho_{00}(X)$ is comparatively large, since the ρ_{00} is localized near the point X^{**}, whereas the ρ_{11} function is extended strongly to the right side, where $L = 0$. This extension to the right side is amplified by the "Morse type" anharmonicity of the potential $U_v(X)$ (Fig. 2).

Hence, for compounds with double-minimum potential and with X_e significantly larger than the critical distance X^{**} the tunnelling probability is small for the proton transition from the zero level of the low-frequency vibration (m=0) and increases with the increase of m. On the contrary, for compounds with $X < X^{**}$, the tunnelling probability from the zero level can be expected to be larger than that from the higher levels.

The result demonstrated in Fig. 4 resembles the Fuke and Kaya effect[21]. These authors have found that the intermolecular stretching vibration, being excited selectively, promotes the proton tunnel transfer in the azaindole dimer, its rate increasing by several times with excitation at low temperature. A similar effect was observed also by Sekiya, Nagashima and Nishimura[24]: the tunnelling splitting in the proton double well potential increases under the excitation of some vibrational modes in tropolone. Both experiments were performed in a gas phase (in a supersonic jet) where environment played no part and the protonic potential is symmetric. Proper interpretations under

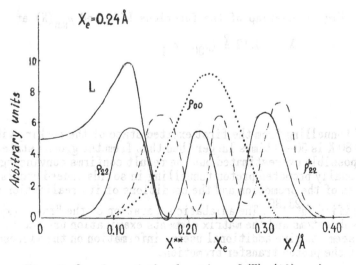

Fig. 6. Overlap of the functions $L(X)$ (18) and $\rho_{mn}(X) = \psi_m(X)\psi_n(X)$ at $X_e = 0.24$ Å (ρ_{00}, ρ_{22}). The broken line ρ_{22}^h refers to the harmonic potential $U_v^h(X)$.

such a condition were given in[28,31]. Our study deals with the same phenomenon in condensed phase.

Any direct measurements of the proton transfer promoted by vibrational excitation in condensed media are unknown to the authors. On the contrary, the hydrogen atom transfer assisted by low–frequency vibrational excitation was directly studied by many authors (see Ref.[22,26] and references there in). Theoretical study of the phenomenon with proton in condensed media can be found in[23,30]. Meyer and Ernst[30], in interpreting the experimental data for the NMR relaxation time for the benzoic acid crystal with H and D isotopes using a quantum jump excitation model, concluded that an adequate interpretation is possible only by applying a two–dimensional model PES. They have found that the pro-

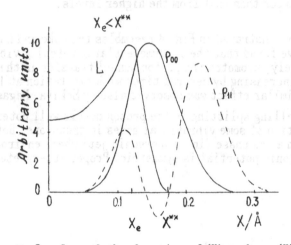

Fig. 7. Overlap of the functions L(X) and ρ_{mn}(X) at

$$X_e = 0.12 \text{ Å } (\rho_{00}, \rho_{11}).$$

bability of tunnelling from the first excited state of the rocking vibration at T = 40 K and 80 K is 50–60 times larger than that from the ground state. These values are possibly overestimated but the result confirms convincingly that the vibrationally promoted proton tunnelling in solids indeed occurs. A detailed mechanism of the promotion and the conditions of its realization were however not clarified in[23,30]. The adiabatic separation of the "fast" (x) and "slow" (X) subsystems and the matrix elements examination used above in our treatment, seems to give additional useful information on the microscopic mechanism of the proton transfer promotion.

In the model, suggested by Shida, Barbara and Almlof[29,31] a detour reaction path is considered along which the potential barrier is, as has been mentioned above, much lower than for the straight path along x. Similarly to[28] the tunnelling splitting and hence the tunnelling probability are increased strongly with vibrational excitation of a heavy atom mode. The potential barrier lowering along the detour path (with the <m|X|m> decrease) is indirectly taken into account by our model also: the "anti-Morse type" anharmonicity plays a significant role in the above proton transfer mechanism. Contrary to our model the quasi-classical model suggested in[29,31] seems to miss a possibility of the proton transfer promotion in the case of harmonic $O...O$ vibrations.

It should be noted that the proton tunnelling probability dependence on m is very sensitive to the shape of PES (in particular to the $V_0(X)$ function shape). For instance, the decrease of the parameter λ in (4) on 2.5% (from 19.5 to 19.0) results in an increase of a total probability transition from the m=2 and m=3 levels by three to four times. This demonstrates once again that our results are mainly qualitative.

Concluding our discussion we would like to stress the following. In strong hydrogen bonded systems a proton tunnelling, occurred from excited states of the "slow" subsystem, happens by several ways simultaneously. The mechanism of such tunnelling transitions is in some respect similar to the mechanism of an appearance of the molecular vibronic spectra. In both cases the transitions, induced in the "fast" subsystem, gives rise to the low-frequency transitions. This effect can be conveniently described with adiabatic approximation, examining the respective matrix elements. A difference between the two cases is that in the proton transfer problem a specificity of the double minimum potential makes, as a rule, the Condon approximation inapplicable due to a sharp fall of the $L(X)$ function. In the vibronic spectra description this approximation is known to be acceptable in most cases.

REFERENCES

1. E. Ady, and J. Brickman, Non-empirical calculations of the intramolecular rearrangement in the dimer of formic acid, Chem.Phys. Lett. 11:302 (1971).
2. K. Loth, F. Graf, and Hs.H. Gunthart, Chemical exchange dynamics and structure of intramolecular $O...H...O$ bridges: an ESR and INDO study of 2-hydroxi- and 2,6-dihydroxy-phenoxyl, Chem.Phys. 13:95 (1976).
3. G. Karlstrom, H. Wennerstrom, B. Jonsson, S. Forsen, J. Almlof, and B. Roos, An intramolecular H bond. *Ab initio* MO calculations of the enol toutomer of malonaldialdehyde, J.Am.Chem.Soc. 97:4188 (1975).
4. J.S. Del Bene, and W.S. Kochenour, Molecular orbital theory of the hydrogen bond. XV. Ring closure and proton transfer in formic acid dimer and β-hydroxyacrolein, J.Am.Chem.Soc. 98:2041 (1976).
5. S. Scheiner, Proton transfer in H bonded systems, J.Am.Chem.Soc. 103:315 (1981).
6. A. Agresti, M. Bacci, and A. Ranfagni, How reliable are calculations on proton tunneling? Chem.Phys.Lett. 79:100 (1981).
7. F. Graf, R. Meyer, T.-K. Ha, and R.R. Ernst, Dynamics of H bond exchange in carboxyle acid dimers, J.Chem.Phys. 75:2914 (1981).
8. S. Scheiner, Proton transfer in hydrogen bonded systems, J.Chem.Phys. 77:4039 (1982).
9. S. Nagaoka, N. Hirota, T. Matsushita, and K. Nishimoto, *Ab initio* calculation on proton transfer in the benzoic acid dimer, Chem.Phys. Lett. 92:498 (1982).
10. J. Bicerano, H.F. Schaefer III, and W.H. Miller, Structure and tunnelling dynamics of malonaldehyde. A theoretical study, J.Am.Chem.Soc. 105:2550 (1983).
11. S. Scheiner, Theoretical study of proton transfers, Acc.Chem.Res. 18:174 (1985).
12. E.A. Hillenbrand, and S. Scheiner, Analysis of the principles governing proton-transfer reactions. Carboxyl group, J.Am.Chem. Soc. 108:7178 (1986).
13. S. Scheiner, P. Redfern, and E.A Hillenbrand, Factors influencing proton

positions in biomolecules, Inter.J.Quant. Chem. 29:817 (1986).

14. T. Carrington Jr., and W.H. Miller, Reaction surface description of inter-molecular H atom transfer in malonaldehyde, J.Chem.Phys. 84:4364 (1986).

15. S. Scheiner, Relationship between the angular characteristics of a hydro-gen bond and the energetic of proton transfer occurring within, J. Mol. Struct. 177:79 (1988).

16. W. Siebrand, T.A. Wildman, and M.Z. Zgierski, Temperature dependence of hydrogen tunnelling rate constants, Chem.Phys.Lett. 98:108 (1983).

17. N.D. Sokolov, M.V. Vener, and V.A. Savel'ev, Tentative study of strong hydrogen bond dynamics. Part II. Vibrational frequency considerations, J. Mol. Struct. 222:365 (1990).

18. M.A. El-Bayoumi, and P. Avoureis, Dynamics of DPTR in the excited state of 7-azaindole H bonded dimer. A time-resolved fluorescence study, J.Chem.Phys. 62:2499 (1975).

19. A. Sarai, Dynamics of proton migration in free base porphines, J.Chem.Phys. 76:5554 (1982).

20. P.F. Barbara, Tunnelling mediated by heavy atoms large amplitude motion, in:"Tunnelling", J. Jortner and B. Pullman, eds, D. Reidel Publ. Co., Dortrect (1986).

21. K. Fuke, and K. Kaya, Dynamic of double-proton-transfer reaction in the excited mode H-bonded base pairs, J.Phys.Chem. 93:614 (1989).

22. B. Prass, J.P. Colpa, and D. Stehlik, Intermolecular H-tunnelling in solid state. Photoreaction promoted by distinct low-energy nuclear fluctuation mode, Chem.Phys. 136:187 (1989).

23. A. Stockli, B.H. Meier, R. Keis, R. Meyer, and R.R. Ernst, Hydrogen bond dynamics in isotopically substituted benzoic acid dimers, J.Chem.Phys. 93:1502 (1990).

24. H. Sekiya, Y. Nagashima, and Y. Nishimura, Electronic spectra of jet-cooled tropolone. Effect of the vibrational excitation on the proton tunnelling dynamics, J.Chem.Phys. 92:5761 (1990).

25. M.Ya. Ovchinnikova, The tunnelling mechanism of the low temperature H-exchange reactions, Chem.Phys. 36:85 (1979).

26. W. Siebrand, T.A. Wildman and M.Z. Zgierski, Golden rule treatment of hydrogen transfer reactions . 2. Applications, J.Am.Chem.Soc. 106:4089 (1984).

27. Z. Smedarchina, W. Siebrand, and T. Wildman, Intramolecular tunnelling exchange of the inner hydrogen atoms in free-base porphyrines, Chem.Phys.Lett. 143:395 (1988).

28. N.Sato, and S. Iwata, Promotion of the proton transfer reaction by the intermolecular stretching mode, J.Chem.Phys. 89:2932 (1988).

29. N. Shida, P.F. Barbara, and J.E. Almloff, A theoretical study of multidi-mensional nuclear tunnelling in malonaldehyde, J.Chem.Phys. 97:4061 (1989).

30. R. Meyer, and R.R. Ernst, Transitions induced in a double minimum system by interaction with quantum mechanical heat bath, J.Chem.Phys. 93:5518 (1990).

31. N. Shida, P.F. Barbara, and J.E. Almlof, A reaction surface Hamiltonian treatment of the double proton transfer of formic acid dimer, J.Chem.Phys. 94:3633 (1991).

32. E.A. Pchenichnov, and N.D. Sokolov, Intermolecular proton transfer in solution, Doklady Akademii Nauk SSSR, 159:174 (1964) (in Russian).

33. E.A. Pshenichnov and N.D. Sokolov, Effect of the hydrogen bond and the solvent on kinetics and mechanism of intermolecular proton transfer, Intern.J.Quant.Chem. 1:855 (1967).

34. J. Brikmann, and H. Zimmermann, Uber der Tunneleffect des Protons im Dop-pelminimumpotential von Wasserstoffbruckenbindungen, Ber. Buns.phys.Chem. 71:160 (1967).

35. R.G. Carbonell, and M. Kostin, Quantum mechanics of transfer reactions, J.Chem.Phys. 60:2047 (1974).

36. P.E. Parris, and R. Silbey, Low temperature tunnelling dynamics in conden-sed media, J.Chem.Phys. 83:5619 (1985).

37. P. Banacky, Quantum–statistical theory of proton transfer reactions. Dynamics of proton motion in an asymmetric double–well potential interacting with a heat bath, Chem.Phys. 109:307 (1986).

38. R. Meyer, and R.R. Ernst, Hydrogen transfer in double minimum potential: Kinetic properties derived from quantum dynamics, J.Chem.Phys. 86:784 (1987).

39. N.D. Sokolov, and V.A. Savel'ev, Dynamics of the hydrogen bond: two–dimensional model and isotope effects, Chem.Phys. 22:383 (1977).

40. S. Nagaoka, T. Terao, F. Imashiro, A. Saika, N. Hirota, and S. Hayashi, An NMR relaxation study on the proton transfer in the hydrogen bonded carboxylic acid dimers, J.Chem.Phys. 79:4694 (1983).

41. J.L. Skinner, and H.P. Trammsdorf, Proton transfer in benzoic acid crystalls: A chemical spin–bozon problem. Theoretical analysis of NMR, INC and optical experiments, J.Chem.Phys. 89:897 (1988).

42. N.D. Sokolov, M.V. Vener, and V.A. Savel'ev, Tentative study of strong hydrogen bond dynamics. Part I. Geometric isotope effects, J.Mol.Struct. 177:93 (1988).

43. B.R. Johnson, New numerical methods applied to solving the one–dimensional eigenvalue problem, J.Chem.Phys. 67:4086 (1977).

THERMAL GENERATION AND MOBILITY OF CHARGE CARRIERS
IN COLLECTIVE PROTON TRANSPORT IN HYDROGEN-BONDED CHAINS

M. Peyrard, R. Boesch and I. Kourakis

Physique non linéaire: Ondes et Structures Cohérentes
Faculté des Sciences, 6 blvd Gabriel, 21000 Dijon, France.

I. INTRODUCTION

The transport of protons in hydrogen-bonded systems is a long standing problem which has not yet obtained a satisfactorily theoretical description. Although this problem was examined first for ice, it is relevant in many systems and in particular in biology for the transport along proteins or for proton conductance across membranes, an essential process in cell life. This broad relevance makes the study of proton conduction very appealing. Since the original work of Bernal and Fowler on ice[1], the idea that the transport occurs through chains of hydrogen bonds has been well accepted. Such "proton wires" were invoked by Nagle and Morowitz[2] for proton transport across membranes proteins and more recently across lipid bilayers[3]. In this report, we assume the existence of such an hydrogen-bonded chain and discuss its consequences on the dynamics of the charge carriers. We show that this assumption leads naturally to the idea of soliton transport and we put a special emphasis on the role of the coupling between the protons and heavy ions motions. The model is presented in section II. In section III we show how the coupling affects strongly the dynamics of the charge carriers and in section IV we discuss the role it plays in the thermal generation of carriers. The work presented in section III has been performed in 1986 and 87 with St Pnevmatikos and N. Flytzanis[4] and was then completed in collaboration with D. Hochstrasser and H. Büttner[5]. Therefore the results presented in this part are not new but we think that they are appropriate in the context of this multidisciplinary workshop because they provide a rather complete (and tractable) example of the soliton picture for proton conduction. Section IV discusses the thermal generation of the charge carriers when the coupling between the protons and heavy ions dynamics is taken into account. The results presented in this part are very recent and will deserve further analysis but they already show that the coupling can assist the formation of the charge carriers.

Since the results presented here consider only the ionic defects along a pre-existing hydrogen bonded chain they give a partial view of the proton transport mechanism. However, since the coupling between the motion of the carriers and the dynamics of the underlying lattice generates a very characteristic response, we hope that these results might suggest some experimental test of the soliton picture for proton transport.

II. THE ANTONCHENKO-DAVYDOV-ZOLOTARIUK MODEL

Since the original work of Bernal and Fowler[1], it is now accepted that, in ice as well as in water, protons are transferred by jumps from one water molecule to another along hydrogen bonds (fig. 1 a). According to this mechanism the charge carriers are H_3O^+ and OH^- ionic defects. However the motion of these defects is not sufficient to explain a permanent proton conductivity since, after one defect as passed, the chain is left in a state that cannot carry charge in the same direction again.

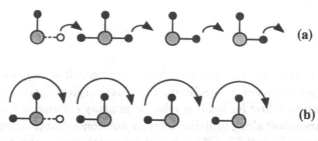

Figure 1: (a) Schematic picture of proton transport according to the Bernal Fowler mechanism. The figure presents the case of an H_3O^+ ionic defect. (b) Motion of a Bjerrum defect across an hydrogen bonded chain. The defect restores the chain in a state that can again carry charge according to the Bernal Fowler mechanism.

The chain has to be restored to its original state by another type of defect which rotates the water molecules, the so-called Bjerrum defects[6] (fig 1 b). In 1978 Nagle and Morowitz[2] extended the ideas of Bernal and Fowler and Bjerrum to membrane proteins, showing the great biological importance of proton transport across an hydrogen-bonded chain. However, in their model, the dynamics of the charge carriers was not described quantitatively. The first model describing this dynamics was proposed by Antonchenko-Davydov-Zolotariuk[7]. It combined a well known soliton model, the ϕ^4 model[8], and the dynamics of the heavy ions. This ADZ model only describes the H_3O^+ and OH^- ionic defects. Other models were introduced later to combine the ionic and Bjerrum defects in a single description[9], or to introduce other degrees of freedom of the heavy ions not included in the original ADZ model[5,10]. However, although this is not generally mentionned, these extended models are dependent on the actual crystal geometry while the ADZ model can be introduced by general arguments independent of the geometry of the chain. Moreover the ADZ model is easily tractable analytically and describes the essential features of the hydrogen bonded chain. This is why, in this report, we have chosen to restrict our discussion to this particular model.

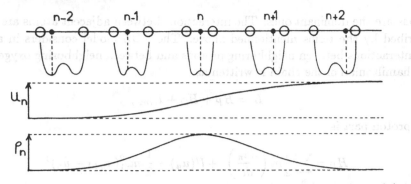

Figure 2: Picture of the Antonchenko-Davydov-Zolotariuk model showing the chain with the proton potentials, and a example of the solution of the equations of motion.

A schematic drawing of the ADZ model is shown in fig. 2. It consists of a chain of protons and heavy ions (which are henceforth called "oxygens" although they can be OH^- or more complicated entities depending on the system). A classical description of the dynamics of this chain could in principle be derived from the hypersurface of potential energy as it can be obtained from ab-initio calculations. However, this potential energy depends on all the variables in the system and cannot be used as such to obtain a model which is analytically tractable. The main idea of all the soliton models is to select in the potential energy the part which is relevant for proton transport and split it into several components relative to the proton sublattice, the oxygen sublattice, and the coupling between the two sublattices. The different terms can be justified if we consider a simple prototype hydrogen-bonded system, the proton-bound water dimer[11] $(H_2O\ H \cdots OH_2)^+$. This system can be characterized by two parameters, the distance X between the oxygens and the position u of the bonding proton (note that, by choosing only these parameters which are the most relevant parameters for the bonding-proton dynamics, we have already greatly simplified the expression of the potential energy of the system). Ab-initio calculations show that, when X has its equilibrium value, there are two energetically degenerate positions for the bonding-proton, close to one oxygen or close to the other. If X is maintained fixed, the transfer of the proton between the two sites requires to overcome a potential energy barrier between the sites so that the potential energy for the proton has the shape of a double well. But if X is allowed to vary, the proton transfer is accompanied by a reduction in the oxygens distance which lowers significantly the potential energy barrier. In some compounds the barrier can even vanish completely[12]. This effect is described in the ADZ model by writting the potential energy of the proton as the sum of a double well potential depending only on the proton position u, and an interaction term that depends both on u and the distance of the two adjacent oxygens. The model must also take into account the existence of a stable equilibrium distance a between neighboring oxygens by adding a term in the potential energy which, for simplicity, is chosen as a harmonic potential for the variable $\rho = a - X$. Moreover the hypersurface of potential energy couples any variable in the chain to any other. It is reasonable to assume that the shortest interactions are the dominant ones. The interactions between adjacent atoms are already described by the terms mentionned above. The ADZ model considers in addition the interactions between neighboring protons and between neighboring oxygen pairs. The hamiltonian of the chain is written as

$$H = H_P + H_O + H_{int} , \qquad (1)$$

The proton part is

$$H_P = \sum_n \frac{1}{2} m \left(\frac{du_n}{dt} \right)^2 + U(u_n) + \frac{1}{2} m \omega_1^2 (u_{n+1} - u_n)^2$$

where the index n designates the unit cells and m is the proton mass. The first term is the kinetic energy term, $U(u_n)$ is the double well potential obtained for fixed X in the ab-inito calculations. It is written as

$$U(u_n) = \epsilon_0 (1 - u_n^2/u_0^2)^2 , \qquad (2)$$

and the last term represents the harmonic coupling with characteristic frequency ω_1 between neighboring proton. In the double well expression $U(u_n)$, ϵ_0 is the height of the potential barrier between the two equilibrium sites situated at the positions $u = \pm u_0$.

The oxygen part is

$$H_O = \sum_n \frac{1}{2} M \left(\frac{d\rho_n}{dt} \right)^2 + \frac{1}{2} M \Omega_0^2 \rho_n^2 + \frac{1}{2} M \Omega_1^2 (\rho_{n+1} - \rho_n)^2 . \qquad (3)$$

M is the oxygen mass, Ω_0 is the frequency of the optical mode corresponding to the oscillations of the distance between adjacent oxygens and Ω_1 characterizes the coupling between neighboring oxygen pairs.

The interaction part is

$$H_{int} = \sum_n \chi \rho_n (u_n^2 - u_0^2) , \qquad (4)$$

where χ measures the strength of the coupling between the proton and hydrogen sublattices. The expression of the interaction potential in the ADZ model is interesting because it is both physically relevant and mathematically convenient. It describes correctly the interaction because if ρ_n increases (i.e. the distance between two adjacent oxygens decreases), the addition of the interaction potential to $U(u_n)$ generates a double well potential with a lower barrier and closer minima. Moreover, the specific form of H_{int} gives equations of motions which, in some cases, reduce to the well known ϕ^4 model and are thus solvable.

The comparison between the results of the ab-initio calculations and the ADZ model shows that this model provides a rather natural description of the hydrogen-bonded chain. It includes indeed some approximations since the hypersurface of potential energy has been severely simplified and because some degrees of freedom of the oxygens are ignored since an oxygen pair is described by a single variable ρ_n. The model does not include an overall translation of the pair, i.e. the acoustic modes of the oxygen sublattice. But, as the main change in the distance between adjacent oxygens is caused by a local optical motion, the approximation is reasonable.

III. Mobility of the Charge Carriers in the ADZ Model

The mobility of the carriers can be determined by investigating the dynamics of the ADZ model. We show in this section that its equations of motion have two solitonlike solutions which correspond to the H_3O^+ and OH^- ionic defects and we discuss the dynamics of these solutions.

The hamiltonian (1) generates a set of coupled differential equations for the u_n and ρ_n that cannot be solved analytically, but, if the nearest neighbor couplings between protons and between oxygen pairs are strong enough, one can use a continuum approximation which replaces a set of functions $u_n(t)$ by the two-variable function $u(x,t)$, and similarly for the $\rho_n(t)$ which are replaced by $\rho(x,t)$. Within this approximation, the original set of coupled differential equations is replaced by two coupled PDE's

$$u_{tt} - c_0^2 u_{xx} - \frac{4\epsilon_0}{mu_0^2} u \left(1 - \frac{u^2}{u_0^2}\right) + \frac{2\chi}{m}\rho u = 0 \,, \tag{5}$$

$$\rho_{tt} - v_0^2 \rho_{xx} + \Omega_0^2 \rho + \frac{\chi}{M}(u^2 - u_0^2) = 0 \,, \tag{6}$$

where $x = na$ is the continuous space variable, $c_0 = a\omega_1$ is the sound speed in the proton sublattice, and $v_0 = a\Omega_1$ is a parameter which characterizes the dispersion of the oxygens optical mode. A charge carrier moving at speed v is described by a permanent profile solution, i.e. a solution which depends only on $\xi = (x - vt)/a$. The equations of motion are therefore reduced to

$$\frac{1}{a^2}(c_0^2 - v^2)u_{\xi\xi} + \frac{4\epsilon_0}{mu_0^2} u \left(1 - \frac{u^2}{u_0^2}\right) - \frac{2\chi}{m}\rho u = 0 \,, \tag{7}$$

$$\frac{1}{a^2}(v^2 - v_0^2)\rho_{\xi\xi} + \Omega_0^2\rho + \frac{\chi}{M}(u^2 - u_0^2) = 0 \,. \tag{8}$$

Analytical solutions of this set of coupled equations are only known in some particular cases. If the coupling between the two sublattices vanishes ($\chi = 0$), Eq. (7) reduces to the ϕ^4 equation[8] which has kinklike permanent profile solutions, while equation (8) is simply linear. In the presence of the coupling, an exact analytical solution can only be obtained for the particular speed $v = v_0$ because, in this case, Eq. (8) gives an expression of ρ as a function of u. Introducing this expression in Eq. (7), we get

$$\frac{1}{a^2}(c_0^2 - v^2)u_{\xi\xi} + \frac{4}{mu_0^2}\left(\epsilon_0 - \frac{\chi^2 u_0^4}{2\Omega_0^2 M}\right) u \left(1 - \frac{u^2}{u_0^2}\right) = 0 \,, \tag{9}$$

which is again the standard ϕ^4 equation with a renormalized barrier ϵ between the two proton sites

$$\epsilon = \epsilon_0 - \frac{\chi^2 u_0^4}{2\Omega_0^2 M} \,. \tag{10}$$

The simplicity of this equation is due to the special form chosen for the interaction

69

term in the hamiltonian. Equation (9) has kinklike solutions

$$u = \pm u_0 \tanh(\xi/L) , \qquad (11)$$

where L measures the kink width and is given by

$$\frac{1}{L^2} = \frac{2}{m\omega_1^2 u_0^2} \left(\epsilon_0 - \frac{\chi^2 u_0^4}{2\Omega_0^2 M} \right) \frac{1}{1 - v^2/c_0^2} . \qquad (12)$$

The corresponding solution in the oxygen sublattice has a bell shape

$$\rho = \rho_0 \text{sech}^2(\xi/L) \quad \text{with} \quad \rho_0 = \chi u_0^2/M\Omega_0^2 . \qquad (13)$$

This solution is shown in fig. 2 with the plus sign for u. The figure shows that the kink in u generates a local reduction in the proton density, which amounts to creating a negatively charged carrier in the chain. This solution corresponds to the OH^- defect in the Bernal Fowler picture. The other solution with a minus sign in u increases the local proton density and it corresponds to the H_3O^+ defect. Both are accompanied by a local reduction in the distance between adjacent oxygens which is associated to a decrease in the effective barrier for the protons. The kink solutions of the ϕ^4 model are not solitons in the strict sense because they don't survive collisions with simply a phase shift. However they are stable solitonlike structures which propagate with a constant shape and speed. Therefore this particular solution for $v = v_0$ suggests that the coupling between the two sublattices assists the proton transport because it makes the jump from one position to the other easier by lowering locally the potential energy barrier. However the investigation of the dynamics of the model for other velocities shows that this is not always the case.

For $v \neq v_0$, an exact analytical solution cannot be obtained and we have to rely on numerical methods. A scheme using an effective hamiltonian has been recently designed to find permanent profile solutions moving at any speed[10], but we can also take advantage of the exceptional stability of the solitonlike solutions we are looking for to use the system itself as an equation solver. The idea is to run a molecular dynamics simulation in which a static solution obtained by energy minimization is forced to move at the desired speed by an external force. A small damping is added to absorb the radiations emitted while the static solution is accelerated. When a steady state is achieved, the external force and damping are gradually removed. This procedure shows that the two velocity domains $v \leq v_0$ and $v > v_0$ are fundamentally different.

(i) for $v \leq v_0$, a permanent profile solution exists and when the external force and damping are removed, it propagates freely at constant speed. An approximate analytical description of this solution can be derived because the shape of the kink in the proton sublattice is only weakly modified by the coupling with the oxygens. Therefore it is well approximated by the solution of Eq. (9) with $v < v_0$ although this equation does not treat Eq. (8) exactly in this case. Then the displacements in the oxygen sublattice are obtained by solving Eq. (8) with a known u solution, *i.e.* by treating the oxygen motions as if they were forced by a given proton kink. The amplitude of the oxygen pulse obtained by this approach is in good agreement with the numerical results.

(ii) for $v > v_0$, there are no permanent profile solutions. The numerical simulations show that, as the kink in the proton sublattice propagates, instead of being accompanied by a localized solution in the oxygen sublattice, it radiates waves in this sublattice. The same approximate analytical treatment as for $v \leq v_0$ shows that Eq. (8) forced by a kink in u moving at velocity $v > v_0$ has no localized solution. The radiation in the oxygen sublattice corresponds to a transfer of energy from the proton kink so that a proton kink launched at a speed $v > v_0$ slows down until its speed reaches v_0 where the radiation stops.

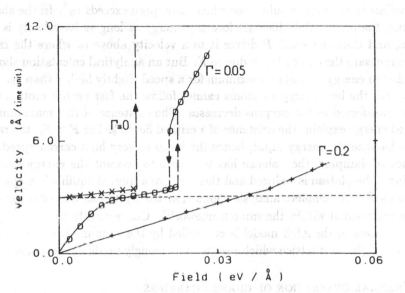

Figure 3: Velocity of the proton kink driven by an external field as a function of the field strength for several values of the damping coeffiecient Γ. The horizontal dotted line indicates the velocity v_0. Note the hysteresis phenomena for $\Gamma = 0.05 \ 10^{14} \ s^{-1}$.

Therefore, whereas for $v \leq v_0$ the coupling between the two sublattices is assisting the motion of the charge carriers, for $v > v_0$ it has an opposite effect. This generates a strong nonlinearity in the mobility of the carriers when they are submitted to an external field. We have determined this mobility by carrying numerical simulations in the presence of an external field. When it drives the ionic defect, the field feeds energy in the system. In a three-dimensional system part of this energy is distributed among many degrees of freedom which are not included in the one-dimensional ADZ model. This energy transfer has been approximately modelled by adding a phenomenological damping term in the equations of motion. Figure 3 shows the equilibrium velocity v_f of the proton kink as a function of the applied field F for different values of the damping coefficient Γ.

Without damping ($\Gamma = 0$) there is an abrupt discontinuity in v_f when F reaches a critical value F_c. For $F < F_c$, the velocity of the solitary wave is always larger than v_0 and increases very slowly with F (plateau in fig. 3). For $F > F_c$, v_f jumps to a value close to c_0 where it is limited by discreteness effects which cause radiation in the proton sublattice[13]. In the presence of damping this abrupt discontinuity is smoothed out but, for intermediate damping, some hysteresis is found in the charge carrier mobility when F is increased to a value larger than the critical value and then decreased. Consequently the numerical results show a very strong nonlinear response of the charge carriers to an external field. This behavior is due to the coupling between the two sublattices and can be understood if one considers the balance between the energy transmitted to the carriers by the external field and the energy they radiate in the oxygen sublattice when their speed exceeds v_0[4]. In the absence of damping, the proton kink does not lose any energy as long as its velocity is smaller than v_0 and thus any small F drives it to a velocity above v_0 where the radiation can compensate the energy input due to F. But an analytical calculation shows that the radiated energy exhibits a maximum for a speed slightly higher than v_0. At very high speed, the heavy oxygens atoms cannot follow the fast proton motion and the energy transferred to the oxygens decreases. The existence of this maximum in the radiated energy explains the existence of a critical field F_c. For $F > F_c$, the radiation cannot balance the energy input, hence the jump to very high carrier speed. In the presence of damping, the balance has to take into account the energy loss due to damping. The plateau is reduced and there is now a limited equilibrium speed when F exceeds F_c. A complete analysis shows that the hysteresis observed numerically can be understood within the same framework[4]. Consequently the mobility of the charge carriers in the ADZ model is controlled by the coupling between the charge carriers and the host lattice which results in a strongly nonlinear response.

IV. THERMAL GENERATION OF CHARGE CARRIERS

The analysis of conductivity experiments in hydrogen-bonded systems generally assume that the number of ionic defects is fixed. At low temperature some of them are trapped by defects or inhomogeneities and they are gradually released when temperature increases. One may ask however if the thermal *creation* of ionic defects is possible. This section examines this question. As for the mobility of the carriers, we show that the coupling between the proton and the oxygen sublattice has a strong influence.

The first step to determine whether thermal creation is possible is to compute the energy of an ionic defect. This is the energy of the solitonlike solutions determined previously and, for $v = v_0$ it is given by

$$E(v_0) = \frac{8}{3\sqrt{2}} \frac{\sqrt{m}}{\sqrt{1 - v_0^2/c_0^2}} \frac{c_0}{a} u_0 \sqrt{\epsilon}(1 + C) , \qquad (14)$$

with

$$C = \frac{4}{5} \frac{v_0^2}{c_0^2} \frac{\chi^2 u_0^2}{mM\Omega_0^2} . \qquad (15)$$

The coupling between the two sublattices appears in two places in this formula.

Table I. Energies E of ionic defects for different parameter sets.

ϵ_0	u_0	ω_1	Ω_0	v_0/c_0	χ	ϵ	C	E
2.0 eV	1 Å	2.21	0.184	0.192	0.1	1.99 eV	$4 \ 10^{-3}$	16.1 eV
0.1 eV	0.8 Å	1.0	0.570	0.1	0.1	0.996 eV	$2 \ 10^{-5}$	0.48 eV
0.18 eV	0.8 Å	1.0	0.570	0.1	1.5	0.987 eV	$6 \ 10^{-3}$	0.48 eV

First it shows up in the term $\sqrt{\epsilon}$ because, in the absence of coupling one would get ϵ_0 instead of ϵ. Since the effective barrier ϵ is lowered by the interaction as indicated by Eq. (10), the coupling contributes to *reduce* the creation energy of a defect. Second the coupling is responsible for the correction term C in Eq. (14). This term represents the energy which exists in the oxygen sublattice due to the local distorsion that accompanies the proton kink. This extra energy contributes to *increase* the creation energy of a defect. Table I lists the energy of an ionic defect for three different parameter sets.

The choice of appropriate parameters for the model is delicate because some of them are not directly accessible to experiments. The first set was introduced by Spatschek, Laedke and Zolotariuk[14] and used later by ourselves[4] and more recently by Nylund and Tsironis[15] for a comparison between model results and experiments. This set gives a defect energy which is extremely high (16 eV) end is certainly not correct. The two other sets have been chosen in order to give equal energies for the defects with weak ($\chi = 0.1$) or strong ($\chi = 1.5$) coupling between the sublattices. The values of ϵ_0 have been chosen to be consistent with the ab-initio results[11] and also with a very extensive analysis of a large number of hydrogen bonded compounds performed by Sokolov et al.[12]. The value of ω_1 has been chosen so that the characteristic width of a defect, *i.e.* its spatial extent L, is if the order of two lattice spacings. The frequency Ω_0 of the oxygens optical mode has been set to 300 cm^{-1}. The parameter v_0 is the most difficult to chose because it is related to the dispersion of this optical mode which is not accessible by a spectroscopy experiment. We have chosen to impose $v_0/c_0 = 0.1$ but it would be interesting to refine this value either by comparison with the results of ab-initio calculations on a system involving at least two oxygen pairs so that the coupling between them can be obtained, or by using dispersion curves determined by neutron diffraction. The coupling constant χ is also a parameter which is not well known, which is why we have considered two cases, $\chi = 0.1$ and $\chi = 1.5$ corresponding respectively to weak and strong coupling. In the weak coupling case, table I shows that ϵ is very close to ϵ_0 and the correction C is very small. In the strong coupling case $\epsilon \approx 0.5 \, \epsilon_0$, while the correction term remains rather small. The global effect is a significant reduction of the energy of the defect.

However, even if the model parameters are not perfectly known, the energy of an ionic defect is of the order of 0.5 eV so that its thermal creation around room temperature is extremely unlikely unless some particular mechanism can intervene to localize thermal energy in the chain. In a nonlinear system like the hydrogen-bonded chain such a mechanism exists; it is the *modulational instability* of a plane wave. Before attempting to study it for the coupled lattices of the ADZ model, let us discuss the idea on a simple case with only the proton sublattice. We start from the equation of motion (5) with $\chi = 0$, $i.e.$

$$u_{tt} - c_0^2 u_{xx} - \frac{4\epsilon_0}{mu_0^2}u\left(1 - \frac{u^2}{u_0^2}\right) = 0,\tag{16}$$

In order to study the build-up of large amplitude solutions we expand around the equilibrium position u_0 using a multiple scale expansion which goes beyond a simple linear expansion

$$u = u_0 + \epsilon u_1(T_0, T_1, T_2, \ldots, X_0, X_1, X_2, \ldots) + \epsilon^2 u_2(T_0, T_1, T_2, \ldots, X_0, X_1, X_2, \ldots),\tag{17}$$

with $T_0 = t$; $T_1 = \epsilon t$ is a slow time which will describe the slow evolution of the solution, $T_2 = \epsilon^2 t$, and in a similar manner $X_0 = x$, $X_1 = \epsilon x$, etc. At order ϵ one gets an evolution equation for u_1 which gives

$$u_1 = A(X_1, T_2)e^{i(kx - \omega t)} + CC,\tag{18}$$

where ω and k are related by the linear dispersion relation of the lattice. In the small amplitude limit, u_1 would simply correspond to the phonon modes of the lattice. But if nonlinearity is taken into account, the higher order terms give a nonlinear Schrödinger equation (NLS equation) for A,

$$i\frac{\partial A}{\partial T_2} + P\frac{\partial^2 A}{\partial X_1^2} + Q|A|^2 A = 0,\tag{19}$$

with $P = (c_0^2 - v_g^2)/2\omega$, $v_g = c_0^2 k/\omega$, and $Q = 24\epsilon_0/\omega u_0^4$. This equation has an exact space-independent solution $A = A_0 \exp(iQA_0^2 T_2)$ which corresponds to a plane wave solution in u_1. However this solution in which the energy is evenly distributed over the chain is unstable. Looking for a perturbed solution $A = [A_0 + A_1(X_1, t_2)]\exp(iQA_0^2 T_2)$, one finds that, if $PQ > 0$, which is the case for the equation that we consider here, the perturbation tends to grow and generates a spontaneous modulation of the wave. A more complete analysis shows that the plane wave tends to break-up into solitary waves, or breather solutions of the NLS equation, in which the energy is concentrated. The same mechanism is also true for the thermal fluctuations. When their amplitude is sufficient to excite the nonlinearities in the system, an energy localization occurs and promotes the formation of kink-antikink pairs.

This mechanism is well known for an equation like (16), but the case of the ADZ model of the hydrogen-bonded chain is more complicated because we must consider the two coupled equations (5) and (6). Nevertheless the same energy localization mechanism exists and it is even made more efficient by the coupling between the two sublattices[16]. A multiple scale expansion performed on the variables u and ρ yields at first order

$$u_1 = A_+(X_1, T_2)e^{i(\omega_+ t - kx)} + A_-(X_1, T_2)e^{i(\omega_- t - kx)} + CC\tag{20}$$

$$\rho_1 = B_+(X_1, T_2)e^{i(\omega_+ t - kx)} + B_-(X_1, T_2)e^{i(\omega_- t - kx)} + CC\tag{21}$$

where the two sets of terms arise from the existence of two branches in the dispersion curve of the diatomic chain, with frequencies ω_+ and ω_-. Each sublattice sees its own dispersion curve and a "shadow" of the dispersion curve of the other sublattice due to the coupling. For instance, the component of amplitude A_- in the proton sublattice is the "shadow" of the component B_- in the oxygen sublattice. Therefore A_- is related to B_- and similarly B_+ can be expressed as a function of A_+ so that the only independent factors are A_+ and B_-. They are solutions of a set of coupled NLS equations

$$i\frac{\partial A_+}{\partial T_2} + P_1\frac{\partial^2 A_+}{\partial X_1^2} + Q_{11}|A_+|^2 A_+ + Q_{12}|B_-|^2 A_+ = 0 \tag{22}$$

$$i\frac{\partial B_-}{\partial T_2} + P_2\frac{\partial^2 B_-}{\partial X_1^2} + Q_{22}|B_-|^2 B_- + Q_{21}|A_+|^2 B_- = 0 \tag{23}$$

The coefficients P_1, P_2, Q_{ij} have lengthy expressions in terms of the model parameters but they can be obtained analytically[16]. Similar systems of coupled NLS equations have been obtained previously for birefringent fibers[17] or coupled lasers beams in aplasma[18,19]. One general feature of these coupled NLS is that modulational instability is more likely to occur, and the growth rate of the instability is larger than for a single NLS. This is also true for the equations derived above for the hydrogen-bonded chain. Instead of the single condition $PQ > 0$ for a single NLS, there are now several sufficient instability conditions for the exact plane wave solution

$$A_+ = A_0 \exp[i(Q_{11}A_0^2 T_2 + Q_{12}B_0^2 T_2)] \tag{24}$$

$$B_- = B_0 \exp[i(Q_{21}A_0^2 T_2 + Q_{22}B_0^2 T_2)]. \tag{25}$$

The first one

$$P_1 Q_{11} A_0^2 + P_2 Q_{22} B_0^2 > 0 \tag{26}$$

is simply the generalization of the instability condition of a single NLS equation. The others have complicated expression, but the net result is that, with the model parameters listed in table I, the instability is always present.

Figure 4 shows the result of a numerical simulation performed with the second parameter set of table I (weak coupling case). In this simulation using molecular dynamics at constrained temperature, a chain containing 256 protons is slowly heated. At low temperature the protons were all on the same side of the double well ($u_n \approx +u_0$) so that the chain had no ionic defect. The black areas which corresponds to domains in which the protons have moved to the opposite position $-u_0$ show that ionic defects have been formed at high temperature.

The tendency for energy localization appears clearly because the large amplitude motions start in a small region of the chain. In this region one can notice that the formation of a large domain in which the protons have switched well is preceeded by a

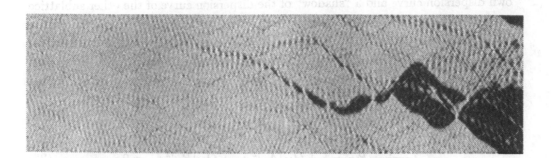

Figure 4: Thermal generation of a pair of ionic defects in the ADZ model. The figure shows the positions of the protons in the chain with a gray scale. Light grey corresponds to protons around the position $+u_0$ and black corresponds to protons around the position $-u_0$. The chain extends along the vertical axis. The temperature extends along the horizontal axis. In this experiment the system was heated with a linear temperature ramp so that the horizontal axis is also proportional to time. An ionic defect appears on this figure as an interface between a clear and a dark region.

sequence of alternating balck and white regions which grow bigger and bigger. These dots correspond to large amplitude oscillations of the protons in a small domain of the chain. A few protons move toward $-u_0$ giving rise to a dark dot, then come back to their original position then move toward $-u_0$ again. This type of motion is typical of a breather mode of the NLS equation. As the temperature is raised, the amplitude of the breather increases while its frequency decreases until it freezes, giving rise to a kink-antikink pair. This type of oscillatory precursor motions before a pair of ionic defects are formed shows that the NLS description, hence the modulational instability mechanism, is the appropriate description of the thermal generation of ionic defects in the hydrogen bonded chain. However, in spite of the enhancement of the energy localization due to the coupling between the proton and oxygen sublattices, the thermal generation of the carriers still requires a high temperature. In the simulation shown in fig. 4, the temperature varies from 500K to 1500K and the formation of a pair of ionic defects is observed around 1200K. Since the energy of the two defects is 0.96 eV (corresponding to T=11,130K if we set $k_B T = 0.96$ eV) the energy localization is responsible a substantial decrease in the generation temperature. Nevertheless, in spite of the enhancement due to the coupling between the two sublattices, the thermal generation of ionic defects around room temperature can be expected to be a rare event.

V. CONCLUSION

The analysis of the different terms of the potential energy of the ADZ model has shown that they provide an approximate description of the hypersurface of potential energy of an hydrogen-bonded chain. Therefore, if one accepts the idea that protons are transported along such a chain, the model leads naturally to solitonlike solutions. These solutions provide a general description of the two types of ionic defects: very narrow solutions correspond to independent proton jumps while broader ones describe a collective proton transport. Only the broad solutions can be expected to have solitonlike properties because, if the kink width is only of the order of the lattice spacing, discreteness effects trap it[13]. This soliton picture could also be obtained with a simpler model assuming that the heavy ions are fixed. However ab-initio calculations or the analysis of vibrational proton frequencies in a large number of hydrogen bonded systems[12] show that the proton motion is accompanied by a rather large distorsion of the heavy ion sublattice. This distorsion is included in the ADZ model and we have shown that it has a very stong influence, both on the mobility and on the thermal generation of ionic charge carriers. Due to energy exchanges between the protons and heavy ions a nonlinear mobility is found. This characteristic behavior could provide an experimental test of the validity of the model. We have also shown that the distorsion of the heavy ion sublattice enhances the modulational instability that can localize the energy and promote the thermal generation of ionic defects. However, even with this enhancement, the thermal generation of defects at room temperature remains a rare event.

The ADZ model is still a fairly simple description of a hydrogen bonded system and it can be improved to describe the Bjerrum defects[9], include acoustic modes for the oxygens[5] or even include anharmonic interaction between the protons to describe the difference in energy between the H_3O^+ and OH_- ions. However, the most urgent research to carry in this domain is the determination of appropriate parameters for the model. They are essential to allow a comparison between theory and experiments and to decide whether the soliton picture is closer to reality than independent proton jumps. Good model parameters could probably be provided by ab-initio calculations on systems big enough to include the dynamics of the heavy ions and to test the degree of cooperativity in the proton motions.

ACKNOWLEDGEMENTS

Part of this work has been supported by the EEC Science program under contract SCI /0229-C (AM)The present paper has been written during the stay of two of us (M.P. and R.B.) at the Center for Nonlinear Studies of the Los Alamos National Laboratory. We thank the Center for its hospitality.

The work presented here has been initiated by a collaboration with Stephanos Pnevmatikos and would not have been done without him. We would like to stress here once more the very important role he played in popularizing the soliton concept in several domains of physics and in bringing together researchers with various interests. He was also a great friend who will be always remembered.

REFERENCES

1. J. D. Bernal and R. H. Fowler, A theory of water and ionic solution, with particular reference to hydrogen and hydroxyl ions, J. Chem. Phys. 1:515 (1933)

2. J. F. Nagle and H. J. Morowitz, Molecular mechanism for proton transport in membranes, Proc. Natl. Acad. Sci. USA, 75:298 (1978)

3. J. F. Nagle, Theory of Passive Proton Conductance in Lipid Bilayers, J. Bioenergetics and Biomembranes, 19:413 (1987)

4. M. Peyrard, St. Pnevmatikos and N. Flytzanis, Dynamics of two-component solitary waves in hydrogen-bonded chains, Phys. Rev. A 36:903 (1987)

5. D. Hochstrasser, H. Büttner, H. Desfontaines and M. Peyrard, Solitons in hydrogen-bonded chains, Phys. Rev. A 38:5332 (1988)

6. N. Bjerrum, Structure and Properties of Ice, Science, 115:385 (1952)

7. V. Ya. Antonchenko, A. S. Davydov and A. V. Zolotariuk, Solitons and proton motion in ice-like structures, Phys. Status Solidi B 115:631 (1983)

8. J. A. Kruhmansl and J. R. Schrieffer, Dynamics and statistical mechanics of a one-dimensional model hamiltonian for structural phase transitions, Phys. Rev. B 11:3535 (1975)

9. St. Pnevmatikos, Soliton dynamics of hydrogen-bonded networks: a mechanism for proton conductivity, Phys. Rev. Lett. 60:1534 (1988)

10. St. Pnevmatikos, A. V. Savin, A. V. Zolotariuk, Yu. S. Kivshar and M. J. Velgakis, Nonlinear transport in hydrogen-bonded chains: Free sclitonic excitations, Phys. Rev. A 43:5518 (1991) and references therein

11. S. Scheiner, P. Redfern and M. M. Szcześniak, Effects of External Ions on the Energetics of Proton Transfers across Hydrogen Bonds, J. Phys. Chem. 89:262 (1985)

12. N. D. Sokolov, M. V. Vener and V. A. Savel'ev, Tentative study of strong hydrogen bond dynamics. Part II. Vibrational frequency considerations, J. Molecular Structure 222:365 (1990)

13. M. Peyrard and M. D. Kruskal, Kink dynamics in the highly discrete sine-Gordon system, Physica D 14:88 (1984)

14. E. W. Laedke, K. H. Spatschek and A. V. Zolotariuk, Phys. Rev. A 32:1161 (1985)

15. E. S. Nylund and G. P. Tsironis, Evidence for solitons in hydrogen-bonded systems, Phys. Rev. Lett. 66:1886 (1991)

16. I. Kourakis, Instabilité modulationnelle dans les systèmes à liaisons hydrogènes, DEA report, University of Dijon

17. C. R. Menyuk, IEEE J. Quantum Electron. 23:174 (1987)

18. B. Gosh. , K. P. Das, Nonlinear interaction of two compressional hydromagnetic waves, J. Plasma Phys. 39:215 (1988)

19. G. G. Luther and C. J. McKinstrie, Transverse modulational instability of collinear waves, J. Opt. Soc. Am. B 7:1125 (1990)

SOLITON MODELLING FOR THE PROTON TRANSFER IN HYDROGEN-BONDED SYSTEMS

St. Pnevmatikos[1,2,†], A.V. Savin[1,3], and A.V. Zolotaryuk[1,4]

[1]Research Center of Crete, P.O.Box 1527, 71110 Heraklion, Crete, Greece
[2]Department of Physics, University of Crete, 71409 Heraklion, Crete, Greece
[3]Institute for Physico-Technical Problems, 119034 Moscow, Russia
[4]Institute for Theoretical Physics, Academy of Sciences, 252130 Kiev, Ukraine

ABSTRACT

The longitudinal and transverse collective dynamics of hydrogen-bonded protons is studied on the basis of a zig-zag model in which the heavy negative ions of the background sublattice are considered to be frozen. The on-site potential for each proton is constructed as an appropriate sum of the Morse potentials. The nearest-neighbor protons are assumed to interact harmonically with each other. Ionic and orientational defect solutions of soliton type have been obtained numerically by using the steepest descent minimization scheme. On the basis of the zig-zag modelling a simple one-dimensional model which belongs to the nonlinear Klein-Gordon family has been used for the studies of the longitudinal proton transfer under external electric fields and damping.

INTRODUCTION

There is a large number of hydrogen-bonded molecular systems, such as ice[1-3], imidazole[4], lithium hydrazinium sulfate[5], hydrogen halides[6,7], protein and other biological macromolecules[8] (for a comprehensive review see Nagle and Tristram-Nagle[9]), which exhibit a considerable electrical conductivity, even though the electron transfer through these systems is hardly supported. The systems of this type consist of hydrogen-bonded chains and the protons in hydrogen bonds are considered to be dominant charge carriers, so that the (protonic) conductivity along the direction of the hydrogen-bonded chains is of several orders greater than in the perpendicular directions. Therefore, the physical mechanism of the protonic conductivity in these cases has a quasi-one-dimensional nature and a regular one-dimensional hydrogen-bonded chain may be adopted as an appropriate model for the theoretical studies[8].

The protonic conductivity is associated with the motion along a hydrogen-bonded chain of the ionic (ionization) and Bjerrum (orientational or bonding) defects[1,8-10]. The former involve translational motions of hydrogen-bonded protons whereas the latter are results of rotations of hydroxyl ions or some other hydroxyl groups. The transport of protons may begin either with the passage of an ionic defect or the passage of an orientational defect, but thereafter the motion of these defects must strictly alternate[8,9].

†Deceased before the completion of this work

Recently, using a well established fact that the motion of the protons in adjacent hydrogen bonds is correlated, so that the proton dynamics in the hydrogen-bonded chain should be considered as a collective process[11-13], different one- and two-component soliton models have been suggested to describe the motion of the ionic and Bjerrum defects[14-29]. The main idea is that the interacting protons move in an on-site background potential with a double-well topology (when only one of the two types of the defects, ionic or orientational, is considered)[14-19,22,23,25,26,29] or with a doubly periodic form (when both the types of the defects are considered simultaneously)[20,21,24,27,28]. The on-site potential is created by the surrounding heavy (negative) ions and when these ions are assumed to be frozen, then only the proton displacements from the top of barriers in the on-site potential are considered resulting in a one-component soliton model[14-17,19,20,22,23,27,28]. The second important idea is the allowance for the heavy ions of the chain to move. In this case the height of the potential barrier associated with the intrabond proton transfer essentially depends on the relative distance between the surrounding heavy ions. On the other hand, any intrabond proton displacement produces a local deformation of the heavy sublattice. Therefore, a number of two-component soliton models described in the continuum limit by coupled nonliner field equations has been proposed in oder to study the two-sublattice soliton dynamics[15,17,18,21,24-26,29]. In some of these models only the coupling of the proton subsystem with an optical mode of the background sublattice has been considered[15] whereas in the other models the protonic chain is coupled only with an acoustic mode of the heavy ion subsystem[17,18,21].

An improved and more general version of the two-component model for the ionic defects has been suggested by Zolotaryuk et al.[25]. In this model the protonic chain is coupled with both the acoustic and optical modes and the coupling is given explicitly from any pair ion-proton potential. In general, this dependence is nonlinear and it involves an kink-antikink asymmetry that means that the positive and negative ionic defects differ in their properties (they have different energies, effective masses, Peierls potential reliefs, mobilities, etc.). But the most important thing that should be emphasized when the heavy ions are allowed to move is the possibility to study the effects of temperature in a correct manner. In this case only the heavy ion sublattice should be considered to contact a heat bath (see Nylund and Tsironis[15], and Savin and Zolotaryuk[29]). The other (proton) sublattice is affected by the heat bath via the heavy sublattice, i.e. the covalent bonds.

Our aim in this paper is to study the dynamics of the ionic and orientational defects in hydrogen-bonded systems in the framework of models with more realistic structure (for example, a zig-zag chain[23,27]) and the parameter values close to the experimental data. Of course, such modelling can be realized and studied only numerically[27] but some simple models of the soliton type can also be extracted from these studies (for instance, a one-component model with a doubly periodic on-site proton potential[28], etc.) and investigated analytically. In this case some important dynamical quantities of the defects can be easily calculated[28] which describe particle-like properties of both the ionic and orientational defects.

A ZIG-ZAG MODEL FOR THE IONIC AND ORIENTATIONAL DEFECTS

A direct way to model the proton transfer in hydrogen-bonded systems like hydrogen halides $(HX)_\infty$ (for example, $(HF)_\infty$, $(HCl)_\infty$, or $(HBr)_\infty$) is to consider a two-dimensional zig-zag network (see Fig. 1) where both the longitudinal and transverse proton motions can be considered simultaneously. Here l is the equilibrium length of the $X - X$ bond between the two neighboring negative ions, α is the angle between these

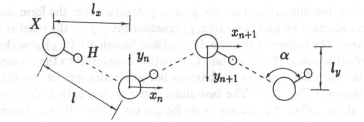

Figure 1: Schematic two-dimensional representation of a zig-zag hydrogen-bonded chain. The large (small) circles represent heavy negative ions (protons).

bonds, and $l_x = l \sin(\alpha/2)$, $l_y = l \cos(\alpha/2)$. In a first crude approximation we assume all the negative ions to be frozen and we focus our attention to the two-dimensional dynamics of protons. The coordinates (x_n, y_n) are measured from the center of the n-th negative ion as shown in Fig. 1.

Two-dimensional on-site proton potential

Consider the proton interaction with the ion subsystem. It is convenient to describe the pair interaction of a proton with a negative ion by the Morse potential $V(r) = D_0\{\exp[-b(r - r_0)] - 1\}^2$ where D_0 and r_0 are the energy and the length of the X – H bond in an isolated molecule HX and b is the phenomenological parameter. We construct the on-site proton potential $U(x, y)$ as the sum of nine Morse potentials in the form $U(x,y) = \sum_{i=-4}^{4} V(r_i)$ where $r_i = [(x - il_x)^2 + (y - (1 - (-1)^i)l_y/2)^2]^{1/2}$.

To define the values of the parameters r_0 and b we consider a one-dimensional on-site proton potential given by the sum

$$
\begin{aligned}
U(u) &= V(\frac{l}{2} - u) + V(\frac{l}{2} + u) - V(\frac{l}{2} - u_0) - V(\frac{l}{2} + u_0) = \\
&= \varepsilon_I \left(\frac{\cosh bu - \cosh bu_0}{1 - \cosh bu_0} \right)^2
\end{aligned} \tag{1}
$$

where u is the displacement of the proton from the midpoint of the distance between the adjacent ions, ε_I and $2u_0$ are the height and width of the barrier in the intrabond potential:

$$
\varepsilon_I = D_0(1 - \cosh bu_0)^2 / \cosh^2 bu_0 , \quad u_0 = \frac{1}{b} \cosh^{-1}\{\frac{1}{2} \exp[b(\frac{l}{2} - r_0)]/2\}. \tag{2}
$$

Here b is the phenomenological parameter, and in the following we denote by $r_1 = \frac{l}{2} - u_0$ and K_{HX} the length and stretching force constant of the covalent bond X–H in the hydrogen-bonded chain $(HX)_\infty$; we have $K_{HX} = 2\varepsilon_I b^2 / \tanh^2(bu_0/2)$. In this section we use the parameter values for the following three hydrogen halides $(HX)_\infty$: $(HF)_\infty$, $(HCl)_\infty$, and $(HBr)_\infty$ given in Anderson et al[30]. In these three cases we have: l=2.50Å, D_0=5.869eV, r_1=0.97Å, K_{HX}=550.3N/m (ω_s=3045.5cm^{-1}) for $(HF)_\infty$; l=3.688Å, D_0=4.433eV, r_1=1.275Å, K_{HX}=431.5N/m (ω_s=2697cm^{-1}) for $(HCl)_\infty$; l=3.927Å, D_0=3.758eV, r_1=1.414Å, K_{HX}=340.3N/m (ω_s=2395cm^{-1}) for $(HBr)_\infty$. From these values we obtain b=3.3148Å^{-1} and r_0=0.9262Åfor $(HF)_\infty$, b=2.7033Å^{-1} and r_0=1.2583Åfor $(HCl)_\infty$, and b=2.65098Å^{-1} and r_0=1.39405Åfor $(HBr)_\infty$.

The resulting two-dimensional on-site proton potential takes the form depicted in Fig. 2. For convenience we plot the $-U(x,y)$ function cutting off the relief at a certain level by a plane. The valleys of the two-dimensional function $U(x,y)$ give the path of the lowest potential energy for a proton. This path corresponds to the proton channel along which the potential relief has two barriers per one lattice cell, i.e. doubly periodic topology as shown in Fig. 3(a). The two-dimensional proton path is depicted in Fig. 3(b). All the plots in Figs. 2 and 3 are made for the crystalline $(HF)_\infty$. The numerical calculations for the other materials $(HCl)_\infty$ and $(HBr)_\infty$ give the similar plots.

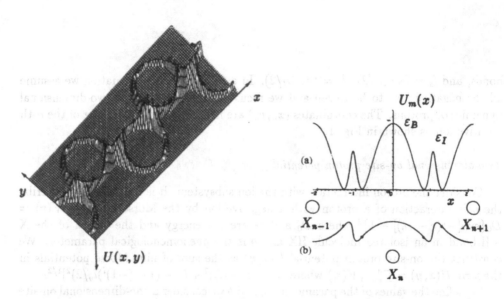

Figure 2: Two-dimensional relief $-U(x,y)$ which is the result of the interaction of one proton H with its nine nearest negative ions X. The parameter values have been taken for $(HF)_\infty^{30}$.

Figure 3: (a) One-dimensional potential relief in the longitudinal direction (along the proton channel). (b) Proton channel (path) $y = f(x)$ of the lowest potential energy $U_m(x) = U(x, f(x)) = \min_y U(x,y)$. The parameter values correspond to $(HF)_\infty^{30}$.

Ionic defects

In oder to find the stationary states of the ionic defects we have to solve the following minimization problem:

$$E_I = \sum_{n=2}^{N-1} [\frac{K_I}{2}(x_{n+1} - x_n)^2 + \frac{K_I}{2}(y_{n+1} - y_n)^2 + U(x_n, y_n)] \rightarrow$$

$$\rightarrow \min : x_2, \ldots, x_{N-1}, y_2, \ldots, y_{N-1} \tag{3}$$

with the boundary conditions $x_1 = l_x - x_0$, $y_1 = l_y - y_0$, $x_N = x_0$, $y_N = y_0$ for a positive ionic defect I_+ and with the conditions $x_1 = x_0$, $y_1 = y_0$, $x_N = l_x - x_0$, $y_n = l_y - y_0$ for a negative ionic defect I_-. Here $K_I = K_{HX}((\omega_a/\omega_s)^2 - 1)/2$ is the proton-proton stretching force constant which characterizes the vibration of the covalent bond X–H in the hydrogen-bonded chain $(HX)_\infty$, ω_s and ω_a are the frequencies of the symmetric and antisymmetric vibrational modes. All these parameter values[30] are summarized in Table 1.

Table 1. Spectroscopic parameter values for the hydrogen halides $(HF)_\infty$, $(HCl)_\infty$, and $(HBr)_\infty$.

$(HX)_\infty$	$\omega_s(cm^{-1})$	$\omega_a(cm^{-1})$	$K_{HX}(N/m)$	$K_I(N/m)$
$(HF)_\infty$	3045.5	3386	550.3	32.5
$(HCl)_\infty$	2697	2741.5	431.5	3.6
$(HBr)_\infty$	2395	2431	340.3	2.6

The minimization problem (3) admits the solutions with two types of symmetry: a solution with an integral center of symmetry, when $l_x/2 - x_{N/2+k} = x_{N/2-k} - l_x/2$, $l_y/2 - y_{N/2+k} = y_{N/2-k} - l_y/2$ and a solution with a half-integral center of symmetry, for which $l_x/2 - x_{N/2+k} = x_{N/2-k+1} - l_x/2$, $l_y/2 - y_{N/2+k} = y_{N/2-k+1} - l_y/2$. The stationary ionic defect with a half-integral center is stable whereas the stationary ionic defect with an integral center is unstable. The energies of the integral- and half-integral-centered ionic defects are given in Table 2. The longitudinal and transverse profiles of these defects are plotted in Fig. 4.

Table 2. Ionic defect energies in the integral- and half-integral states.

$(HX)_\infty$	$E_{I_\pm}(n)(eV)$	$E_{I_\pm}(n+1/2)(eV)$
$(HF)_\infty$	0.783	0.373
$(HCl)_\infty$	1.644	0.204
$(HBr)_\infty$	1.255	0.167

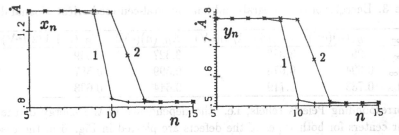

Figure 4: Profiles of the static stable (curve 1) and unstable (curve 2) ionic defects in $(HF)_\infty$.

Orientational defects

In oder to find the static states of the orientational defects we have to solve the following minimization problem:

$$E_B = \sum_{n=3}^{N-2} [\frac{K_{B_1}}{2}(\rho_{n,1} - \rho_1)^2 + \frac{K_{B_2}}{2}(\rho_{n,2} - \rho_2)^2 + U(x_n, y_n)] \rightarrow$$

$$\rightarrow \min : x_3, \ldots, x_{N-2}, y_3, \ldots, y_{N-2} \quad (4)$$

with the conditions $x_1 = x_2 = \pm x_0$, $x_N = x_{N-1} = \mp x_0$, $y_1 = y_2 = y_{N-1} = y_N = y_0$ for a positive and a negative orientational defect $B_\pm$. Here $\rho_{n,1} = [(l_x + x_{n+1} - x_n)^2 + (l_y - y_{n+1} - y_n)^2]^{1/2}$ is the distance between the n−th and $(n+1)$−th protons, $\rho_1 = [(l_x^2 + (l_y - 2y_0)^2]^{1/2}$, $\rho_{n,2} = [(2l_x + x_{n+2} - x_n)^2 + (y_{n+2} - y_n)^2]^{1/2}$ is the distance between the n−th and $(n+2)$−th protons, $\rho_2 = 2l_x$, and K_{B_1}, K_{B_2} are the proton-proton stretching force constants which characterize the dipole-dipole interaction ($K_{B_2} = K_{B_1}(\rho_1/\rho_2)^3$). For $(HF)_\infty$ K_{B_1}=41N/m, for $(HCl)_\infty$ K_{B_1}=5.1N/m, and for $(HBr)_\infty$ K_{B_1}=4.5N/m.

The minimization problem (4) admits the solutions with two types of symmetry: a solution with an integral center of symmetry, when $x_{N/2+k} = -x_{N/2-k}$, $y_{N/2+k} = y_{N/2-k}$, and a solution with a half-integral center of symmetry, for which $x_{N/2+k} = -x_{N/2-k+1}$, $y_{N/2+k} = y_{N/2-k+1}$. The defect with an integral center is stable for $(HF)_\infty$, and is unstable for $(HCl)_\infty$ and $(HBr)_\infty$, while the defect with a half-integral center is stable for $(HCl)_\infty$ and $(HBr)_\infty$, and is unstable for $(HF)_\infty$. The profiles of the static orientational defects are plotted in Fig. 5 for the case of $(HF)_\infty$. Their energies in the integral- and half-integral states are given in Table 3.

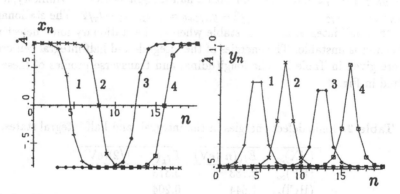

Figure 5: Profiles of the static stable (curves 2,4) and unstable (curves 1,3) orientational defects in $(HF)_\infty$.

Table 3. Energies of the integral- and half-integral-centered orientatioanal defects.

$(HX)_\infty$	$E_{B_+}(n)$(eV)	$E_{B_+}(n+1/2)$(eV)	$E_{B_-}(n)$(eV)	$E_{B_-}(n+1/2)$(eV)
$(HF)_\infty$	3.297	3.323	3.327	3.359
$(HCl)_\infty$	0.694	0.678	0.599	0.587
$(HBr)_\infty$	0.733	0.719	0.644	0.633

The corresponding Peierls reliefs, i.e. the dependence of the energy of the defects on their centers for both types of the defects are plotted in Fig. 6 in the case of the crystalline $(HF)_\infty$.

A SIMPLE ONE-DIMENSIONAL TWO-KINK MODEL

The two-dimensional modelling for the proton transfer considered in the previous section can be used as a basis to introduce a simple one-component chain model known in the dislocation theory as the Frenkel-Kontorova (FK) model[31]. Usually the on-site potential in this model is chosen to be a periodic function of the sine/cosine form. As was shown in the previous section (see Fig. 3(a)), the potential energy function along the longitudinal proton transfer has two barriers per one period of the zig-zag chain. Therefore, the one-dimensional FK model with a doubly periodic on-site potential of

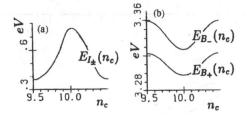

Figure 6: Peierls potential reliefs in $(HF)_\infty$ for the (a) ionic $(E_{I_\pm}(n_c))$ and (b) orientational $(E_{B_\pm}(n_c))$ defects.

the general form[20] can be introduced in order to describe the dynamics of both the ionic and orientational defects *simultaneously*, i.e. in the framework of the same model.

Thus, the proton dynamics in this model is governed by the Hamiltonian written in the dimensionless form as follows[20,21,24,28,32].

$$H = \varepsilon_0 \sum_n \left[\frac{1}{2}\left(\frac{du_n}{d\tau}\right)^2 + \frac{1}{2}(u_{n+1} - u_n)^2 + \Omega_0^2 V(u_n)\right] \tag{5}$$

where u_n is the dimensionless proton displacement from the midpoint in the X–X bond and Ω_0 is the dimensionless frequency of small-amplitude oscillations of protons at the local minima $u = 4\pi n \pm u_0$, $n = 0, \pm 1, ...$, $0 < u_0 < 2\pi$, of a doubly periodic (with period 4π) function $V(u)$ shown in Fig. 7 (curve 1). In this case the potential $V(u)$ is normalized in such a way that $V''(4\pi n \pm u_0) = 1$ where the prime denotes differentiation. In other words, the parameter Ω_0 measures the ratio of the height of the energy barriers in the on-site proton potential to the proton-proton intersite coupling. This coupling appears as a result of the cooperativity of the hydrogen bond and in the continuum limit it determines the velocity c_0 of small-amplitude waves in the proton sublattice. For numerical calculations and notations it is convenient to use the scaled time $\tau = t/t_0$ with $t_0 = l/c_0$ (l is shown in Fig. 1) as well as the energy unit $\varepsilon_0 = m(c_0/4\pi)^2$.

Equations of motion in the dimensionless form can be written as

$$\frac{d^2 u_n}{d\tau^2} = u_{n+1} - 2u_n + u_{n-1} - \Omega_0^2 \frac{dV}{du_n} + f - b\frac{du_n}{d\tau} \tag{6}$$

where $f = 4\pi Fl/\varepsilon_0^2$ and $b = t_0\Gamma$, with F an external force and Γ a phenomenological damping coefficient. When a constant external force F is applied, the degeneracy of the ground (vacuum) states for each proton is removed, as shown in Fig. 7 (see curve 2). In this case the equilibrium positions are displaced to the right (as shown in Fig. 7) for $F > 0$, and to the left for $F < 0$. For sufficiently small F there is also an infinite set of equilibrium positions for the biased potential

$$U(u) = \Omega_0^2 V(u) - fu \tag{7}$$

which are roots of the equation

$$\Omega_0^2 \frac{dV}{du} = f. \tag{8}$$

The roots of Eq. (8) for which $d^2V/du^2 > 0$ determine stable equilibrium positions and they correspond to the local minima of the biased potential $U(u)$, while the roots with

85

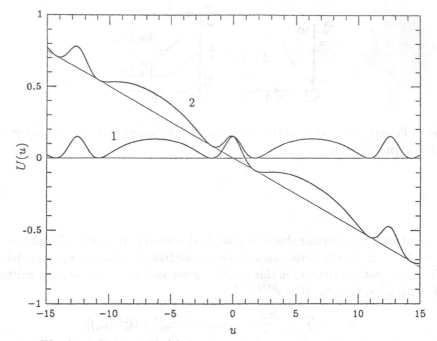

Figure 7: The biased potential (7) for $f = 0$ (curve 1) and for $f = 0.1$ or $F = 0.3\text{eV}/\text{Å}$(curve 2).

$d^2V/du^2 < 0$ give unstable equilibrium points, i.e. the local maxima of the potential (7).

Let u_1^* and u_2^* ($u_1^* < u_2^*$) be any pair of adjacent stable roots of Eq. (7). Then for small external forces ($|f| \ll 1$) they can be found explicitly:

$$u_1^* = -u_0 + \Omega_0^{-2}f \quad \text{and} \quad u_2^* = u_0 + \Omega_0^{-2}f. \tag{9}$$

Due to the periodicity of the function $V(u)$ all the stable equilibrium positions can be represented by the two sets of points $4\pi n \pm u_i^*$ for $i = 1, 2$ and $n = 0, \pm 1, \dots$.

In general, the difference-differential equation (6) can be solved only numerically. For analytical investigations we apply the continuum approximation, substituting n by x and $u_n(\tau)$ by $u(x, \tau)$. Then Eq. (6) can be transformed to the (linearly) damped nonlinear Klein-Gordon equation with an external force:

$$u_{\tau\tau} - u_{xx} + \Omega_0^2 \frac{dV}{du} + bu_\tau = f. \tag{10}$$

Note that the continuum approximation (displacive limit[33]) can be applied if Ω_0 is sufficiently small. In this case the discreteness effects can be neglected and the Peierls potential relief disappears. Also, for the studies of particle-like properties of the ionic and orientational defects the continuum limit is very useful. We introduce the moving (with velocity s) coordinate $\xi = x - s\tau$ and look for the solutions of Eq. (10) with a kink/antikink profile. We denote these solutions, which in general are not required to be constant in time, by $u_K(x, \tau)$ where $K = I_\pm$ or $B_\pm$. As in the previous section I_+ stands for a positive ionic defect (antikink), I_- for a negative ionic defect (kink), B_+ for a positive orientational (Bjerrum) defect (antikink), and B_- for a negative orientational defect (kink). The boundary conditions at any time τ can be written in the form

$$u_{I_-}(-\infty, \tau) = u_{I_+}(\infty, \tau) = 4\pi n + u_1^*, \quad u_{I_+}(-\infty, \tau) = u_{I_-}(\infty, \tau) = 4\pi n + u_2^*,$$

$$u_{B_-}(-\infty,\tau) = u_{B_+}(\infty,\tau) = 4\pi n + u_2^*, \; u_{B_+}(-\infty,\tau) = u_{B_-}(\infty,\tau) = 4\pi(n+1) + u_1^*, \quad (11)$$

where $n = 0, \pm 1, \dots$. We should also impose the condition that the spatial derivative of the kink profile $u_K(x,\tau)$ is a function localized in space, i.e.

$$u_{K;x}(\pm\infty,\tau) = 0. \quad (12)$$

A number of physical quantities associated with the free ($f = b = 0$) ionic and orientational kink/antikink solutions such as the energy, momentum, effective mass, width, and charge of both types of the defects can be defined and calculated[28,34]. Following the physical prerequisities of the hydrogen bond a two-parameter on-site proton potential

$$V(u;\alpha,\beta) = \frac{2}{1-\alpha^2}\left(\frac{\varphi}{1-\beta\varphi}\right)^2 \quad (13)$$

with $\varphi = \cos(u/2) - \alpha$ and $\alpha = \cos(u_0/2)$ can be introduced[28]. Its form has been depicted in Fig. 7 (see curve 1) for the parameter values $\alpha = 0.666$, $\beta = 1.251$, $\Omega_0 = 0.358$, and $u_0 = 1.684$ which have been chosen to satisfy the continuum approximation of Eq. (6). In this potential α controls the relative widths of both barriers as well as the distance between the two minima in each lattice cell. The second parameter β controls the relative heights of the two barriers. These heights are given by

$$V(4\pi n) = 2R_0^2(1+\alpha\beta-\beta)^{-2} \text{ and } V(4\pi(n+\tfrac{1}{2})) = 2R_0^{-2}(1+\alpha\beta+\beta)^{-2} \quad (14)$$

where $n = 0, \pm 1, \dots$, $0 \le \alpha < 1$, $-(1+\alpha)^{-1} < \beta < (1-\alpha)^{-1}$ and $R_0 = [(1-\alpha)/(1+\alpha)]^{1/2}$. The on-site potential form (13) is reduced to the well known sine-Gordon case[34] for $\alpha = \beta = 0$ and to the double sine-Gordon case[35] when only $\beta = 0$. The kink/antikink solution of Eq. (10) with $f = b = 0$ in the case of the potential (13) can been written analytically[28]. The velocity spectrum for these solutions is the interval $-1 < s < 1$.

In the presence of an external field and damping ($f \ne 0$ and $b > 0$) the kink velocity is no more a free parameter. For a constant force f it takes some fixed value s_K, $K = I_\pm$ and $B_\pm$, uniquely determined by the external (constant) force F and damping Γ. The effects of both F and Γ (or f and b) are cancelling each other, and as a result of this cancellation a uniform motion of the kink (or antikink) occurs similarly to Stoke's law. Therefore, one can study the mobility and other particle-like quantities of such carriers as the ionic and orientational defects in hydrogen-bonded chains.

In order to introduce some important physical quantities in the presence of external fields and damping we write the equation of balance between the energy dissipated and the energy provided by the field. Differentiating the expression (cf. (5))

$$E(\tau) = \varepsilon_0 \sum_n \left[\frac{1}{2}\left(\frac{du_n}{d\tau}\right)^2 + \frac{1}{2}(u_{n+1} - u_n)^2 + U(u_n)\right] \quad (15)$$

for the total energy and using Eq. (6), we find the following balance equation:

$$\frac{dE}{d\tau} + \varepsilon_0 b \sum_n \left(\frac{du_n}{d\tau}\right)^2 = 0 \; . \quad (16)$$

The balance equation (16) describes an uniform motion of kinks/antikinks. Therefore, in the continuum limit the solutions with constant profile can be written as $u_K = u_K(\xi)$ where $\xi = x - s_K\tau$, $x = n$, and $s_K = v_K/c_0$. In this limit, the first two terms in the energy (15) are constant in time and the differentiation of the third term gives

$$\sum_n \frac{dU(u_n)}{dt} = -s\Delta U_K \quad (17)$$

where

$$\Delta U_K = \int dU(u_K) = U[u_K(\infty)] - U[u_K(-\infty)], \tag{18}$$

so that $\Delta U_{I_\pm} = \pm[U(u_1^*) - U(u_2^*)]$ for the ionic kinks/antikinks and $\Delta U_{B_\pm} = \pm[U(u_2^*) - U(u_1^* + 4\pi)]$ for the orientational kinks/antikinks. Therefore, Eq. (16) is transformed to

$$bs_K \int (\frac{du_K}{d\xi})^2 d\xi = \Delta U_K \tag{19}$$

where the profile of the kink/antikink $u_K(\xi)$ depends on its velocity s_K, so that this relation may be considered as an implicit equation for finding the kink velocity s_K. It follows from this equation that for $F > 0$, when the protons move to the right, the kink (negative defect) moves to the left and the antikink (positive defect) moves to the right.

Only in the two particular cases[22,36,37] the kink profiles can be given analytically and therefore Eq. (19) can be written in an explicit form. In the general case we can calculate the integral in Eq. (19) only approximately, using the profile for the free kinks (when $f = b = 0$). We expect this profile to be not very different from the kink profile in the case of $f \neq 0$ and $b \neq 0$, at least for small values of f. As a result, we find the following expression for the kink velocity:

$$s_k = s_K(f) = \Phi_K(f)[1 + \Phi_K^2(f)]^{-1/2} \tag{20}$$

where the function $\Phi_K(f)$ is defined by

$$\Phi_K(f) = \frac{2^{-3/2}\Delta U_K}{b\Omega_0 A_K}, \quad A_K = \int_{\Lambda_K} [V(u)]^{1/2} du, \tag{21}$$

where Λ_K denotes the integration substrate which for $K = I_\pm$ is $[0, u_0]$ and for $K = B_\pm$ is $[u_0, 2\pi]$.

Having the velocity-force characteristic (20) established, we are able to define the mobilities of the defects as

$$\mu_K = |v/\mathcal{E}| = \mu_0 |s_K/f| \tag{22}$$

where $\mathcal{E}$ is an external electric field ($F = e\mathcal{E}$), $v_K = s_K c_0$, $\mu_0 = 4\pi el/mc_0$ and the velocity s_K is given by (20). In the limit $f \to 0$ we have

$$|\Delta U_K| = 2R_K |f| + O(f^2) \tag{23}$$

where $R_K = u_0$ for $K = I_\pm$ (ionic defects) and $R_K = 2\pi - u_0$ for $K = B_\pm$ (orientational defcts). Therefore, in limit $f \to 0$, the expressions (20) – (23) yield the following exact formula for the kink mobility:

$$\mu_K(f \to 0) = 2^{-1/2}\mu_0 R_K / b\Omega_0 A_K. \tag{24}$$

One of the other important physical quantities is the kink momentum given by

$$P_K(\tau) = -\frac{\varepsilon_0}{c_0} \int u_{K;x}(x, \tau) u_{K;\tau}(x, \tau) dx. \tag{25}$$

In the case of a constant external force and damping, Eq. (19) can be used to find the momentum of a uniformly moving kink. As a result, we find

$$P_K = \frac{\varepsilon_0 \Delta U_K}{c_0 b}. \tag{26}$$

The case of an alternative external field is rather more complicated but in this case we are also able to establish some particle-like properties of both the types of defects. Let us assume the external force f in (6) to be a function of time τ. The varying field $f(\tau)$ clearly causes the "vacuum" motion $\varphi(\tau)$ of the displacement field $u(x,\tau)$ in the vicinities of the equilibrium points $4\pi n \pm u_0$ where $n = 0, \pm 1, \ldots$. Then, following Olsen and Samuelsen[38] we assume the varying force $f(\tau)$ to be sufficiently small in order to represent the field $u(x,\tau)$ as the sum of the pure kink/antikink part $u_K(x,\tau)$ and of the vacuum part $\varphi(\tau)$, i.e. we can write

$$u(x,\tau) = u_K(x,\tau) + \varphi(\tau). \tag{27}$$

In this case the local minima $4\pi n \pm u_0$ at any time τ are displaced by $\varphi(\tau)$, so that $u(x,\tau)$ satisfies the boundary conditions

$$u_x(\pm\infty,\tau) = 0 \quad \text{and} \quad u_\tau(-\infty,\tau) = u_\tau(\infty,\tau). \tag{28}$$

The equation of motion for the vacuum part $\varphi(\tau)$ can be easily established. Indeed, substituting (27) into Eq. (10) and using the boundary conditions (28) we find the linearized equation

$$\varphi_{\tau\tau} + b\varphi_\tau + \Omega_0^2\varphi = f. \tag{29}$$

To study the dynamics of the defects in external varying fields we use a dynamical equation for a momentum of the defects. To find this equation we represent, by using (27), the total momentum (25) as the sum $P = P_K - (\varepsilon_0/c_0)d_K\varphi_\tau$ where $d_K = u(\infty,\tau) - u(-\infty,\tau)$. Next, multiplying Eq. (10) by $u_x = u_{K;x}$ and integrating it by using the boundary conditions (28) as well as Eqs. (27) and (29), we find the following equation for the kink/antikink momentum $P_K(\tau)$:

$$\frac{d}{d\tau}P_K + bP_K + (\varepsilon_0/c_0)d_K\Omega_0^2\varphi = 0. \tag{30}$$

A general solution of this equation is written as

$$P_K(\tau) = P_K(0)e^{-b\tau} - \frac{\varepsilon_0}{c_0}d_K\left[\int_0^\tau f(\tau')e^{b(\tau'-\tau)}d\tau' - \varphi_\tau + \varphi_\tau(0)e^{-b\tau}\right]. \tag{31}$$

In order to obtain the dependence of the velocity of the defect s_K on time we suppose the velocity-force characteristic (20) to be valid also for varying fields f (adiabatic approximation). Then, from the expressions (20), (21), and (26) we find the dependence of the defect velocity s_K on its momentum:

$$s_K = P_K[2(2\varepsilon_0\Omega_0 A_K/c_0)^2 + P_K^2]^{-1/2}. \tag{32}$$

For small velocities this dependence has a linear behavior, i.e.

$$s_K = dx_K/d\tau = (c_0/2^{3/2}\varepsilon_0\Omega_0 A_K)P_K \tag{33}$$

where $x_K = x_K(\tau)$ denotes the position of the kink/antikink center.

Some particular cases of the solution (31), (32), or (33) can be considered. Let, for instance, the external force be absent, i.e. $f(\tau) = 0$, and the initial condition have the form: $P_K(0) = 0$, $\varphi(0) = \varphi_0$, and $\varphi_\tau(0) = 0$. Then, using the solutions of Eq. (29) with $f(\tau) = 0$ we find from the expression (31) the dependence of the defect momentum P_K on time:

$$P_K(\tau) = -\frac{\varepsilon_0}{c_0}d_K\varphi_0\Omega_0^2\frac{\sin(\Omega\tau)}{\Omega}e^{-b\tau/2}, \quad \Omega^2 = \Omega_0^2 - (b/2)^2. \tag{34}$$

For small velocities (see (33)) we arrive at the following dependence of the velocity s_K on time:

$$s_K = -2^{-3/2}\varphi_0 \frac{d_K}{A_K}\Omega_0 \frac{\sin(\Omega\tau)}{\Omega} e^{-b\tau/2}. \qquad (35)$$

The other interesting case is that of an external harmonic force, i.e. $f(\tau) = f_0 \cos(\Omega_e\tau + \delta_0)$ where Ω_e and δ_0 are the frequency and initial phase of the ac-driving field. In this case under the initial conditions $\varphi(0) = 0$ and $\varphi_\tau(0) = 0$ from the general expression we find

$$P_K(\tau) = P_K(0)e^{-b\tau} + \frac{\varepsilon_0}{c_0}d_K f_0 \{ \frac{1}{b^2 + \Omega_e^2}[e^{-b\tau}(b\cos\delta_0 + \Omega_e\sin\delta_0) -$$

$$- b\cos(\Omega_e\tau + \delta_0) - \Omega_e\sin(\Omega_e\tau + \delta_0)] + [(\Omega_e^2 - \Omega_0^2)^2 + b^2\Omega_e^2]^{-1/2} \times \qquad (36)$$

$$\times [e^{-b\tau/2}\left(\Omega_0^2\cos(\delta + \delta_0)\frac{\sin(\Omega\tau)}{\Omega} + \Omega_e\sin(\delta + \delta_0)(\cos(\Omega\tau) - \frac{b\sin(\Omega\tau)}{2\Omega})\right) -$$

$$-\Omega_e\sin(\Omega_e\tau + \delta + \delta_0)]\}$$

where the phase δ is given by $\tan\delta = b\Omega_e/(\Omega_e^2 - \Omega_0^2)$. In the particular case when the damping can be neglected, i.e. $b = 0$, from (36) after integration, the relation $x_K = x_K(\tau)$ can be easily obtained. Let the initial soliton velocity be zero and $\Omega_e \neq \Omega_0$ (non-resonance case). Then, $\Omega = \Omega_0$, $\delta = 0$, and the behavior of the soliton center is realized under the action of the harmonic force as follows

$$x_K(\tau) = x_K(0) + (\varepsilon_0/M_K c_0^2)d_K f_0 \{|\Omega_e^2 - \Omega_0^2|^{-1} \times$$

$$\times \left[\left(\frac{\Omega_e}{\Omega_0}\sin(\Omega_0\tau) - \sin(\Omega_e\tau)\right)\sin\delta_0 + (\cos(\Omega_e\tau) - \cos(\Omega_0\tau))\cos\delta_0\right] + \qquad (37)$$

$$+ \Omega_e^{-2}\left[(\Omega_e\tau - \sin(\Omega_e\tau))\sin\delta_0 - 2\sin^2\frac{\Omega_e\tau}{2}\cos\delta_0\right]\}$$

where $M_K = 2^{3/2}(\varepsilon_0/c_0^2)\Omega_0 A_K$ is the rest mass of the kinks/antikinks. It follows from (37) that the motion of a defect includes an oscillating term plus a drifting term with the velocity proportional to $\sin\delta_0$ (note that the constant component of the external force is absent in this case). Such a result means a peculiar phasing of oscillations of the vacuum state caused by the external force and reflects the particle-like property of the soliton state. Indeed, a particle, under the action of a varying force, according to the equation of motion $mdv/dt = A_0\cos(\omega t + \delta)$ acquires the constant component of its velocity equal to $(A_0/\omega)\sin\delta$. This result is in good agreement with numerical calculations[39].

CONCLUSIONS AND DISCUSSION

We have presented in this paper a theoretical study of a quasi-one-dimensional hydrogen-bonded system. This study takes into account a zig-zag structure of the system and uses the realistic parameter values known for $(HF)_\infty$, $(HCl)_\infty$, and $(HBr)_\infty$ in the literature[30]. Similarly to the previous studies[27], the on-site potential for the hydrogen-bonded protons in these zig-zag chains has been constructed as the sum of the two-dimensional Morse potentials which represent the pair potentials for a proton and are centered at *nine* nearest sites of the chain. Summing over more than nine pair potentials has been proved not to change practically the resulting on-site proton potential and this is why the number *nine* was chosen under this modelling. Defining the on-site proton potential in this way we were able to study the collective proton dynamics on the two-dimensional plane. Note that the nearest-neighbor proton-proton interactions in the present paper have been chosen in a more reasonable way than in our previous work[27].

Modelling the on-site potential relief in the two dimensions by using the Morse potential we have found it to form the zig-zag trajectory (channel) for the longitudinal motion of hydrogen-bonded protons. Along this channel, where the potential energy of the protons takes the lowest values, the relief takes the form of a periodic function (as shown in Fig. 3(a)) with two barriers per one period. Therefore, the well known one-dimensional Frenkel-Kontorova model[31] with a doubly periodic on-site potential can be effectively adopted and many results known for nonlinear Klein-Gordon equations may be used for the studies of the collective proton dynamics. Some physical quantities such as the energy, effective mass, and mobility of both the ionic and orientational defects can be analytically calculated if the continuum approximatiom is adopted. Of course, this is a crude approximation because for the realistic parameter values, as was shown in the present paper, the Peierls potential relief has been proved to exist. Therefore, the discreteness effects should be taken into account. In general, for the one-dimensional FK model these effects are well studied. Concerned to the studies of the proton transfer the migration of the defects should be studied in the framework of the *zig-zag* model with the *two-dimensional* on-site proton potential. The motion of heavy ions, which allows to study the temperature effects in a correct manner, should be also taken into account in the way as it was done previously[29] for the one-dimensional two-component model. All these studies are now in progress and they will be the subject of our subsequent papers.

ACKNOWLEDGMENTS

Both of the authors (A.V.S. and A.V.Z.) are indebted for hospitality to the Research Center of Crete that enabled this work to be completed.

REFERENCES

1. P.V. Hobbs, "Ice Physics" Clarendon, Oxford (1974).

2. "Physics of Ice", N. Riehl, B. Bullemer, and H. Engelhardt, eds., Plenum, New York (1969).

3. "Physics and Chemistry of Ice", E. Whalley, S.J. Jones, and L.W. Gold, eds., Royal Society of Canada, Ottawa (1973).

4. A. Kuwada, A.R. McGhie, and M.M. Labes, J. Chem. Phys. **52**: 1438 (1970).

5. V.H. Schmidt, J.E. Drumheller, and F.L. Howell, Phys. Rev. **B4**: 4582 (1971).

6. M. Springborg, Phys. Rev. Lett. **59**: 2287 (1987); Phys. Rev. **B38**: 1438 (1988).

7. R.W. Jansen, R. Bertocini, D.A. Pinnick, A.I. Katz, R.C. Hanson, O.F. Sankey, and M. O'Keeffe, Phys. Rev. **B35**: 9830 (1987).

8. J.F. Nagle and H.J. Morowitz, Proc. Natl. Acad. Sci. U.S.A. **75**: 298 (1978); J.F. Nagle, M. Mille, and H.J. Morowitz, J. Chem. Phys. **72**: 3959 (1980); E.W. Knapp, K. Schulten, and Z. Schulten, Chem. Phys. **46**: 215 (1980).

9. J.F. Nagle and S. Tristram-Nagle, J. Membrane Biology **74**: 1 (1983).

10. N. Bjerrum, K. danske Vidensk. Selsk. Skr. **27**: 1 (1951); Science **115**: 385 (1952).

11. T. Mitani, G. Saito, and H. Vrayama, Phys. Rev. Lett. **60**: 2299 (1988).

12. B. Brzezinski, G. Zundel, and R. Kramer, J. Phys. Chem. **91**: 3077 (1987).

13. G.P. Johari and S.J. Jones, J. Chem. Phys. **82**: 1019 (1985); S.F. Fischer, G.L. Hofacker, and M.A. Ratner, J. Chem. Phys. **52**: 1934 (1970).

14. J.H. Weiner and A.Askar, Nature **226**: 842 (1970).

15. V.Ya. Antonchenko, A.S. Davydov, A.V. Zolotaryuk, phys. stat. solidi (b) **115**: 631 (1983); E.W. Laedke, K.H. Spatschek, M. Wilkens, Jr., and A.V. Zolotaryuk, Phys. Rev. **A32**: 1161 (1985); M. Peyrard, St. Pnevmatikos, and N. Flytzanis, Phys. Rev. **A36**: 903 (1987); H. Weberpals and K.H. Spatschek, Phys. Rev. **A36**: 2946 (1987); H. Lan and K. Wang, Phys. Lett. **A139**: 61 (1989); E.S. Nylund and G.P. Tsironis, Phys. Rev. Lett. **66**: 1886 (1991).

16. Y. Kashimori, T. Kikuchi, and K. Nishimoto, J. Chem. Phys. **77**: 1904 (1982).

17. S. Yomosa, J. Phys. Soc. Jpn. **51**: 3318 (1982); **52**: 1866 (1983).

18. A.V. Zolotaryuk, K.H. Spatschek, and E.W. Laedke, Phys. Lett. **A101**: 517 (1984); A.V. Zolotaryuk, Teor. Mat. Fiz. **68**: 415 (1986) [Sov. Theor. Math. Phys. **68**: 916 (1987)]; J. Halding and P.S. Lomdahl, Phys. Rev. **A37**: 2608 (1988); D. Hochstrasser, H. Büttner, H. Desfontaines, and M. Peyrard, Phys. Rev. **A38**: 5332 (1988); H. Desfontaines and M. Peyrard, Phys. Lett. **A142**: 128 (1989).

19. St. Pnevmatikos, Phys. Lett. **A122**: 249 (1987).

20. A.V. Zolotaryuk, Kink models in biophysics, in: "Biophysical Aspects of Cancer", J. Fiala and J. Pokorný, eds., Charles University, Prague (1987), pp. 179-188.

21. St. Pnevmatikos, Phys. Rev. Lett. **60**: 1534 (1988); G.P. Tsironis and St. Pnevmatikos, Phys. Rev. **B39**: 7161 (1989).

22. A. Gordon, Physica **B146**: 373 (1987); **B150**: 319 (1988); **B151**: 453 (1988); L.N. Kristoforov and A.V.Zolotaryuk, phys. stat. solidi (b) **146**: 487 (1988).

23. E.S. Kryachko, Solid State Commun. **65**: 1609 (1988); O.E. Yanovitskii and E.S. Kryachko, phys. stat. solidi (b) **147**: 69 (1988).

24. A.V. Zolotaryuk and St. Pnevmatikos, Two-component kink dynamics of hydrogen-bonded chains, in: "Singular Behavior and Nonlinear Dynamics", St. Pnevmatikos, T. Bountis, and Sp. Pnevmatikos, eds., World Scientific, Singapore (1989), vol. 2, pp. 384-410; St. Pnevmatikos, G.P. Tsironis, and A.V. Zolotaryuk, J. Molec. Liquids **41**: 85 (1989).

25. A.V. Zolotaryuk and A.V. Savin, Proton conductivity in systems with hydrogen bonding: a new application of the soliton theory, in: "Solitons and Applications", V.G. Makhankov, V.K. Fedyanin, and O.K. Pashaev, eds., World Scientific, Singapore (1989), pp. 429-433; A.V.Zolotaryuk and St. Pnevmatikos, Two-sublattice nonlinear dynamics of kink-bearing models, in: "Nonlinear World", A.G. Sitenko, V.E. Zakharov, and V.M. Chernousenko, eds., Naukova Dumka, Kiev (1989), vol. 1, pp. 211-214; A.V. Zolotaryuk, St. Pnevmatikos, and A.V. Savin, Physica D **44** (1991).

26. E.S. Kryachko, Chem. Phys. **143**: 359 (1990).

27. A.V. Savin, A two-dimensional soliton model for proton transport in hydrogen-bonded molecular chains, in: "Nonlinear World", A.G. Sitenko, V.E. Zakharov, and V.M. Chernousenko, eds., Naukova Dumka, Kiev (1989), vol. 1, pp. 165-168; St. Pnevmatikos, A.V. Savin, and A.V. Zolotaryuk, Longitudinal and transverse proton collective dynamics on a two-dimensional multistable substrate, in: "Nonlinear Coherent Structures" (Lecture Notes in Physics, vol. 353), M. Barthes and J. Léon, eds., Springer, Berlin (1990), pp. 83-92.

28. A.V. Zolotaryuk and St. Pnevmatikos, Phys. Lett. **A143**: 233 (1990); St. Pnevmatikos, A.V. Savin, A.V. Zolotaryuk, Yu.S. Kivshar, and M.J. Velgakis, Phys. Rev. **A43**: 5518 (1991).

29. A.V. Savin and A.V. Zolotaryuk, Two-sublattice solitons in hydrogen-bonded chains with dynamical disoder, in: "Nonlinearity and Disorder", F.Kh. Abdullaev, A.R. Bishop, and St. Pnevmatikos, eds., World Scientific, Singapore (1991).

30. A. Anderson, B.H. Torrie, and W.S. Tse, J. Raman Spectroscopy **10**: 148 (1981).

31. J. Frenkel and T. Kontorova, J. Phys. USSR **1**: 137 (1939).

32. St. Pnevmatikos, "Solitons: Part II. Topological Anharmonic Waves" (in Greek), Crete University, Heraklion (1991).

33. J.A. Krumhansl and J.R. Schrieffer, Phys. Rev. **B11**: 3535 (1975).

34. J.F. Currie, J.A. Krumhansl, A.R. Bishop, and S.E. Trullinger, Phys. Rev. **B22**: 477 (1980).

35. C.A. Condat, R.A. Guyer, and M.D. Miller, Phys. Rev. **B27**: 474 (1983); R.M. DeLeonardis and S.E. Trullinger, Phys. Rev. **B27**: 1867 (1983).

36. E. Maggari, Z. Phys. B **550**: 137 (1984).

37. J.C. Kimball, Phys. Rev. **B21**: 2104 (1980).

38. O.H. Olsen and M.R. Samuelsen, Phys. Rev. **B28**: 210 (1983).

39. St. Pnevmatikos, A.V. Savin, A.V. Zolotaryuk, Yu.S. Kivshar, and M.J. Velgakis, Nonlinear transport in hydrogen-bonded chains: solitons under external fields and damping (to be published).

SOLITON PROPAGATION THROUGH INHOMOGENEOUS HYDROGEN-BONDED CHAINS

Yuri S. Kivshar

Institute for Low Temperature Physics and Engineering
47 Lenin Avenue, SU-310164 Kharkov (U.S.S.R.)

Present address: Departamento de Física Teórica I
Facultad de Ciencias Físicas, Universidad Complutense
E-28040 Madrid (Spain)

1 Introduction

Since the basic work [1] (see also [2]) on the dynamics of hydrogen-bonded chains, various models using the same basic idea have been proposed (see, e.g., Refs.[3]-[10]). The main idea is that protons move in double wells due to hydrogen bonds with a heavy-ion lattice (oxygen lattice) which is deformable. The local distortions of the oxygen lattice can lower the activation barrier for the protons and thus promote their motion. In order to describe this effect models must include two coupled degrees of freedom at each lattice site, so that the proton sublattice supports topological solitons (kinks) while, in the absence of coupling, the equations for the oxygen sublattice are linear ones. The models proposed in Refs.[1]-[10] seem to be an effective description of the proton mobility in hydrogen-bonded chains, and they may also play an important role in interpreting certain biological processes in such systems[11].

To describe the proton mobility in hydrogen-bonded systems, it is also necessary to estimate the influence of inhomogeneities which always exist in real systems. Such impurities may sufficiently change the steady-state motion of solitons because they may attract or repulse solitons so that a soliton *may be trapped by the inhomogeneities*. Moreover, propagating in an inhomogeneous medium, a soliton emits radiation in the form of linear waves. This radiation is usually small for one-component soliton-bearing systems because the on-site potential supporting the topological solitons (kinks) yields a gap in the spectrum of linear excitations. Due to the gap, the soliton emission stipulated by inhomogeneities is exponentially small in the soliton velocity (see, e.g., Ref.[12]). However, for an attractive impurity, the kink soliton may excite an impurity mode (see, e.g., Refs.[13],[14]), a localized oscillating state at the impurity site, which also may be important in the energy exchange mechanizm of the kink scattering [15].

There are two basic models for hydrogen-bonded systems which have difference even at the level of their linear spectra. One type of the models corresponds to the interaction between the optical branch, resulting from the hydrogen subsystem, which takes into account the proton motion in the effective double-well potential, and the acoustic branch, which describes oscillations of the heavy-ion oxygen subsystem [4]-[10]. The other type of the models takes into account effectively interactions between different hydrogen-bonded chains, assuming that the heavy-ion oxygen subsystem is described as the other optical branch [1],[2], the effective gap being due to the inter-chain interactions. In the present paper I will deal with a model of the first type. It is clear that in such a case emission of a soliton moving with a variable velocity is not small: it is easy for it to generate linear waves in the oxygen subsystem, because the latter has no gap. Therefore, the soliton emission in hydrogen-bonded models has to play an important role in estimating the proton mobility in real systems.

The paper considers the soliton scattering by inhomogeneities in the hydrogen-bonded systems and it is based on my original paper[16]. Section 2 presents the model of an inhomogeneous hydrogen-bonded chain which takes into account a local change of the parameters of the oxygen subsystem. In Section 3 I discuss importance of radiative effects accompanying the soliton scattering and in Section 4 I demonstrate that the soliton may be trapped by an inhomogeneity due to radiation-induced losses. Section 5 deals with the soliton propagation in a disordered hydrogen-bonded chain. Using the independent-scattering approach I estimate the propagation length of the soliton which losses its energy due to emission of acoustic waves in the oxygen subsystem. Section 6 concludes the paper.

2 Model

The model, similar to that considered in Refs.[1],[4], consists of a diatomic chain of proton and oxygen atoms (see, e.g., Fig.1 of Ref.[6]). Such a model includes all degrees of freedom for the oxygen atoms, unlike the original model investigated in the pioneering paper[1]. The Hamiltonian consists of three parts. The proton part is given by

$$H_p = \sum_n [\frac{1}{2}m(\frac{dp_n}{dt})^2 + \frac{1}{2}k_p(p_{n+1} - p_n)^2 + U(p_n)], \qquad (1)$$

with the on-site potential

$$U(p) = \frac{1}{4}U_0[1 - (p/p_0)^2]^2. \tag{2}$$

The proton coordinate p_n is the displacement of the hydrogen from the middle of the bond in a static case. The oxygen Hamiltonian is written as

$$H_O = \sum_n [\frac{1}{2}M(\frac{dq_n}{dt})^2 + \frac{1}{2}k_q(q_{n+1} - q_n)^2], \tag{3}$$

where q_n describes the displacement from the equilibrium position. In the original model proposed in [1] (see also [2]) only optical vibrations were considered, whereas in the models considered in Refs.[4], [8] acoustic vibrations of the oxygen sublattice were taken into account.

The interaction Hamiltonian is in the standard form (see, e.g., Ref.[1]),

$$H_{int} = \frac{1}{2}W_0 \sum_n (q_{n+1} - q_n)[1 - (p_n/p_0)^2]. \tag{4}$$

The physical content of this interaction is the lowering of the double potential barrier due to the oxygen displacement. In principle, analytical results may be obtained for a more general form of the interaction potentials (see, e.g., Refs.[2],[4], and [9]).

In the continuum limit, which is valid for the long-wave approximation, when the variable $na \approx z$ is considered as continuous, the Hamiltonian (1)-(4) yields the motion equations,

$$u_{tt} - u_{xx} - u(1 - u^2) = -\alpha uv_x, \tag{5}$$

$$v_{tt} - s^2 v_{xx} = \alpha\beta uu_x, \tag{6}$$

where

$$u = p/p_0, v = q/q_0, \quad s^2 = s_0^2/c_0^2, \tag{7}$$
$$c_0^2 = k_p a^2/m, \quad s_0^2 = k_q a^2/M,$$
$$\alpha = (\frac{aq_0 W_0 \omega_0}{U_0 c_0}), \quad \beta = (\frac{m}{M})(\frac{p_0}{q_0})^2, \quad \omega_0^2 = \frac{U_0}{mp_0^2},$$

a being the lattice spacing. In Eqs.(5) and (6) I have used the dimensional variables,

$$\tau = \omega_0 t, \quad z = \frac{\omega_0}{c_0}x. \tag{8}$$

The parameter α stands for an effective coupling between the hydrogen and oxygen sublattices.

In the absence of impurities, the homogeneous equations (5) and (6) support the steady-state propagation of the kink soliton in the hydrogen subsystem together with its "shadow" in the oxygen subsystem,

$$u(x,t) = \pm \tanh(\xi/l(V)), \ \xi = x - Vt, \tag{9}$$

$$\frac{dv(\xi)}{d\xi} = \frac{\alpha\beta}{2(s^2 - V^2)\cosh^2(\xi/l(V))}, \tag{10}$$

where the parameter $l(V)$,

$$l^2(V) = \frac{2(1 - V^2)(V^2 - s^2)}{(V^2 - s^2 + \alpha^2\beta)}, \tag{11}$$

has a sense of the soliton width.

The simple analysis demonstrates that the kink solution (9)-(11) exists provided

$$0 < V^2 < s^2 - \alpha^2\beta, \tag{12}$$

or

$$s^2 < V^2 < 1, \tag{13}$$

and it is proved that in the region (12) the soliton is a stable solution [4].

The standard model for the hydrogen-bonded chain briefly presented above assumes homogeneous parameters along the chain and, as a result, the motion of soliton excitations is possible in the steady-state regime (see, e.g., Refs.[2] and [5]). In this paper I will investigate the influence of local inhomogeneities (impurities) of the oxygen subsystem on the soliton motion. To obtain equations of motion in the inhomogeneous case, I assume that in the vicinity of the impurity the parameters of the oxygen subsystem are changing, so that

$$M \to M + a\Delta M\delta(z) = M(1 + \epsilon_M\delta(x)), \tag{14}$$

$$k_p \to k_q + a\Delta k_q\delta(z) = k_q(1 + \epsilon_q\delta(x)), \tag{15}$$

where

$$\epsilon_M \equiv a\left(\frac{\omega_0}{c_0}\right)\frac{\Delta M}{M}, \ \epsilon_q \equiv a\left(\frac{\omega_0}{c_0}\right)\frac{\Delta k_q}{k_q}. \tag{16}$$

The equation of motion taking into account the influence of the impurity may be derived directly from the Hamiltonian (1)-(4) using Eqs.(14),(15),

$$u_{tt} - u_{xx} - u(1 - u^2) = -\alpha u v_x, \tag{17}$$

$$v_{tt} - s^2 v_{xx} = \alpha\beta u u_x - \epsilon_M\delta(x)v_{tt} + s^2\epsilon_q[\delta(x)v_x]_x. \tag{18}$$

To conclude this section, I will note that the more general case which describes the change of the *all* parameters of the model, i.e. the hydrogen as well as oxygen subsystems, has been considered in my original paper [16].

3 Emission of acoustic waves by soliton

During the scattering by impurities a soliton changes its velocity and it generates radiation [12]. The total energy emitted by the soliton is stipulated by spectral properties of the linear system because the radiation is usually small-amplitude one. For example, the radiation is exponentially small in the case of the Klein-Gordon models for $V^2 \ll 1$ (see details, e.g., in the review paper [12]). Due to the radiative losses, a soliton moving from infinity and scattering by an attractive impurity may be captured by it. The radiation-induced threshold velocity for such a process is known [12]. In the hydrogen-bonded chains two subsystems are coupled, and the oxygen sybsystem usually has no gap. It means that radiation-induced effects in the two-component hydrogen-bonded chains play much more important role then in one-component ones. For example, the thershold velocity of the radiation-induced capture of the soliton by a local attractive inhomogeneity is not small in the two-component models, and such an effect is very important for transport properties of hydrogen-bonded systems [17].

To demonstrate some features of radiation-induced effects in the model under consideration, I will calculate the spectral density and the total energy emitted by the soliton scattering by the local impurity. In the zeroth approximation I have to consider emission generated by the solution (9),(10) with $\xi = x - Vt$, assuming, for simplicity, the case of a slowly moving kink, when $V^2 \ll s^2 - \alpha^2 \beta, s^2$.

To simplify the calculations, firstly let us consider the equation

$$v_{tt} - s^2 v_{xx} = F(x,t), \qquad (19)$$

which is, in fact, Eq.(18), where $F(x,t) \equiv \alpha u u_x - \epsilon_M \delta(x) v_{tt} + s^2 \epsilon_q [\delta(x) v_x]_x$. If $F(x,t)$ is a localized function of x, the calculations of the energy emitted under the action of such a force is standard. Let us represent the wave field emitted by this force in the form,

$$v(x,t) = \frac{1}{2\pi} \int_{-\infty}^{\infty} dk\, v_k(t) e^{ikx}, \qquad (20)$$

and similarly for the force $F(x,t)$, its Fourier amplitude is $F_k(t)$. The equation of motion for the Fourier amplitude $v_k(t)$ can be brought into the form,

$$\frac{da_k(t)}{dt} = -iska_k(t) + F_k(t), \qquad (21)$$

where

$$a_k(t) \equiv \frac{dv_k(t)}{dt} - iskv_k(t). \qquad (22)$$

The proper energy of waves in the oxygen susbsystem may be presented as

$$E_{em} = \frac{1}{2\beta} \int_{-\infty}^{\infty} dx(v_t^2 + s^2 v_x^2) = \int_{-\infty}^{\infty} dk\, \mathcal{E}(k), \qquad (23)$$

where

$$\mathcal{E}(k) = \frac{1}{2\beta} | \int_{-\infty}^{\infty} dt \int_{-\infty}^{\infty} dx F(x,t) e^{-ik(x-st)} |^2. \qquad (24)$$

The result (23),(24) is a basis to calculate radiation-induced losses of the soliton due to emission in the oxygen subsystem.

If $F(x,t)$ stads for the r.h.s. of Eq.(18), then emission will be generated only due to the second and third terms. Substituting these terms into Eq.(24), we may obtain for $V^2 \ll s^2 - \alpha^2\beta$ the spectral density in the form,

$$\mathcal{E}(k) = \frac{1}{8}\pi^2\alpha^2\beta(\frac{kl_0}{V})^4 \frac{(\epsilon_M V + \epsilon_q s)^2}{\sinh^2(\pi k s l_0/2V)}. \qquad (25)$$

where $l_0 \equiv l(V=0)$.

The total emitted energy may be calculated according to Eq.(23) as the following $(V > 0)$,

$$E_{em} = \frac{4\pi}{15} \frac{\alpha^2\beta V}{s^5 l_0}(\epsilon_M V + \epsilon_q s)^2. \qquad (26)$$

As can be seen from Eq.(26), the emission is determined by the soliton velocity and it is not exponentially small as for one-component models.

4   Soliton trapping by impurity

The emitted energy (26) allows us to calculate the threshold velocity for the soliton to be captured by the attractive impurity. Indeed, let us consider a soliton scattering by the impurity which starts to move from infinity, where its velocity is V_0. In the case of the attracltive impurity (see details in Ref.[16]), radiative losses calculated above may result in the capture of the soliton. The threshold velocity for such an effect may be calculated equating the kinetic energy of the soliton, $E_k = \frac{1}{2}m_{eff}V_0^2$, $m_{eff} = \frac{2}{3}l_0(1+\frac{\alpha^2\beta}{2s^4})$ being the effective mass of the kink, and the total emitted energy due to the scattering which is defined by Eq.(26). The result is

$$(V_0)_{thr} = \frac{8\pi}{15} \frac{\alpha^2\beta\epsilon_q^2}{s^3 l_0 m_{eff}}, \qquad (27)$$

where I have taken into account that the term $\sim \epsilon_M$ is of the second order in the impurity amplitude. Therefore, for $V_0 \leq (V_0)_{thr}$ the soliton will be captured by the impurity due to radiative losses, and this effect does not depend on the effective change of the oxygen mass M in the vicinity of the impurity.

5 Soliton propagation in a disordered chain

Let us consider now a disordered hydrogen-bonded chain with local impurities. When the concentration ρ of the impurities is low, i.e. in the dilute limit, the average distance between two nearest-neighboring impurities is larger than the soliton size. In this limit the scattering of a soliton by many impurities can be approximately treated independently. In other words, I don't take into account interference effects. The transmitted soliton for the i-th impurity is the incident one for the (i+1)-th impurity. Then I can write the following relation

$$E_{i+1} = E_i T_i(E_i), \tag{28}$$

where $T_i(E_i)$ is the soliton transmission coefficient of the i-th impurity. Then,

$$\Delta E_{i+1} = E_{i+1} - E_i = -E_i[1 - T_i(E_i)] = -E_i R_i(E_i), \tag{29}$$

where $R_i(E_i)$ is the reflection coefficient. Since on the average there are $(\Delta x)\rho$ impurities in the interval between x and $x + \Delta x$, from Eq.(29) the following differential equation can be derived,

$$\frac{dE}{dx} = -\rho E(x) R[E(x)]. \tag{30}$$

The soliton reflection coefficient $R(E)$ may be defined as a ratio of the energy, emitted by the soliton during its scattering in the direction opposite to its propagation direction. In the case when the soliton emits radiation in another (oxygen) subsystem, for the reflected energy I may use the total energy emitted by the soliton in this subsystem. Then, the value $R(x)E$ has the sense of the emitted energy E_{em} defined above as the soliton radiation-induced losses during its scattering by the inhomogeneity. If, instead of E, I will put in Eq.(30) the kinetic energy of the soliton defined as $E_k = \frac{1}{2}m_{eff}V_0^2$, where m_{eff} is the soliton effective mass, which takes into account a renormalization due to the oxygen subsystem (see definition above), then Eq.(30) gives rise the law of radiation-induced damping of the soliton velocity in the case when impurities are installed into the oxygen subsystem only,

$$m_{eff}\frac{dV}{dx} = -\rho B(\epsilon_M V + \epsilon_q s)^2, \tag{31}$$

where

$$B = \frac{4\pi}{15}\frac{\alpha^2 \beta}{s^5 l_0}. \tag{32}$$

The solution of Eq.(32) with the initial condition $V(x = 0) = V_0$ has the form

$$V(x) = \frac{\epsilon_M V_0 - \epsilon_q s C x}{\epsilon_M (1 + C x)}, \tag{33}$$

where

$$C = \frac{4\pi}{15} \frac{\rho \epsilon_M \alpha^2 \beta (\epsilon_M V_0 + \epsilon_q s)}{s^5 l_0 m_{eff}}. \tag{34}$$

Equations (33),(34) allow us to estimate the propagation length L of the soliton in the disordered system,

$$L \sim \frac{\epsilon_M V_0}{\epsilon_q s C}, \tag{35}$$

where C is defined in Eq.(34). It is important to note that, according to Eq.(35), L tends to infinity for $\epsilon_M \to 0$, i.e. for a pure isotopic disorder.

6 Conclusions

In conclusion, I have studied analytically the soliton motion in a hydrogen-bonded chain with impurities. It has been shown that the soliton may be captured by the attractive impurity due to emission of linear waves into the oxygen subsystem. The emission of the soliton in a disordered hydrogen-bonded chain results in a fast decreasing of the soliton velocity. I have calculated the decay of the transmission coefficient of the soliton in such a case and I have determined the propagation length of it in a disordered hydrogen-bonded system using the so-called independent-scattering approach. The results obtained in the paper may be useful to describe the effect of disorder on the hydrogen mobility in hydrogen-bonded systems.

I would like to thank Michel Peyrard, Stephanos Pnevmatikos (it is very regrettable that he tragically died while I was working on this problem), and Alex Zolotaryuk for valuable discussions.

References

[1] V.Ya.Antonchenko, A.S.Davydov, and A.V.Zolotaryuk, phys. status solidi (b) 115, 631 (1983).

[2] A.V.Zolotaryuk, Teor. Mat.Fiz. 68, 415 (1986).

[3] S.Yomosa, J.Phys. Soc. Jpn. 51, 3318 (1982).

[4] A.V.Zolotaryuk, K.H.Spatschek, and E.W.Ladke, Phys. Lett. A 101, 517 (1984).

[5] E.W.Ladke, K.H.Spatschek, M.Wilkens, Jr., and A.V.Zolotaryuk, Phys. Rev. A 20, 1161 (1985).

[6] M.Peyrard, St.Pnevmatikos, and N.Flytzanis, Phys.Rev. A **36**, 903 (1987).

[7] St.Pnevmatikos, Phys. Lett. A **122**, 249 (1987).

[8] D.Hochstrasser, H.Bütner, H.Desfontaines, and M.Peyrard, Phys. Rev. A **36**, 5332 (1988).

[9] St.Pnevmatikos, Phys.Rev.Lett. **60**, 1534 (1988).

[10] J.Halding and P.S.Lomdahl, Phys.Rev. A **37**, 2608 (1988).

[11] J.F.Nagle, M.Mille, and H.J.Morovitz, J.Chem.Phys. **72**, 3959 (1980).

[12] Yu.S.Kivshar and B.A.Malomed, Rev.Mod.Phys. **61**, 793 (1989).

[13] T.Fraggis, St.Pnevmatikos, and E.N.Economou, Phys. Lett.A **142**, 361 (1989).

[14] O.M.Braun and Yu.S.Kivshar, Phys.Rev.B **43**, 5419 (1990).

[15] Zhang Fei, Yu.S.Kivshar, and L.Vázquez, submitted to Phys.Rev. B (1991).

[16] Yu.S.Kivshar, Phys.Rev. A **43**, 3117 (1990).

[17] St.Pnevmatikos, A.V.Savin, A.V.Zolotaryuk, Yu.S.Kivshar, and M.J.Velgakis, Phys.Rev. A **43**, 5518 (1990).

THE CONCEPT OF SOLITON-CARRIER COLLECTIVE VARIABLE FOR PROTON

TRANSFER IN EXTENDED HYDROGEN-BONDED SYSTEMS: OVERVIEW

Eugene S. Kryachko [§+] *and Vladimir P. Sokhan* [‡+]

[§] *University of Munich, D-8000 Munich 2, FRG*
[+] *Institute for Theoretical Physics, Kiev, 252130, USSR*
[‡] *University of Constance, D-7750 Constance 1, FRG*

PREFACE

It seems now to be the right time to take a overlook at the application of the concept of soliton[1,2] in the theory of hydrogen bonding, and the theory of anomalous proton conductivity in liquid water and ice particularly. The important progress continues to be made in this area at a substantial pace. The hot example is just these Proceedings. A number of problems which are nearly as old as the object under study itself (e.g., liquid water) remains open, and the new problems continue to arise and develop. On the other hand, this area is now about ten years old, and already many problems have been solved and the key to the general theory of proton transfer we guess has been posed.

At what point in the development of this new area our overlook be started? And the review be written about? Is anything in the hydrogen bonding theory related to solitons? More than a decade ago no one believed in that. But, in 1982-83 there appeared three papers[3-5] (see also[6]) which gave the second (or maybe third) "breath" to the theory of collective proton transfer in extended hydrogen-bonded systems. And that was exactly the solitonic "breath". Does it look like completely new? "Yes" and "no".

Let start with "no". Could we speak about or even think of solitons in hydrogen-bonded systems if the latter ones have no the nice property what we already called the coooperativity of hydrogen bonding? Certainly not! In our case the cooperativity is the cornerstone in applying the solitonic doctrine.

The cooperativity of hydrogen bonding has different facets especially in liquid water and ice (see[7] and refs. therein) which organization in a rather extended hydrogen-bonded network obeys the Bernal—Fowler-Pauling rules[8]. The energetic one goes back to Frank and Wen[9] who used this terminology to explain the stability of the H-bonded structure of liquid water. The non-additivity of the binding energy of water oligomers[10] frames the energetic facet. The charge facet of cooperativity is determined by non-additivity of the total dipole moment of water clusters[11]. The striking pattern of the other one, the so-called spectroscopic facet, is the well-known drastic change in the harmonic force constant of stretching vibrations[12]. For liquid water, under a passage from gas phase to the liquid or amorphous, or ice, it consists of about -300 cm^{-1} (the red shift of OH-stretching vibration). This red shift cannot be explained in terms of dimer-like conformations with corresponding change of the harmonic force constant and $\Delta\nu$ of about $-60 \div -80$ cm^{-1}, but becomes well understood by using the concept of cooperativity[13].

If we are talking about a proton transfer, the cooperativity means in fact an existence of some collective variables, no matter of how we nickname them, of the number much less the total number of all degrees of freedom of the system in question. For example, these fewer variables describe the concerted proton transfer in the systems with two or more hydrogen bonds observed by Scheiner[14] in his computer screen.

Theory of solitons is the mathematical subject dealing with the specific solutions of nonlinear Euler-Lagrange equations of motion of completely integrable systems. Physics as well as chemistry always include mathematics pragmatically to use as a language, or "tool". And this is exactly that case. The aforementioned collective variables are framed by the solitonic Ansatz in such "funny" way that they can be often handled analytically that advertised the people to make the business in this area. That is why the beauty of the concept of soliton in view of the hydrogen bonding theory is in its analyticity as well as in the physical transparency, although once again, it is worth to notice that the theory of solitons is not a physical theory, it is only the "tool" to investigate the collective variables. How to choose them is the problem of physics, and the physical intuition guides us how properly to write down a Hamiltonian.

1. SOLITONIC CONCEPT OF PROTON TRANSFER (BABY VERSION)

A baby version of the solitonic model for proton transfer has been delivered the decade ago[3-5]. Its parents limited protons to move in one-dimensional channel formed by the molecules of water which indeed looks like a rather good "caricature" of real liquid water or ice (1D ice)[15-17]. To make the next step, that is, to write Hamiltonian involving the solitonic-carrier collective variables, one needs primarily to define what we are going to describe by them. Let we call them as bond defects.

1.1. Bond Defects Revisited

Bond defects can be defined as breakdowns of either Bernal-Fowler-Pauling rule: A number of protons, occupying a chosen H-bridge, is equal to: [i] zero ("empty" bond) or two ("double" bond) (a case of three or more protons is inconceivable); [ii] oxygen possesses single or three neighboring protons.

Taking a look at these breakdowns' definitions, one can immediately give the examples of bond defects. A bond defect, corresponding to the former definition, is called an orientational defect, because one can organize the "empty" or "double" H-bond by rotating a single OH-bond in the Bernal-Fowler configuration out or to a given H-bond, respectively. The simplest model of orientational defect was suggested by Bjerrum[18]: L-defect corresponds to a single "empty" ("leer") H-bond without proton, O...O; D-defect to a single "double" ("doppelt") H-bond, O-H...H-O. The orientational defects of both types originate from rotating a single OH-bond in the Bernal-Fowler configuration around the bisector of $O_1...O...O_2$, through the formation of DL-pair. It is clear that L-defect is characterized by a deficiency of positive charge, that is why it carries a certain negative charge, say $-e_B$, while the D-defect is positively charged, for its charge being $+e_B$.

Another bond defect refers to the ionic one. As in the above case, one can fairly easy give the simplest candidate for the ionic defect. The H_3O^+ ion, i.e., the oxygen atom with three neighboring protons, represents itself a positive ionic defect (call it P-defect) with charge $+e_I$, and OH$^-$ a negative one (or N-defect) granted by the charge $-e_I$.

Now we are quite prepared ourselves to write the scenario of proton migration in liquid water and how it contributes to the anomalous conductivity in terms of given simple models of bond defects. Assuming the H_3O^+ - OH$^-$ pair is formed by the thermal fluctuations in the H-bonded network and applying an external electric field, one can break it down that results in migrating P- and N-defects separately, in the opposite directions along hydrogen bonds, until they reach the electrodes. One can provide a conductivity process to be continuous by suggesting that at the moment of arrival of ionic defects at electrodes, the H-bonded pathway, or "channel" is turned by means of generating orientational defects. Migrating and reaching the edges of the protonic channel, these defects give a start to the subsequent step of proton-transfer, and so on. Therefore, the proton-transfer process rather schematically consists of two stages, for each stage being responsible for carrying the definite charge which unifying at the edges, results in the total charge of proton, by the absolute value, i.e., $e = e_B + e_I$, what was expected[19].

Concluding this Subsection it should be noted that the picture given above is rather simplified, mainly due to the rough and schematic model of bond defects applied which neglects a cooperativity. As a result, the distortion of the network of H-bonds in the vicinity of bond defects are poorly described because the definition of localized bond defects assumes tacitly that the surrounding H-network remains the Bernal-Fowler-Pauling one. This contradicts to the numerical calculations[20-22] of the lattice relaxation taking place under either defect to be organized, and the experimental data of the positive volume extention in the its region[23], too.

Now reminding that we already have in hands a one-dimensional channel of water molecules, how the generalization of the definition of bond defects to take the H-bonding cooperativity into account is going on. Two geometrical variables play the basic role. The first one is the angle $\Theta_i \equiv 1/2 \sphericalangle O_{i+1} - O_i - O_{i-1}$ describing a rotation of the proper OH-bond dipole belonging to the i-th water molecule in the interior part of a channel. Due to the symmetry of the model system chosen, the own potential energy $U(\Theta_i)$ of the proper i-th OH-bond dipole has two equivalent minima at $\Theta_i = \pm \Theta_e$. The $2\Theta_e$-th fluctuation jumping, classically over the barrier $U_B^{(0)} = U(\Theta_i = 0)$, from one minimum at $\Theta_i = +\Theta_e$ to the equivalent at $\Theta_i = -\Theta_e$, correlated with associated transition of neighboring OH-bond dipoles owing to the presence of the potential energy term,

$$\frac{x^{(0)}}{2} \left(\Theta_i - \Theta_j \right)^2 \tag{1}$$

generates the orientational defect. We hope that the term in Eq.(1) or something like it describes a cooperativity between the jumps of i-th and j-th dipoles. In Eq.(1) one assumes that the orientational coupling constant, $x^{(0)}$, is certainly non-vanishing. This term allows to overcome the aforementioned difficulties of the Bjerrum model corresponding to $x^{(0)} = 0$. In the latter case, the DL-pair generated at the i-th site decays into quasi-isolated D- and L-defects as time is going on, due to the subsequent OH-bond reorientations, that finally results in the defected chain, comprising of two subchains. Each subchain involves its own D- or L-defect. They differ from each other by the directions (topology) of the OH-bonds at their "edges". For the left one, involving the positive D-defect, its left-side-edge OH-bond occupies the minimum at $\Theta = +\Theta_e$, while the opposite one, at $\Theta = -\Theta_e$. The reverse arragement of minima of the edge OH-bonds is peculiar for the right subchain with the negative L-defect. Generalizing that property on non-vanishing $x^{(0)}$, one can define the orientational defects as those structures in the chain which satisfy the following boundary conditions[24,25]:

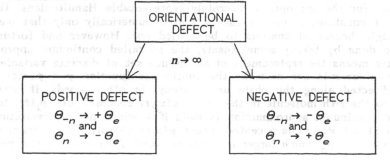

107

Therefore, the orientational defect represents itself a certain transition state between these limiting topology of OH-bonds. For the word "transition" means a discrete set of OH-bond angles, $\{\Theta_i\}$, realizing rotations of OH-bond dipoles from one limiting angle $(\pm\Theta_e)$ to the opposite one $(\mp\Theta_e)$.

The second variable, a co-length r of OH-bond, is related to ionic defects. We assume that a proton can migrate one-dimensionally, along a single H-bond, O-H...O, and that process is described by a unique reaction coordinate r directed along the O...O axis. We also assume that the i-th proton moves in the effective potential field $U(r_i)$, symmetrical relative to $r_i = 0$, the centre of the O...O bond, and posessing two equivalent minima at $r_i = \pm r_0$. Here $r_0 = (R/2 - r_e)$, r_e is the equilibrium length of the OH-bond (the equilibrium co-length is r_0), and R that of the O...O bond. These minima are separated by the energy barrier with height $U_B^{(i)} = U(r_i = 0)$. A migration of the i-th proton is no doubt coupled with those of other protons in the chain, due to the following term,

$$\frac{x^{(i)}}{2}\left(r_k - r_I\right)^2, \qquad (2)$$

for instance, with the non-vanishing (i)onic coupling constant $x^{(i)}$. Generalizing the schematic model of the ionic defect such as H_3O^+ and OH^-, on a domain of non-vanishing coupling constants, one can very similarly to the case of the orientational defects, give the definition of ionic defect as the definite structures in our quasi-one-dimensional chain, obeying the following "edge" conditions[5,6,26]:

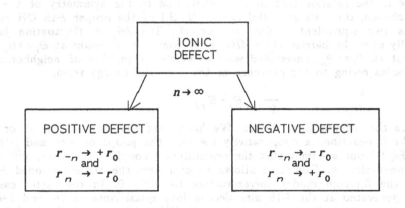

Pictorially, the ionic defect is such one-dimensional structure of water molecules whose OH-bonds are "lengthened" discretely (that is described by the set $\{r_i\}$) between two "edge" OH-bonds with the equilibrium lengths but the different oxygen atom property. Therefore, we can suggest that the collective excitations (or structures) of the orientational and ionic defect types described by the sets of corresponding variable, $\{\Theta_i\}$ or $\{r_i\}$, do exist. As we learned, the rigorous way to elucidate their behavior is to solve the Euler-Lagrange equations of motion for the appropriate discretely representable Hamiltonians. Unfortinately, these equations of motion can be solved numerically only that makes hard a thorough theoretical analysis to be carried out. However and fortunately, it could be done by taking some Ansatz, the so-called continuum approximation. The latter means the replacement of an infinite set of discrete variables $\{\varphi_i(t)\}$ where $\varphi = \Theta$ or r in our case, by the continuous function $\varphi = \varphi(x,t)$ with the x-axis directed along the chain under study. In other words, if $x = x_i$ corresponds to the i-th molecule in the chain, $\varphi(x_i, t)$ becomes just $\varphi_i(t)$. It is clear that the continuum approximation is valid if a width of the excitation region $2\Delta x$ around $x = x_0$ (x_0 is a centre) where $\varphi(x, t)$ sharply differs from the equilibrium values $\pm\varphi_e$, much larger a lattice constant L (relative to the x-th axis).

1.2. Somnething More About Continuum-Discrete Competition

Assume here that we could write down the quite reasonable Hamiltonian for our discrete chain. A passage grom discrete Hamiltonian to its continuum analogue can be made by using, for example, the following relation

$$\varphi_{i+1}(t) \equiv \varphi(x_{i+1}) = \varphi(x_i + L, t)$$

$$= \varphi_i(t) + L\partial\varphi/\partial x + O\left((\partial\varphi/\partial x)^2\right) \tag{3}$$

and neglecting further the last term in *rhs* of Eq. (3). For this reason, the potential energy term

$$\frac{x}{2}\left(\varphi_{i+1} - \varphi_i\right)^2 \tag{4}$$

[compare with Eqs. (1) and (2)] transforms to

$$\frac{x}{2}L^2\left(\partial\varphi/\partial x\right)^2 \tag{5}$$

Let assume for a while that we have already got the soliton-like solution $\varphi(x, t)$ of the Euler-Lagrange equation within the continuum approximation,

$$\frac{\partial^2\varphi}{\partial t^2} - C_0^2\frac{\partial^2\varphi}{\partial x^2} = -\frac{dU}{d\varphi} \tag{6}$$

Here $U(\varphi)$ is the on-site potential, and C_0^2 comes from the numerical factor in Eq. (5).

How we should come back to the discrete case? Or, in other words, what is the inverse procedure to construct $\{\varphi_i(t)\}$ from $\varphi(x, t)$? At the first glance, the answer seems to be very simple: substituting discrete values of x into $\varphi(x, t)$ results in $\{\varphi_i(t)\}$. However, it is not that case, because when the continuum approximation is chosen, the coupling term (4) is linearized, that is, it is suggested that the terms under $O((\partial\varphi/\partial x)^2)$ in *rhs* of Eq. (3) are considerably smaller than $L\partial\varphi/\partial x$. But the main difficulty still remains if we even take into account the above terms in a somewhat way. It consists in different symmerties inherent for solutions. Within the continuum approach neglecting the terms under $O((\partial\varphi/\partial x)^2)$ together with the common kinetic energy term the term in Eq. (5) leads to that Eq. (6) becomes Lorentz covariant. Hence, if $\varphi(x - x_0)$ stands for its static solution obeying a "half" of Eq. (6),

$$C_0^2\frac{\partial^2\varphi}{\partial x^2} = \frac{dU}{d\varphi} \tag{7}$$

a travelling solitary wave $\varphi(x, t)$ is merely $\varphi(x - x_0 - Vt)$. The latter describes the static configuration, $\varphi(x - x_0)$, propagating along the chain in the positive direction with a constant velocity $V > 0$. Dislike this, the translational symmetry becomes discrete and forbids to take values of the displacement of soliton which are not integral multipliers of the lattice constant L. So, the caution is of great necessity when we speak about the defect in a discrete chain in terms of its continuum prototype because the reverse step is not yet clear so far.

1.3. Hamiltonian and Kink-Defects in Baby Version

We guess that the reader still patiently expects a written Hamiltonian. Now we are ready to write it for the subblattice of proton with mass m,

$$H = \sum_{n=-N}^{N}\left[\frac{m}{2}\left(\dot{\varphi}_n\right)^2 + U\left(\varphi_n\right)\right] + \sum_{n,m=-N}^{N}V_{nm}\left(\varphi_n - \varphi_{n-m}\right), \tag{8}$$

where $\dot{\varphi}_n \equiv d\varphi_n/dt$.

H seems to be of a rather general form which includes all possible Hamiltonians for our system and reproduce its image H' resulting, in turn, in Eq. (7). In Eq. (8) the summation ranges over all sites labelled by n, however, that over m in the elastic strain term is not rarely cutted on the first and/or second neighbors. This term can be chosen to have the Hook's-law form,

$$V_{n,n-m} = \frac{1}{2} A_m (\varphi_n - \varphi_{n-m})^2, \quad m = 1, 2; \tag{9}$$

although speaking generally, A_i can depend on $\varphi_n - \varphi_{n-i}$, too, e.g.[27],

$$A_i = A_i^{(0)} + \frac{1}{3} A_i^{(1)} (\varphi_n - \varphi_{n-i}) + \frac{1}{3 \times 4} A_i^{(2)} (\varphi_n - \varphi_{n-i})^2. \tag{9'}$$

Some baby versions of H preferred to choose the on-site potential term in sine form to ascribe a complete integrability for the system under study. In our treatment it has the well-known φ-four form. Within the continuum approximation, H transforms to the following ones,

$$H^{(0)'} = \frac{1}{2L} \int_{-\infty}^{\infty} dx \left\{ m r_e^2 (\dot{\theta})^2 + m r_e^2 C_{00}^2 (\partial \theta / \partial x)^2 + 2 U_B^{(0)} \left[1 - (\theta / \theta_e)^2 \right]^2 \right\} \tag{10a}$$

$$H^{(i)'} = \frac{1}{2L} \int_{-\infty}^{\infty} dx \left\{ m (\dot{r})^2 + m C_{i0}^2 (\partial r / \partial x)^2 + 2 U_B^{(i)} \left[1 - (r / r_0)^2 \right]^2 \right\}. \tag{10b}$$

Some comments are needed now to estimate parameters involved in $H^{(0)'}$ and $H^{(i)'}$. We assume that OH-bond dipoles located at different water molecules interact with each other as point-like dipoles. The equilibrium dipole moment μ_e of a single OH-bond at the equilibrium distance $r = r_e$ corresponds to the effective charge q_H of the "proton". Beyond $r = r_e$, the bond dipole moment behaves to a good accuracy as a sine function[28,26]. The values of parameters obtained in such a way[24-26] are given in table 1.

Table 1. Properties of orientational and ionic defects in zig-zag chain $(H_2O)_\infty$

Orientational defect	Ionic defect
$r_e = 1.0$ Å, $\quad 2\theta_e = 109.5°$	
$R = 2.76$ Å (ordinary ice)	$R = 2.5$ Å (symmetric ices)
$q_H = 0.6 \, e^{15}$, $\mu_e = 2.85$ D	$q_H = 0.85 \, e$, $\left. d\mu/dr \right\|_{r_e} = 4.083$ D Å^{-1} [28,29]
$U_B^{(0)} = 7.68$ kcal mol^{-1}	$U_B^{(i)} = 10.2$ kcal mol^{-1}
	$U_B^{(i)}{}_{expt} = 9.5 \pm 0.7$ kcal mol^{-1} [30]
$C_{00} = 4.24 \times 10^6$ cm s^{-1}	$C_{i0} = 65.6 \times 10^5$ cm s^{-1}
$E_f^{(0)} = 15.7$ kcal mol^{-1}	$E_f^{(i)} = 23.74$ kcal mol^{-1}
$E_f^{(0)}{}_{expt} = 15.6 \pm 0.9$ kcal mol^{-1}	$E_f^{(i)}{}_{expt} = 22 \pm 3$ kcal mol^{-1}

Deriving the Euler-Lagrange equation of motion from Eqs. (10a-b) and taking the boundary conditions into account, one gets the typical solitonic solutions,

$$\theta(x,t) = \mp \theta_e \tanh \left(\sqrt{2 U_B^{(0)} / \left[m \left(1 - (V/C_{00})^2 \right) \right]} \, \frac{x - x_0^{(0)} - Vt}{r_e C_{00} \theta_e} \right) \tag{11a}$$

$$r(x,t) = \mp r_0 \tanh\left(\sqrt{2U_B^{(i)} / \left[m\left(1 - (V/C_{i0})^2\right)\right]} \frac{x - x_0^{(i)} - Vt}{C_{i0}\theta_e}\right) \qquad (11b)$$

for the orientational and ionic defects, respectively. Their formation energies are presented in table 1.

"Project" the kink-soliton solutions, Eqs. (11a-b), onto our discrete zig-zag chain $(H_2O)_\infty$ of water molecules. This "projection" procedure ended Subsection 1.1. Consider the case of orientational defect at the rest $(V = 0)$. Instead of $\Theta(x, t = 0)$ we have now the discrete distribution $\{\Theta_i \equiv \Theta(x_i, t = 0)\}$ of angles of OH-bonds constituting, roughly speaking, the discrete orientational defect. The OH-bond of the central water molecule and two neighboring ones refer to "weak", or "broken" H-bonds using the well-known "yes–no" chemical terminology. Even more, the central OH-bond is bifurcated. In the paper[25] we estimated the harmonic force constant of stretching vibrations of this bond via the model developed elsewhere[13] and arrived at the conclusion that it appears to be a good candidate for the "free" OH-group which sounds similarly with that by Guérre[31] from the "peroxide" viewpoint (see also[32] and refs. therein). In our case, such "free" OH-bond is not isolated in the sense that there exists a semi-continuum transition from this bond to the usual, or "strong" ones, and this transition ensemble of OH-bonds is just what we call the orientational kink-defect. One thing here becomes more transparent. It concerns the discreteness-continuum competition. You see that the "free" OH-group lies exactly at the top of the barrier if we "project" the continuum solution onto the discrete reality. So, whether does it mean that a "free" OH-group appears to be unstable? Not certainly because the continuum approximation leaves behind the treatment other, N-1, degrees of freedom N protons, $N \to \infty$ and singles out the only one, the collective variable. These N-1 degrees of freedom are driven by the solitonic collective variable, and apparently are responsible for the local behavior of OH-bonds forming the kink–defect and hence performing some vibrations in the appropriate local potentials which can differ drastically from the original one chosen in the Θ-four form in Eq. (10a), for instance.

2. SOLITONIC CONCEPT OF PROTON TRANSFER (REAL THING)

2.1. Solitons and Hydrogen Bonding

How we should draw the demarcation line between the soliton model of collective proton migration along hydrogen bridges and that of any other type of nonlinear excitation? This doubt still remains. In other words, where is something (it must be) in the nonlinear equations of motions resulting in a soliton (or quasisoliton in a non-integrable case[33]) solution, or in the total Hamiltonian which ensures the well-known specifity of hydrogen bonding.[12,34] One can conclude that this specifity comes from the potential energy part of the total Hamiltonian which if we are talking about hydrogen bonding, should be essentially of three–body nature,[12,34] and hence, the Hamiltonian is two-component. That is why a building block of any extended hydrogen-bonded systems is merely a hydrogen bridge A-H...B with rather specified atoms A and B.

Consider a quasi-one-dimensional infinite chain $(A_i-H_i...A_{i+1})_\infty$ of homoconjugated hydrogen-bonded blocks A-H...A where A can be the oxygen atom, for instance. The potential energy U is approximated by two terms: the intra-block, U_{intra}, and that between any couple of blocks, U_{inter}. The essential feature of U_{intra}, as we have already mentioned, is that U_{intra} is nonseparable, i.e., it cannot be decomposed into the sum of terms describing the motions of the hydrogen atom H, and the heavier atoms separately. and the final term which couples H_i with $A_i...A_{i+1}$. The most adequate analytical candidates for U_{intra} are the double-Morse and Lippincott-Schroeder potentials.[12,34-35] Both absorb many specific features of hydrogen bonding. In particular, for symmetric hydrogen bonds U_{intra} interpreted as a potential curve of proton transfer if A ($= B$) is an atom, takes a single- or double-minimum form depending on R, the distance between the A-th atoms in the single block. For this reason, there exists a definite bridge length R_{thr} at which U_{intra} collapses from the double well to a single well. It is worth to notice here that the total Hamiltonians of the chain $(A\text{-}H\ldots)_\infty$ with the double-Morse and Lippincott-Schroeder potentials are no longer integrable. At least, the contrary is not proved. Hence, there is no guarantee that the kink-soliton exists there at all.

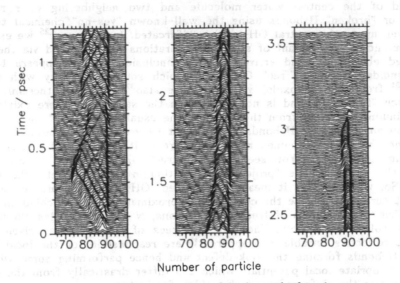

Figure 1. Time evolution of kink–antikink pair.

Just to emphasize the point what does the hydrogen bonding reality dictate us, in figure 1 we plot the numerically generated kink-antikink pair for the two-component chain $(A\text{-}H\ldots)_\infty$ with the Lippincott-Scroeder potential.[36] At the initial moment of time $t = 0$, the pair is generated at the 88-th hydrogen-bonded block. Wavy lines in figure describe the phonon modes of the heavier subsystem.

Almost instantaneously, this pair delivers other pairs in its milieu. At nearly $t = 0.5\times10^{-12}$ s, the ensemble of kink-antikink pairs transforms to a single and broader one embodied about 12-16 blocks. As time advances, this new pair becomes more localizable that finally results in that kink and antikink annihilate each other, and the pair is dead. In the case considered the lifetime of quasisolitonic pair is finite, about 3 ps. This lifetime is rather long and enough to make about 300 stretching vibrations of the A-H type. This example tell us about certain difference between our quasisoliton and those obtained via commonly used potentials.

2.2. Lippincott–Schroeder Kink: Few Results

Unfortunately, the space to present the results about the formation of quasi kink-defect in the $(H_2O)_\infty$ chain with double-Morse[37] and Lippincott-Schroeder potentials[36] in detail is rather limited. That is why the most preferable way we choose is to go through the discussion of the Lippincott-Schroeder quasikink simulated numerically leaving for readers the double-Morse one which is fortunately analytically resolved.[37]

So, let us embark with the Lippincott-Schroeder kink. Consider a one-dimensional infinite chain of water molecules ($A \equiv OH$) of the ordinary ice geometry with potential energy, U_{intra}, for the main building block of the Lippincott-Schroeder type[35]

$$U_{LS}(\varphi, R) = U_1(\varphi, R) + U_2(\varphi, R) + U_3(R) + U_4(R), \tag{12}$$

where R is the distance between two oxygens in the building block O–H...O, and φ is the displacement of proton from the hydrogen bridge midpoint. The first two terms in right-hand side of (12),

$$U_1(\varphi, R) = D_0 \left\{ 1 - \exp\left[- \frac{\alpha(R/2 + \varphi - \varphi_0)^2}{(R/2 + \varphi)} \right] \right\} \tag{13}$$

and

$$U_2(\varphi, R) = - D_0 \exp\left[- \frac{\alpha(R/2 - \varphi - \varphi_0)^2}{(R/2 - \varphi)} \right], \tag{14}$$

represent the proton potential in the field of two neighboring oxygens, and

$$U_3(R) + U_4(R) = A \exp(-bR) - C/R^6 \tag{15}$$

approximates the oxygen-oxygen interaction and includes the van-der-Waals attraction as well as the electrostatic repulsion. The fitted values of the parameters of the potential $U_{LS}(\varphi, R)$ are summarized in table 2.

Table 2. Set of model parameters used in numerical simulations for U_{intra} (Lippincott-Schroeder potential) and U_{inter}. $\tilde{A}_1^{(0)}$ is the harmonic term in series expansion of interaction between heavier atoms.

A	=	8.07×10^6	$kcal\, mol^{-1}$
b	=	4.76	$Å^{-1}$
C	=	295.5	$Å^6\, kcal\, mol^{-1}$
D_0	=	118	$kcal\, mol$
α	=	4.59	$Å^{-1}$
φ_0	=	0.95	$Å^{-1}$
$A_1^{(0)}$	=	24.2	$kcal\, mol^{-1}\, Å^{-2}$
$\tilde{A}_1^{(0)}$	=	156	$kcal\, mol^{-1}\, Å^{-2}$

The potential energy curve for the migration of a hydrogen atom along the O–H...O bond, depending of the value of distance between oxygens, R, may have one or two stable minima. The barrier height, U_b^{LS}, and the position of minima, $\pm \varphi_0^{LS}$, for the "ordinary ice" value of $R = 2.76\,Å$ are given in table 3. The distance at which the double-well potential collapses into a single well is defined by the condition $\partial^2 U_{LS}/\partial\varphi^2|_{\varphi=0} = 0$ which leads to equation for R

$$\frac{\alpha}{16} R^4 - \frac{\alpha}{2} \varphi_0^2 R^2 - \varphi_0^2 R + \alpha \varphi_0^4 = 0 \,,$$

with solution $R_0 = 2.537\,Å$. The property of double-to-single-well collapse of the potential (12) is one of the remarkable properties specifying hydrogen bonding, especially for systems with strong hydrogen bonds.[12,34] No other commonly used potential posesses it, except the double-Morse potential.[37] All the parameters fitted result in the proton potentials being in the agreement with the *ab initio* ones.[14]

The second part of the total potential energy, U_{inter}, is determined by the cutting off the dipole-dipole interaction between the point-like OH-bond dipoles belonging to different building blocks. Its series expansion yields the requi-

Table 3. Properties of proton in equilibrium configuration for Lippincott-Schroeder potential with $L = 2.76$ Å

U_b^{LS}	φ_0^{LS}	$\hat{U}_b^{LS}$	$\varphi_0(L)$
in $kcal\,mol^{-1}$	in $\mathring{A}$	in $kcal\,mol^{-1}$	in $\mathring{A}$
10.59	±0.355	10.36	±0.348

red form, given by equations (9) and (9'). The parameter $A_1^{(0)}$ was fitted to reproduce adequately the ab initio quantum chemical data. Evidently, the parameter $A_1^{(0)}$ depends on L. However, this dependence is rather weak in the interval of reasonable lattice constants,[26] and so can be neglected. Moreover, U_{inter} is also determined by the intermediate-range interaction between the heavier atoms which in the case of a chain consisting of water molecules is quite well-studied and rather accurately approximated. Accepting the similar Hook's law form for this part, one can estimate the corresponding $\tilde{A}_1^{(0)}$ for the heavier subsystem (see table 2).

Repetition of the unit cell generates the chain, in which one oxygen serves as the right-hand oxygen in a one cell and a left-hand in the next. The chain considered contains N unit cells, with typical value of 100 for N.

In dynamical simulation we start with some initial configuration and then the coupled set of equation of motion for protons in the Lippincott-Schroeder chain:

$$m\ddot{\varphi}_n - A_1^{(0)}\left(\varphi_{n+1} - 2\varphi_n + \varphi_{n-1}\right) = -\alpha D_0\left\{\left[3 - \frac{2\varphi_0}{L/2 + \varphi_n} - \left(\frac{\varphi_0}{L/2 + \varphi_n}\right)^2\right]\right.$$

$$\times \exp\left[-\frac{\alpha(L/2 + \varphi_n - \varphi_0)^2}{L/2 + \varphi_n}\right] - \left[3 - \frac{2\varphi_0}{L/2 - \varphi_n} - \left(\frac{\varphi_0}{L/2 - \varphi_n}\right)^2\right]\exp\left[-\frac{\alpha(L/2 - \varphi_n - \varphi_0)^2}{L/2 - \varphi_n}\right]\right\} \quad (16)$$

together with corresponding system of equations for heavier subsystem and with applied periodic boundary conditions was solved. Futher details of numerical procedure used are now given elsewhere.[36]

Results of such simulation, presented in figure 1 shows dynamical behaviour of the kink pictorially. More detailed picture can be obtained by studying the structure of the kink. A kink, or strictly speaking, a quasikink, is the definite solution (if there be one) of the set of coupled algebraic equations:

$$\frac{\partial}{\partial \varphi_n}\left[U(\varphi_n) + \sum_{m=-N}^{N} V_{nm}(\varphi_n - \varphi_{n-m})\right] = 0$$

$$n = -N, -N+1, \ldots, N-1, N$$

satisfying fixed boundary conditions

$$\varphi_{\mp n}(0) = \pm\varphi_0, \qquad \dot{\varphi}_{\mp n}(0) = 0, \quad (17)$$

given initially ($n \to \infty$).

114

Imposing the kink-type boundary conditions makes a conflict among the favourable positions of H-atoms at $\pm\varphi_0$ and positions, determined by the strain term, which tends to keep any pair of atoms at the equilibrium spacing L from each other. This conflict becomes resolved above some threshold energy E_{thr} which correspond to the lowest admissible energy static kink configuration $\{\varphi_{n,\alpha}^s(0)\}$. The upper local minima correspond to the energies

$$E_\alpha \equiv U_{total}\left(\{\varphi_{n,\alpha}^s\}\right), \qquad E_1 \equiv E_{thr},$$

and can be arranged according to increasing energy, $E_1 \equiv E_{thr} \leq E_2 \leq E_3 \ldots$

The kink shape drawn in figure 2 is obtained by simulating numerically the Hamiltonian H of the type (8) where $U(\varphi)$ was chosen in the Lippincott–Schroeder form, for one of the possible values of the lattice spacing that corresponds, in the hydrogen bonding theory terminology, to intermediate (as in ice Ic or Ih at the ordinary conditions) hydrogen bonds ($L = 2.76$ Å) with the frozen (R is fi-

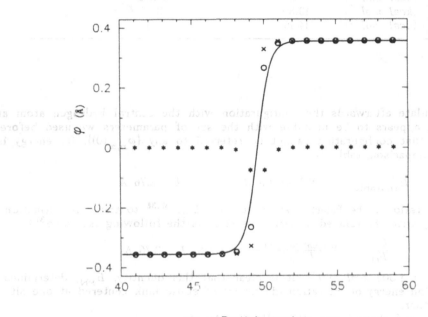

Figure 2. The static kink profile.

xed at the equilibrium lattice constant L) and relaxed heavier subchain. In this figure the positions of H-th atoms, $\{\varphi_n^s(0)\}$, are circled, and they are framed by a solid line which is a kink solution of φ-four approximation of Lippinott-Schroeder potential in continuum approximation with parameters U_b^{LS} and $\varphi_0(L)$ presented in table 3. Table 4 gives the lowest-energy static kink configurations for the chain, as well as kink formation energy E_{thr} and other properties of discrete kink solution for this value of L.

We would like to notice here that within the continuum limit, the kink-like transition state passes over the top of the barrier U_b. One sees that in contrast to the continuum-limit kink, no atoms located at the top of barrier (see, for example, the rows 3 and 4 in table 4). So, the effect of discreteness is very pronounced (see also figures 1 and 2 in[38]). Futhermore, our numerical simulations show that all discrete kinks with the central atom at the top, appear to be unstable for the set of parameters we used.

Table 4. Properties of one- and two-component static kinks

	Heavier subsystem frozen	Heavier subsystem relaxed
E_{thr} ($kcal\ mol^{-1}$)	5.666	3.254
E_{PN} ($kcal\ mol^{-1}$)	7.864	1.003
$\varphi_{\pm1}^{LS}$ (Å)	±0.329	±0.240
$\varphi_{\pm2}^{LS}$ (Å)	±0.354	±0.345
$\Delta\omega_{\pm1}$ (cm^{-1})	−188	−539
$\Delta\omega_{\pm2}$ (cm^{-1})	−6.6	−50
$\Delta\omega_{\pm3}$ (cm^{-1})	−0.23	−3.1
$\Delta_{\pm1}$ ($kcal\ mol^{-1}$)	0.407	1.665
$\Delta_{\pm2}$ ($kcal\ mol^{-1}$)	10.801	9.975
$\Delta_{\pm3}$ ($kcal\ mol^{-1}$)	11.203	11.51
$U_{\pm1}$ ($kcal\ mol^{-1}$)	7.966	3.072
$U_{\pm2}$ ($kcal\ mol^{-1}$)	13.89	12.74
$U_{\pm3}$ ($kcal\ mol^{-1}$)	14.15	14.39

We simulate afterwards the configuration with the central hydrogen atom at the top. It appears to be unstable with the set of parameters we used before. Therefore, that configuration cannot be referred to any $\{\varphi_{n,\alpha}(0)\}$. Its energy is (see, for comparison, table 4)

$$E_{unstable} = 4.257\ kcal\ mol^{-1}, \qquad L = 2.76\ Å$$

Hence, the ratio of the Peierls-Nabarro barrier E_{PN} [37,38] to the kink formation energy E_{thr} with the relaxed heavier subsystem is the following (see also [37]):

$$\frac{E_{PN}}{E_{thr}} \equiv \frac{E_{unstable} - E_{thr}}{E_{thr}} = 0.308, \quad L = 2.76\ Å$$

where $E_{PN} = 1.003\ kcal\ mol^{-1}$. It is clear that this quantity, E_{PN}, determines the activation energy of migration (diffusion) of static kink centered at one site to the another.

All mentioned above frames the following Ansatz:

$$\varphi_{i\alpha}(t) = r_{i\alpha}(t) + \varphi_{i,\alpha}^{s}(0), \quad i = -N, ..., N \tag{18}$$

where α enumerates the admissible, static kink configurations lying underneath the Peierls—Nabarro barrier. Within this Ansatz, the collective variable, $\{\varphi_{i,\alpha}^{s}(0)\}_i$, ensures the time-independent background for the local dynamical variables $\{r_{i\alpha}(t)\}_i$. The latters describe local "phonon" modes of H-atoms vibrating in local potentials organized by α-th discrete static kink,

$$U_{i,\alpha}^{s}(\{r_n\}) \equiv U(r_i + \varphi_{i,\alpha}^{s}(0)) + \sum_{m=-N}^{N} V_{im}\big((\varphi_{i,\alpha}(0) - \varphi_{i-m,\alpha}(0)) + (r_i - r_{i-m})\big).$$

It is evident that the local n-site potential U_n^s (α is omitted) lost already the initial symmetric double-well property. In figure 3 we depict profiles of $U^s(r_n)$'s obtained for the chain with the Lippincott-Schroeder potential. All these potentials become asymmetric. To estimate the effect of the shape change quantatively we introduce the parameter of asymmetry Δ and the barrier height U_b^n (referred to the lowest well)

$$\Delta_n \equiv U^s(-\varphi_0 + \varphi_n^s(0)) - U^s(+\varphi_0 - \varphi_n^s(0)),$$

$$U_b^n \equiv U^s(\varphi_n^s(0)) - U^s(\varphi_0 - \varphi_n^s(0)).$$

In table 4 we present their values and also the harmonic frequency shift $\Delta\omega_n \equiv \omega_n - \omega_0$, where

$$\omega_n = \left[\left(d^2 U_s(r_n)/dr^2 \big|_{r = \varphi - \varphi^s(0)} \right) / 2m \right]^{1/2},$$

obtained for the exact Lippincott-Schroeder potential for $L = 2.76$ Å.

Local "phonon" modes are certainly coupled with the kink-carrier collective variable, including the collective one. This coupling provides the energy transfer between all modes. In particular, if the collective variable travels along the chain, the profiles of local proton potentials become time-dependent. This is well illustrated in figure 4, where the "microscopic" view of the part of kink evolution (see figure 1) is presented. Here the series of local proton positions at consequent time points marked by small filled circles, and open circles cor-

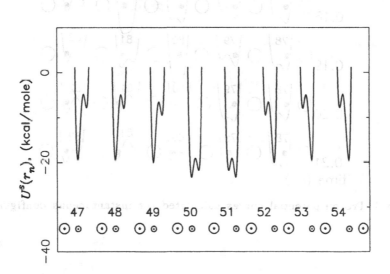

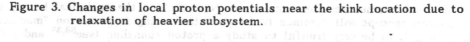

Figure 3. Changes in local proton potentials near the kink location due to relaxation of heavier subsystem.

respond to local positions of the heavier subsystem. One can see the changes of local proton potential of 80th particle, going from the left well to the right one not over the potential barrier, but gliding in deep valley, created by accompanied movement of the heavier subsystem.

4. SOMETHING LIKE CONCLUSIONS

Our contribution has to be written in such a way which reflects the scenario by which the solitonic doctrine penetrated and still penetrates in the hydrogen bonding medium, accompanied by the emphasis of the following points: where does the hydrogen bonding play the part in the solitonic mechanism of proton transfer and how much is its contribution?

For extended hydrogen-bonded systems, the cooperativity frames a collective coordinate variable which carries the kink-defect. The rest "sea" of other degrees of freedom excluded in a "funny" way by the continuum approximation is so huge that still remains mainly beyond the solitonic business. In our opinion, that is exactly the reason why the people border the solitonic mechanism of proton transfer which is not yet properly formulated from the usual ones employing the traditional methods of theoretical physics. The account of other local degrees of freedom washes out this border line although the analytical beauty and physical transparency of the soliton concept still remains and will remain, we hope. However, some corrections in the soliton concept in view of the hydrogen bonding theory have to be made because it is really hard to believe in the complete integrability of hydrogen-bonded systems. Perhaps, it will

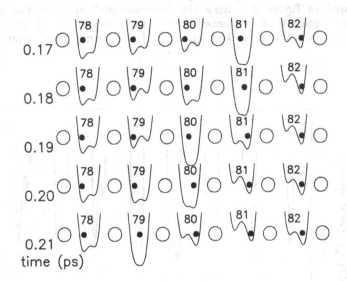

Figure 4. Proton potential curves calculated for instantaneous configuration.

require cumbersome computer simulations. Nevertheless, the analytical beauty of the soliton concept will continue to be attractive owing to instanton "machinery" making it to be very fruitful to study a proton tunneling (see[38,39] and refs. therein).

Acknowledgements

One of the authors, E. S. K., thanks G. Careri, P. L. Christiansen, J. P. Delvin, Y. Marechal, M. Mezei, J. F. Nagle, St. Pnevmatikos , G. E. Walrafen, E. Whalley, and G. Zundel for the fruitful discussions, and The Alexander von Humboldt Foundation and EEC Project "Science 0229-C" for the financial support. V. P. S. gratefully acknowledges the hospitality of the University of Constance. This work was supported in part by the Deutsche Akademische Austauschdienst.

References

1   A. Baroni, F. Esposito, C. J. Magee, and A. C. Scott, *Riv. Nuovo Cim.*, **1**: 227 (1971); A. C. Scott, F. Y. F. Chu, and D. W. McLaughlin, *Proc. IEEE*, **61**: 1443 (1972); A. C. Scott, *The Sciences*, March/April: 29 (1990).
2   A. R. Bishop, and T. Schneider, eds., "*Solitons in Condensed Matter Physics*", Springer, Berlin, (1978); S. E. Trullinger, V. E. Zakharov, and V. L. Pokrovsky

"*Solitons*", North–Holland, Amsterdam, (1986); A. Bishop, D. Campbell, P. Kumar, and S. Trullinger *"Nonlinearity in Condensed Matter"*, Springer, Berlin, (1987).

3   S. Yomosa, *J. Phys. Soc. Japan*, **51**: 3318 (1982).

4   Y. Kashimori, T. Kikuchi, and K. Nishimoto, *J. Chem Phys.* **77**: 1904 (1982).

5   V. Ya. Antonchenko, A. S. Davydov, and A. V. Zolotariuk, *Phys. Stat. Sol.* (b), **115**: 631 (1983).

6   A. S. Davydov, *"Solitons in Molecular Systems"*, 2nd edition, Kluwer, Dordrecht, (1991).

7   Y. Maréchal, *La Recherche*, **20**: 482 (1989).

8   J. D. Bernal, and R. H. Fowler, *J. Chem. Phys.*, **1**: 515 (1933); L. Pauling, *J. Am. Chem. Soc.*, **57**: 2860 (1935).

9   H. S. Frank and W.-Y. Wen, *Disc. Faraday Soc.*, **23**: 72 (1957).

10   D. Hankins, J. W. Moskowitz, and F. H. Stillinger, *J. Chem. Phys.*, **53**: 4544 (1970); J. E. Del Bene and H. A. Scheraga, *Ibid.*, **58**: 5296 (1973); E. Clementi, *"Determination of Liquid Water Structure. Coordination Numbers for Ions and Solvation for Biological Molecules"*, Lecture Notes in Chemistry, vol. **2**, Springer-Verlag, Berlin (1976).

11   E. S. Campbell and M. Mezei, *J. Chem. Phys.*, **67**: 2338 (1977); *Mol. Phys.*, **41**: 883 (1980); E. S. Campbell and D. Belford, *Thoret. Chim. Acta*, **61**: 295 (1982).

12   G. S. Pimentel and A. L. MacClellan, *"The Hydrogen Bond"*, Freeman, San Francisco (1960).

13   E. S. Kryachko, *Int. J. Quantum Chem.*, **30**: 495 (1986).

14   S. Scheiner, *Acc. Chem. Res.*, **18**: 174 (1985).

15   S. Scheiner and J. F. Nagle, *J. Phys. Chem.*, **87**: 4267 (1983).

16   R. Lochmann and Th. Weller, *Int. J. Quantum Chem.*, **25**: 1061 (1984).

17   A. Karpfen and P. Schuster, *Can. J. Chem.*, **63**: 809 (1985).

18   N. Bjerrum, *K. Dan. Vidensk. Selsk. Mat.-Fys. Medd.*, **27**: 3 (1953); *Science* **115**: 385 (1952).

19   D. Eisenberg and W. Kauzmann, *"The Structure and Properties of Water"*, Oxford University Press, New York (1969).

20   J. D. Dunitz, *Nature*, **197**: 860 (1963).

21   N. V. Cohan, M. Cotti, J. V. Iribarne, and M. Weissmann, *Trans. Faraday Soc.*, **58**: 490 (1962).

22   D. Eisenberg and C. A. Coulson, *Nature*, **199**: 368 (1963).

23   R. K. Chan, D. W. Davidson, and E. Whalley, *J. Chem. Phys.*, **43**: 2376 (1965).

24   E. S. Kryachko, *Chem. Phys. Lett.*, **141**: 346 (1987).

25   O. E. Yanovitskii and E. S. Kryachko, *Phys. Stat. Sol.* (b), **147**: 69 (1988).

26   E. S. Kryachko, *Solid State Commun.*, **65**: 1609 (1988).

27   St. Pnevmatikos, N. Flytzanis, and A. R. Bishop, *J. Phys.* C: Solid State Phys., **20**: 2829 (1987).

28   D. D. Klug and E. Whalley, *J. Chem. Phys.*, **83**: 925 (1985).

29   D. D. Klug and E. Whalley, *J. Chem. Phys.*, **81**: 1220 (1985).

30   W. B. Collier, G. Ritzhaupt, and J. P. Devlin, *J. Phys. Chem.*, **88**: 363 (1984).

31   P. A. Giguère and M. Pigeon-Gosselin, *J. Raman Spectr.*, **17**: 341 (1986); P. A. Giguère, *J. Chem. Phys.*, **87**: 4835 (1987).

32   G. E. Walrafen, M. S. Hokmabadi, and Y. C. Chu, *Vibrational and Collision-Induced Raman Scattering from Water and Aqueous Solutions*, in: *"Hydrogen-Bonded Liquids"*, J. C. Dore and J. Teixeira, eds., Kluwer, Dordrecht (1991).

33   M. K. Ali and R. L. Somorjai, *J. Phys.* A: Math. Gen., **12**: 2291 (1979).

34   P. Schuster, G. Zundel, and C. Sandorfy, eds., *"The Hydrogen Bond: Recent Developments in Theory and Experiments"*, North Holland, Amsterdam (1976).

35   E. R. Lippincott, and R. Schroeder, *J. Chem. Phys.*, **23**: 1099 (1955); G. R. Anderson, and E. R. Lippincott, *Ibid.*, **55**: 4077 (1972).

36   E. S. Kryachko and V. P. Sokhan, *Proc. Roy. Soc. London* (in press).

37   E. S. Kryachko, *Chem. Phys.*, **143**: 359 (1990).

38   E. S. Kryachko, *Recent Developments in Solitonic Model of Proton Transfer in Quasi-One-Dimensional Infinite Hydrogen-Bonded Systems"*, in: "Electron and Proton Transfer in Chemistry and Biology", A. Müller *et al.*, eds., Elsevier, Amsterdam (1991).

39   E. S. Kryachko, M. Eckert, and G. Zundel, *J. Mol. Str.* (in press).

DYNAMICS OF NONLINEAR COHERENT STRUCTURES IN A HYDROGEN-BONDED CHAIN

MODEL: THE ROLE OF THE DIPOLE-DIPOLE INTERACTIONS

I. Chochliouros and J. Pouget

Laboratoire de Modélisation en Mécanique (associé au CNRS)
Université Pierre et Marie Curie
4 Place Jussieu, 75252 Paris, Cédex 05, France

Abstract : The problem of the energy transport along H-bonded chains has recently attracted a renewal of interest due to the close connection with basic biological processes. In our study we consider a lattice model consisting of two one-dimensional harmonically coupled sublattices corresponding to the oxygens and protons, the two sublattices being coupled. We also introduce the **dipole-dipole interactions** which are present as a result of the existence of **microscopic dipoles** created by the proton motion. This kind of interaction may affect the response of the **nonlinear excitations** propagating along the chain. We are looking for a solution for which the heavy-ion sublattice can be considered as "frozen", that is the oxygens "stay" at rest and do not participate in motion. A **phi-6 equation** is found, which admits nonlinear excitations of **solitary wave type**. We have different classes of localized solutions for the proton motion and analytical expressions of these types are also given. A parallel study of the problem based on the **potential** is additionally presented, in a way to discuss a physical interpretation of the previous mathematical results. The proton motion is affected by the influence of the dipole interactions and the **proton conductivity** becomes much easier. Numerical simulations are presented for special cases. Finally, possible further extensions of the work are envisaged.

1. INTRODUCTION

Nonlinear equations and their solutions are very important in many areas of physics, chemistry and biology. Special interest has been generated in the concepts of **solitary waves (solitons)**. These excitations can spread in molecular systems at comparatively large distances without changing their shape (profile), energy, momentum, charge and other particle-like properties.
Since the fundamental work of Antonchenko *et al.* [1], and Zolotaryuk *et al.* [2], there has been a great interest concerning the soliton dynamics in one-dimensional hydrogen-bonded systems. These studies provided an effective description of the **proton mobility** in H-bonded chains and also gave a means to investigate and improve previously presented works [3]. The problem of the **proton transport** along H-bonded chains was always very important, especially in the recent years when scientists have tried to use it for the interpretation of certain biological processes [4]. Previous works on **electric conductivity** of ice [5,6,13] have provided a variety of interesting results. In this paper we are trying to examine the influences

of the dipole-dipole interactions on protonic conductivity in H-bonded chains. Simultaneously we propose an analytical study of the nonlinear dynamics of the proton motion.

Our one-dimensional lattice model is an improved form of the **Antonchenko-Davydov-Zolotaryuk (ADZ) model** [1,9] for H-bonded chains. Analytical solutions in the continuum approximation for a certain solitary wave velocity have been found by Antonchenko *et al.* [1]. Other works [7,8,14,15] have provided a variety of numerical results in a greater range of velocities. The important point of our approximation is the **contribution of the dipole interactions** due to the proton motions. The electric dipoles along the chain can exercise a remarkable influence on the response of the system and the proton conductivity becomes much easier. The discrete system is not always convenient for algebraic manipulations. For this reason our study is faced with the continuum approximation of the microscopic model. This approximation leads us to a Φ^6 equation which admits many localized solutions, each one according to several selections and conditions concerning the variables involved. The role of the ionic and the orientational defects remains of the same importance as appeared in previous works [5,7], in order to explain the protonic conductivity. We must notice that the range of the study can be very easily extended, but this will be the aim of further works.

2. CONSTRUCTION OF THE MODEL

Our model for a H-bonded chain consists of two one-dimensional interacting sublattices of **harmonically coupled** protons (mass m) and heavy-ions (hydroxyl groups, mass M) as shown in Fig.1.

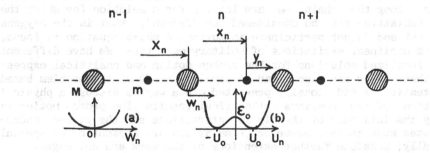

Fig. 1 : One-dimensional lattice model for a hydrogen-bonded diatomic chain. (a) harmonic potential for heavy ions and (b) double-well potential for protons.

In this model the protons alternate within the chain with the heavy-ions. The sublattices are embedded in each other in such a way that every proton is in a symmetric **double-well** potential created by the pair of nearest-neighbour heavy-ions which are usually called as "oxygens".

If we consider a normal state of the chain each proton is connected with an oxygen atom by a covalent bond, on the one hand, and by a hydrogen bond on the other hand. A proton may be transferred along the chain in the form of **ionic defects** (hydroxonium H_3O^+ and hydroxyl OH^-) or **orientational (Bjerrum) defects** due to the rotation of the molecules. The motion of a positive or a negative ionic defect is caused by the **successive jumps** of protons over the barrier. After the motion of an ionic defect from the one end to other, the chain can be led back to the initial state due to the action of the Bjerrum defects. When a proton is moved from the one well to the other, then the two bonds exchange their positions. As mentioned above, the proton potential energy curve has the form of a well with two minima corresponding to the two equilibrium states of a proton. The onsite double-well potential can be approximated [6] by

$$U(y_n) = \epsilon_0 \left(1 - y_n^2/y_0^2\right)^2 .$$ (1)

The potential is schematically presented in Fig.1. In Eq.(1), y_n denotes the displacement of the n-th proton with respect to the center of the oxygen pair in which it is located, ϵ_0 is the height of the potential barrier and y_0 is the distance between the local maximum and one of the minima of the potential-well.

It has been found [8] that the effect of the coupling between protons and oxygens is the reduction of the effective **potential barrier** that protons have to overcome to jump from one molecule to the other and makes their motion easier.

3. EQUATIONS OF THE MODEL

We consider the total Hamiltonian of the system which is expressed as

$$H_{tot} = H_p + H_o + H_{int} + H_{dd} .$$ (2)

Where the Hamiltonian of the proton subsystem is given by

$$H_p = \sum_n \left[\frac{1}{2}m\dot{y}_n^2 + U(y_n) + \frac{1}{2}m\omega_1^2 (y_{n+1} - y_n)^2\right] .$$ (3)

The first of the terms denotes the kinetic energy of the protons, the second term is introduced due to the double well potential, and the third term represents the harmonic coupling with the characteristic frequency ω_1 between neighbouring protons.

In the oxygen part of the Hamiltonian H_o we consider only the relative displacement w_n between two oxygens in a pair, because a possible variation of the O-O distance can modulate the double-well undergone by the protons. For this reason w_n is the oxygen degree of freedom that is normally expected to play an important role in proton motion. Due to this restriction we consider a **coupling** between proton motions and an optical mode of the heavy-ion sublattice. The Hamiltonian H_o is written as

$$H_o = \sum_n \left[\frac{1}{2}M\dot{w}_n^2 + \frac{1}{2}M\Omega_0^2 w_n^2 + \frac{1}{2}M\Omega_1^2 (w_{n+1} - w_n)^2\right] .$$ (4)

The first term denotes the kinetic energy of the oxygens, the second term stands for the coupling between oxygens of the same cell, while the third term represents the harmonic coupling betweeen neighbouring oxygen pairs and also introduces in the model the dispersion of an optical mode (Ω_0 and Ω_1 are characteristic frequencies of the optical mode).

The third term H_{int} of the total Hamiltonian becomes from the **dynamic interaction** between protons and heavy-ions. The physical content of this kind of interaction is the **lowering** of the double potential barrier due to the displacements of the heavy-ions. Further work has been done by Zolotaryuk [10] and interesting mathematical results have been obtained. The interaction is written as in ADZ model [1,8]

$$H_{int} = \sum_n \delta w_n \left(y_n^2 - y_0^2\right) \quad . \tag{5}$$

In Eq.(5), δ measures the strength of the coupling between the two sublattices, and determines the amplitude of the distortion in the oxygen sublattice.

The Hamiltonian H_{dd} arises from the dipole interactions. The existence of electric dipoles in the chain, because of the electric charges, leads us to account for the **mutual interaction between the dipoles**. The dipole-dipole energy is a function of the distance r and the angle between the dipole vector and the distance vector $\vec{r}$ [11]. In the present case, we assume that the distance r between neighbouring dipoles does not depend on the lattice displacement and all vectors of the dipole moment $\vec{P}_n$ are in the same direction (they are all alligned). We can find that due to this assumption the expression for H_{dd} becomes

$$H_{dd} = \bar{\beta} \sum_n P_n P_{n+1} \quad . \tag{6}$$

($\bar{\beta}$ is a constant which may account for the enviroment of the chain). For the dipole moment induced by the proton motion we consider that it must be zero when proton is at either position of the oxygens or when the proton is found at the middle of the distance joining the oxygen pair, where the interactions are opposite. In Fig.2 we present schematically this law for the dipole.

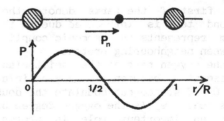

Fig. 2 : Microscopic electric dipole as a function of the proton position with respect to the nearest heavy ion positions

A simple form for the dipole moment P to approximate the above structure has been proposed by Whalley [12] and consists of a **polynomial of the third degree**, that is

$$P_n = \bar{\alpha}(x_n - X_n)(x_n - X_{n+1})\left[x_n - \frac{1}{2}(X_n + X_{n+1})\right] \quad . \tag{7}$$

($\bar{\alpha}$ is a constant). The absolute positions x_n and X_n are shown in Fig.1 and we have the expressions

$$x_n = n\ell_0 + \frac{1}{2}\ell_0 + y_n , \quad X_n = n\ell_0 + w_n , \quad X_{n+1} = (n+1)\ell_0 + w_{n+1} \quad . \tag{8}$$

It is noted that ℓ_0 is the lattice spacing. From the Hamiltonian (1) and after simple calculations we can find the equations of motion of the system which form a **set of coupled nonlinear differential-difference equations** expressed as follows

$$\frac{d^2 y_n}{dt^2} = \omega_1^2 (y_{n+1} - 2y_n + y_{n-1}) + \left(4\epsilon_0/my_0^2\right)\left(1 - y_n^2/y_0^2\right)2y_n - (2\delta/m)w_n y_n$$

$$-\bar{\beta}\,\frac{\partial P_n}{\partial y_n}\,(P_{n+1} + P_{n-1})\,, \tag{9a}$$

$$\frac{d^2 w_n}{dt^2} = \Omega_1^2 (w_{n+1} - 2w_n + w_{n-1}) - \Omega_0^2 w_n - (\delta/m)\left(y_n^2 - y_0^2\right)$$

$$-\bar{\beta}\,\frac{\partial P_n}{\partial w_n}(P_{n+1} + P_{n-1}) - \bar{\beta}\,\frac{\partial P_{n-1}}{\partial w_n}(P_n + P_{n-2})\,. \tag{9b}$$

These equations cannot be solved analytically. For this reason any attempt of further analysis becomes extremely difficult. Our study is focused on a very important case in which the **heavy-ion sublattice** can be considered as "frozen". Now the motion of the oxygens is not so remarkable as that of the protons. This kind of approximation of the problem is not a mathematical simplification but it has also an important physical signification due to **inertia** reasons. The oxygen sublattice cannot easily follow the very fast proton motion, especially for large velocities. We can suppose for this reason that a solution could involve only the proton displacement, while oxygens "stay" at rest and do not contribute to the motion. In the atomic lattice y is of the order of few Å and as w is very small compared to y, it can be considered almost about zero. As a result of this assumption, the total Hamiltonian is reduced to the sum of the Hamiltonians H_p and H_{dd} because the other terms which directly depend on w_n can be considered very small and practically equal to zero.

In order to obtain a solution we propose the units E_0 for energy, t_0 for time and ℓ_0 for length. We can find the derived units $m_0 = E_0 t_0^2/\ell_0^2$ for mass and $f_0 = \epsilon_0/\ell_0$ for force. Then we introduce the dimensionless parameters $\tilde{m} = m/m_0$, $\epsilon = \epsilon_0/E_0$, $\chi_1 = -\bar{\beta}\,\bar{\alpha}^{-2}\ell_0^6/E_0\tilde{m}$. We can additionally consider $u_n = y_n/\ell_0$, $\tilde{H} = H/m\ell_0^2$. After that, the expression for the energy of the system reduced to the proton motion becomes

$$\tilde{H} = \sum_n \left[\frac{1.2}{2}\dot{u}_n + \frac{1}{2}\omega_1^2(u_{n+1} - u_n)^2 + \frac{1}{4}G_0 u_0^2\left(1 - u_n^2/u_0^2\right) - \chi P_n P_{n+1}\right]\,, \tag{10}$$

where the electric dipole is merely given by

$$P_n = u_n\left(u_n^2 - \frac{1}{4}\right)\,. \tag{11}$$

The above expression does not correspond directly to the vector of the dipole moment P_n but is a very convenient notation for the resolution of the equation of motion in which all magnitudes are non-dimensionalized.

Now the equation of the discrete system describing the proton motion reduces to

$$\ddot{u}_n = \omega_1^2(u_{n+1} - 2u_n + u_{n-1}) + G_0 u_n\left(1 - u_n^2/u_0^2\right)$$

$$+ \chi\left(3u_n^2 - \frac{1}{4}\right)\left[u_{n+1}\left(u_{n+1}^2 - 1/4\right) + u_{n-1}\left(u_{n-1}^2 - 1/4\right)\right]\,. \tag{12}$$

It has been previously set $G_0 = 4\epsilon_0/\ell_0^2 u_0^2$ and $\chi = \chi_1 E_0/m\ell_0^2$.

4. CONTINUUM APPROXIMATION OF THE DISCRETE SYSTEM

Further step to our analysis consists of considering the **continuum approximation**. We consider terms up to the second order supposing that the contribution of higher order terms is not significant, as it can be usually done in such type of process. The equation of the proton motion can be written as

$$u_{tt} - \left[1 + \chi(3u^2 - 1/4)^2\right]u_{xx} - \alpha_1 u + \beta_1 u^3 - \gamma_1 u^5 = 0 , \tag{13}$$

where we have set

$$\alpha_1 = G_0 + \chi/8 , \quad \beta_1 = G_0/u_0^2 + 2\chi , \quad \gamma_1 = 6\chi . \tag{14}$$

What is important to notice here is the **dependence** of all three parameters α_1, β_1, γ_1 on χ, which χ is also present in the factor accompanied u_{xx}. Because of this last remark further investigation becomes extremely difficult. We try to simplify the procedure and we propose the following hypothesis : we set $1 + \chi(3u^2 - 1/4) = \tilde{C}_0$ as a function of u. This function has two minima. If we consider a certain value of χ we can find the mean value $\langle \tilde{C}_0 \rangle$ and the next step is to set $\langle \tilde{C}_0 \rangle$ as the coefficient of u_{xx}. This really rough approximation has significance when there is no great difference between the greatest and the lowest values of the function. This occurs if the coefficient χ is considered small enough such that $|\chi| \ll 1$. We now set $\langle \tilde{C}_0 \rangle = \omega^2$ and the equation for study becomes

$$u_{tt} - \omega^2 u_{xx} - \alpha_1 u + \beta_1 u^3 - \gamma_1 u^5 = 0 . \tag{15}$$

The next step is to consider the change of variables $u = \alpha U$, $t = \gamma T$, $x = \beta X$. Then we select $\alpha = \sqrt{\beta_1/\gamma_1}$, $\gamma = \sqrt{\gamma_1/\beta_1}$, $\beta = \omega\sqrt{\gamma_1/\beta_1}$ (it must be $\gamma_1 > 0$, $\beta_1 > 0$). We also set $A = \alpha_1 \gamma^2$. The equation for study takes on the following form

$$U_{TT} - U_{XX} = AU - U^3 + U^5 . \tag{16}$$

We are looking for **localized solutions** with constant profile, moving at a characteristic velocity v, that is for solutions $U = U(\xi)$ where $\xi = X - vT$. We do not dwell on the algebraic manipulations for finding out the different classes of solutions.

5. DIFFERENT TYPES OF LOCALIZED SOLUTIONS

The equation of the proton motion presents a symmetry, so when U is a solution then -U is also a solution. We can set $U_1 = U(\xi \to -\infty)$ and $U = U_2 (\xi \to +\infty)$. We distinguish **different types of solutions** as we try to approach U_2 beginning by U_1, or the inverse.
It is $U_1 = U_2 = 0$ for the solution of the type I (pulse). The corresponding mathematical expression becomes

$$U = \pm \frac{U_m}{\left[1 + P sinh^2 (\Omega\xi/2)\right]^{1/2}} . \tag{17}$$

Where we have considered $U_m^2 = \pm 4A/\left(\sqrt{1 - 4A/A_0} \pm 1\right)$ and also

$P=2\sqrt{1- A/A_0}/\left(\sqrt{1- A/A_0} \pm1\right)$. The sign (+) corresponds to supersonic waves ($|v|>1$) and for this case it must be considered $0<A<(3/16)$. The sign (-) corresponds to subsonic waves ($|v|<1$) and for this case it must be necessarily considered $A<0$. In both cases we have $A_0=(3/16)$ and $\Omega^2=4A/(v^2-1)$.

The solution of the type III presents a kink. In this case we have two opposite non zero values ($U_1=U_0$, $U_2=-U_0$) and the expression is given as

$$U = \frac{\pm U_0\, tanhz}{\left[1 + P(1- tanh^2 z)\right]^{1/2}} .$$ (18)

Now it is $z=(1/2)\Omega\xi$, $P=U_0^2/\left(2U_0^2- 3/2\right)$, $\Omega^2=4U_0^2\left(U_0^2- 1/2\right)/(v^2-1)$. The definition for U_0 is given as $\left(U_0^{\pm}\right)^2=(1\pm\sqrt{1- 4A})/2$, where as it has been explained above, the sign (+) corresponds to supersonic waves (it must be additionally $A<(3/16)$ and $A\neq0$), and the sign (-) corresponds to subsonic waves (it must be additionally $A<(1/4)$ with $A\neq0$, $A\neq(15/64)$).

A solution of the type IV corresponds to a case in which it is set $U_1=U_2\neq0$. We can find that

$$U = \pm\frac{U_0}{\left[P(tanh^2 z - 1)+ tanh^2 z\right]^{1/2}} .$$ (19)

Where it is $P=U_0^2/\left(2U_0^2- 3/2\right)$, $\Omega^2=4U_0^2\left(U_0^2- 1/2\right)/(v^2-1)$. For the constant value U_0 we have $\left(U_0^{\pm}\right)^2=(1\pm\sqrt{1- 4A})/2$. Again the sign (+) corresponds to supersonic waves (for their definition now it must be $(3/16)<A<(1/4)$). The sign (-) corresponds to subsonic waves. (Now we have to consider $A<(1/4)$ with $A\neq0$, $A\neq(15/64)$).

Finally, a particular case (type II) occurs, for which it is $A=3/16$ and the solution represents a kink describing a transition from the state $U_1\neq0$ to the state $U_2=0$. We have (only in case of a supersonic wave $|v|>1$) the expression

$$U = \pm\frac{1}{\left[k_1\, exp(\pm\Omega\xi) + 4/3\right]^{1/2}} .$$ (20)

(Where k_1 is a constant and $\Omega=+(1/2)\sqrt{3/(v^2-1)}$).

6. STUDY OF THE POTENTIAL

The Eq.(16) can be derived from the following Hamiltonian

$$H = \int \left[\frac{1}{2}U_T^2+ \frac{1}{2}U_x^2+ \Psi(U)\right]dx ,$$ (21)

where we have defined

$$\Psi(U) = \frac{1}{2}AU^2- \frac{1}{4}U^4+ \frac{1}{6}U^6 .$$ (22)

We notice that in the Hamiltonian (21) the first term denotes the kinetic energy, the second term stands for the linear interaction (elastic) and the third part is a **potential** of the Φ^6- **type**. The properties of the potential (22) play a crutial role in the existence of different classes of solu-

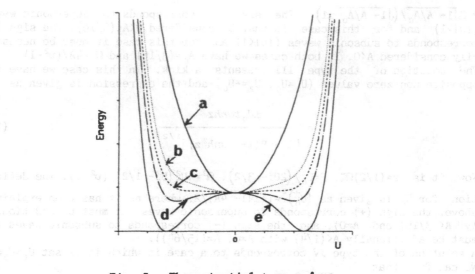

Fig. 3 : The potential ψ as a function of the proton displacement U
for different values of A : (a)
A>1/4, (b) A=1/4, (c) A=3/16, (d)
0<A<3/16 and (e) A<0.

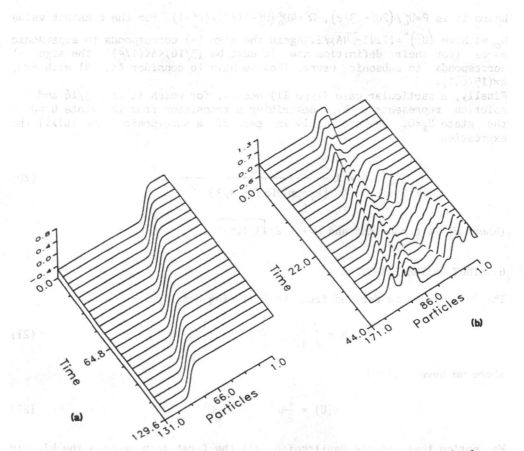

Fig. 4 : Numerical simulations of the lattice model for two particular
excitations, (a) stable kink (solution of type III) and unstable (b)
pulse (solution of type I).

tions. The potential (22) is dessigned in Fig.3 for different values of A. For A>1/4 there is only one minimum for U=0. For A such that (3/16)<A<(1/4) the potential possesses two symmetric metastable minima and two maxima, for U=0 we have the stable minimum. The limited case occurs for A=3/16 and the potential has three stable minima for U=$\pm\sqrt{3}/2$ and U=0. For this case a solution of the type II can exist. When it is 0<A<3/16, the curve has now two nonzero stable minima whereas the minimum at U=0 is metastable. For this situation pulse-like and kink solutions of types I and II can exist. Finally, for A<0, the potential is similar to a symmetric double-well potential with two symmetric stable minima at a nonzero value of U and U=0 is a maximum of the potential.

7. NUMERICAL SIMULATIONS

We present numerical simulations corresponding to the solutions of the types I and III. We study how a localized solution is evolved in time and in order to do so we consider a certain number of particles in each case. The numerical simulations are performed directly by means of the set of discret equations (12). The kink-like solution of the type III corresponds to a proton displacement from the state $-U_0$ to the state $+U_0$ which are wells of the Φ^6 potential. As it can be seen in Fig.4a, this kind of solution is remarkably **stable**. The pulse-like solution of the type I is shown in Fig.4b and presents an **instability**. We remark that after a short lapse of time, the pulse splits into two other pulses travelling in opposite directions accompanied with rather large perturbations behind them, something which makes difficult further investigation.

8. CONCLUSIONS

In this work we have studied how the proton motion is affected by the dipole-dipole interactions, which produce an influence on the electric field of the system and due to this action the proton conductivity becomes much easier. We notice that the description of the **ionic and orientational defects associated with the protonic conductivity** remains of the same validity as in Φ^4 case, studied in other works [8]. In our study we have found several possibilities for the solution describing the proton motion according to the values of the parameters, since the resulting equation possesses **stable, unstable and metastable steady states**. The problem of the continuum approximation can be still investigated as there are also **other possible cases** which can be studied more detailed. The initial introduced model can be extended and enriched by the introduction of an **external electric field** applied on protons (something which can be done in technical applications) and also the introduction of **damping** (in our 1-D model the transfer of energy to extra degrees of freedom, because the real system is 3-D, is modeled by phenomenological damping terms). Further step consists of considering a **rotational motion of the dipoles**, especially if we deal with nonlinear atomic chain. It would be also interesting to examine how the response is modulated by the introduction of **second nearest-neighbouring interactions** and more. From our first results it is found that there are remarkably interesting mechanisms, which could probably explain some of the most fascinating phenomena taking place in biology and in physics.

REFERENCES

[1] V. Ya. ANTONCHENKO , A. S. DAVYDOV and A. V. ZOLOTARYUK, *Phys. Status Solidi B* **115**, 631 (1983).

[2] A. V. ZOLOTARYUK, K. H. SPATSCHEK, and E. W. LAEDKE, *Phys. Lett. A* 101, 517 (1983).

[3] J. D. BERNAL and R. H. FOWLER, *J. Chem. Phys.* 1, 515 (1933).

[4] J. F. NAGLE, M. MILLE, and H. J. MOROWITZ, *J. Chem. Phys.* 72, 3959 (1980).

[5] N. BJERRUM, *Science* 115, 385 (1952).

[6] R. JANOSCHEK, in *The Hydrogen Bond : Recent Developments in Theory and Experiments*, edited by P. Schuster, G. Zundel and C. Sandforty (North Holland, Amsterdam, 1976), p. 165.

[7] E. W. LAEDKE, K. H. SPATSCHEK, M. WILKENS, Jr. and A. V. ZOLOTARYUK, *Phys. Rev. A* 32, 1161 (1985).

[8] M. PEYRARD, St. PNEVMATIKOS, and N. FLYTZANIS, *Phys. Rev. A* 36, 903 (1987).

[9] A. S. DAVYDOV, *Solitons in Molecular Systems* (Dreidel Publishing Company, Dordrecht, Holland 1985).

[10] A. V. ZOLOTARYUK, *Theor. Math. Phys.* 68, 916 (1986).

[11] G. P. JOHARI and S. J. JONES, *Journal de Chimie Physique*, 82, N°11/12, 1019, 1985.

[12] E. WHALLEY, *Journal of Glaciology*, Vol. 21, 13 (1978).

[13] L. ONSAGER, in *Physics and Chemistry of Ice* edited by E. Whalley, S. J. Jones and L. W. Crold (Royal Society of Canada, Ottawa, 1973) pp. 7-12.

[14] D. HOCHSTRASSER, H. BUTTNER, H. DESFONTAINES and M. PEYRARD, *Phys. Rev. A* 38, 5332 (1988).

[15] H. WEBERPALS and K. H. SPATSCHEK, *Phys. Rev. A* 36, 2946 (1981).

THERMALLY EXCITED LATTICE SOLITONS

N. Theodorakopoulos and N.C. Bacalis

Theoretical and Physical Chemistry Institute
National Hellenic Research Foundation
Vas. Constantinou 48, GR - 116 35 ATHENS, Greece

Introduction

This lecture deals with the thermal properties of lattice solitons. Lattice solitons[1] arise in a wide variety of physical contexts, where there is a balancing of nonlinearity and dispersion. Localized nonlinear excitations may possibly be relevant to some biophysical properties, e.g. proton transport[2] and DNA "premelton"[3] dynamics. It is therefore important to have a full understanding of a mathematically tractable model, which includes the essential features of interatomic forces: a short range repulsion and a weak restoring force.

The Toda lattice is such a model. It has long been known to support pulse-like soliton excitations[1]. Furthermore, it is one of the few Hamiltonian systems with the property of complete integrability[4]. This property makes it possible to describe not only dynamics, but statistical mechanics as well, in terms of the the various excitations present at finite temperatures[5-9]. In short, by investigating the properties of the Toda lattice, it is possible to probe into fundamental aspects of nonlinearity.

One of the first questions that can be asked about a nonlinear system at finite T is how many thermal solitons it will support. Muto *et al.* [10,11] carried out a large-scale *numerical simulation* of Toda lattice dynamics. Using an ingenious soliton counting technique, they were able to determine that the fraction of the total degrees of freedom occupied by solitons is proportional to $T^{1/3}$ in the limit of low temperatures.

Now this is the type of result that every respectable *theory of soliton statistics* should be able to check. Indeed, we will present two such calculations, one approximate and one exact, which both predict a $T^{1/3}$ dependence. They are both based on the idea that a soliton has to share the available phase space with the the other excitations present (phonons and solitons). The exact version makes use of the Bethe-Ansatz (BA) thermodynamics of the Toda lattice[5-9], and thus implicitly depends on the exact integrability of the particular model. The approximate version (whose result differs only in the value of the prefactor) only assumes the presence of a non-topological soliton. We thus expect it to be somewhat more generally applicable.

Definitions, Notation

In the following we shall give a concise account of the theoretical background on the dynamics and statistical mechanics of the Toda chain, to the extent that is necessary in order to introduce concepts and notation used in this work. For details the reader is referred to the original literature[1,4-9].

The Hamiltonian

$$H = \sum_{i=1}^{N} \left\{ \frac{p_i^2}{2m} + \frac{a}{b} e^{-b(y_{i+1}-y_i)} + a(y_{i+1}-y_i) \right\} \tag{1}$$

describes a chain of N atoms with equal mass m, coordinates y_i and momenta p_i. The potential consists of a nearest-neighbor exponential repulsive interaction of range b^{-1} and an attractive linear part, mathematically equivalent to an external force of strength a, which is responsible for holding the particles together in the lattice.

The dynamical evolution of particle motion in the infinite Toda lattice is characterized by complete integrability and separability; due to the latter property, two distinct types of dynamical behavior, according to the position of the eigenvalues in the spectrum of the operator associated with the particular nonlinear problem are revealed by the inverse spectral transform (IST)[4]: (a) extended, subsonic, dispersive [phononlike] excitations arising from the continuum part, and (b) localized, supersonic, nondispersive [solitonlike] excitations arising from the discrete part of the spectrum of the associated linear operator. The dispersion curves are displayed in Fig.1.

Scattering of any number of excitations is a completely elastic process and can be described by pairwise additive phase shifts[1,4]. An example is the phase shift[12]

$$\delta(q,\alpha) = \begin{cases} 4\,tan^{-1}\left(tan(q/4)\,coth(\alpha/2)\right), & \text{if } \pi \geq q \geq 0; \\ 4\,tan^{-1}\left(tan(q/4)\,tanh(\alpha/2)\right), & \text{if } -\pi < q < 0, \end{cases} \tag{2}$$

of a phonon of wavevector q due to the presence of a forward moving soliton which streches over $1/\alpha$ lattice constants (the lattice constant is taken equal to unity). The expressions for other phase shifts are given in Ref.1 and elsewhere in the literature[12].

Soliton statistics (approximate)[13]

In a harmonic crystal phonons do not interact. If we impose periodic boundary conditions on a chain of N atoms, the allowed q-values are given by $qN = 2\pi l$, where l is an integer. In the Toda lattice, the presence of a soliton causes a phonon phase shift (Eq. 2). The allowed q-values are now given by $qN + \delta(q,\alpha) = 2\pi l$; alternatively one may speak of a change in the phonon density of states given by $\delta R(q;\alpha) = -(1/2\pi)d\delta/dq$.

If we assume that this is the only mechanism responsible for phase-space sharing (and this is clearly an approximation, since solitons interact with other solitons (multisoliton effects) and nonlinear phonons also interact with other nonlinear phonons) it is possible to estimate the average number of solitons as follows: For a given soliton configuration $\{n(\alpha)\}$, where the occupation number $n(\alpha)$ of each state $\{\alpha\}$ is held fixed, the free energy F of the total (i.e. soliton plus phonon) system is given by

$$-\beta F(\{n(\alpha)\}) = \left(\sum_q lnZ(q) + \sum_\alpha lnZ(\alpha) \right), \tag{3}$$

where $Z(q) = 1/\beta g\omega(q)$, $Z(\alpha) = (1/n(\alpha)!)exp(-\beta n(\alpha)E(\alpha))$, $g = \hbar\omega_o/k_B T_o$, $\omega_o = \sqrt{ab/m}$, $T_o = (a/b)/k_B$, $\beta = T_o/T$, $E(\alpha) = sinh(2\alpha) - 2\alpha$ is the energy (measured in units of a/b), of a soliton with spatial extent $1/\alpha$ and $\omega(q) = 2|sin(q/2)|$ the frequency (in units of ω_o) of a phonon with wavevector q. In transforming the sum over q (Eq. 3) to an integral, we use a phonon density of states reduced by a total amount of

$$\Delta R(q) = \sum_\alpha \delta R(q; \alpha) . \tag{4}$$

It is now possible to minimize F with repect to all $\{n(\alpha)\}$. This yields the average (strictly speaking, most probable) values of $n(\alpha)$:

$$\overline{n}(\alpha) = e^{-\beta(E(\alpha)+\Delta E(\alpha))} , \tag{5}$$

where

$$- \beta \Delta E(\alpha) = ln(2\beta g \, sinh\alpha \, e^{-\alpha}) \tag{6}$$

effectively renormalizes the energy of a single soliton. The total number of solitons is given by

$$N_s = \sum_\alpha \overline{n}(\alpha) = N \frac{8}{\pi} \beta \int_0^\infty d\alpha \, \alpha \, sinh^2\alpha \, e^{-\alpha} e^{-\beta E(\alpha)} , \tag{7}$$

where, in the second step, the density of states for the *free* soliton gas, i.e.[14] $R_o^{sol}(\alpha) = (1/2\pi g)4\alpha sinh\alpha$ has been used.

Evaluation of the integral to leading asymptotic order yields the number of solitons as a fraction of the total degrees of freedom:

$$n_s = N_s/N = \frac{2}{\pi}(3/4)^{1/3}\Gamma(4/3) \, \overline{T}^{1/3} , \tag{8}$$

where $\overline{T} = 1/\beta$. The fraction n_s is plotted as a function of the third root of the reduced temperature $\overline{T}$ in Fig.3, where it is compared with the exact result. (cf. below.)

Soliton statistics (exact)

A complete description of equilibrium statistical mechanics can be given in terms of solitons, phonons and their interactions, or, equivalently, in the framework of the Bethe-Ansatz[5-9]. Thermodynamic properties can be expressed in terms of the distribution of momenta in k-space. The densities of occupied and unoccupied states are denoted by $\rho(k)$ and $\rho_h(k)$ respectively, and the ratio $\rho(k)/\rho_h(k) = exp(-\beta\epsilon(k))$ defines a quasiparticle energy. It has been shown[6] that the function ΔE (Eq. 6) can be interpreted, over large regions of phase space, as a "thermal" correction to the energy of a free soliton.

At zero temperature, the function $\rho(k)$ has a singularity in the classical limit, $\hbar \to 0$; particle states occupy all k values up to $k_F = 2\sqrt{ma/b}/h$; single particle excitations can be shown to correspond exactly to solitons or phonons, according to whether they are particle- or hole-like; the dispersion curves[5] and other dynamical properties[6-9] derived within the BA framework are identical with those of the IST (cf. above).

At finite temperatures, a direct interpretation of the BA quasienergies is less clear[15,16]. There is no singularity in either the $\rho(k)$ or the $\epsilon(k)$ curves. In fact, it has been shown[16] that in the space of the original quantum numbers defined in the BA context, the Fermi sea literally "evaporates" as soon as we heat the [classical] system. Nonetheless, it has been established[7,9] that the classical limit of the BA description is exactly equivalent to an interacting soliton/phonon gas formalism (i.e. a scheme in which solitons and phonons retain their identities, defined at T=0). *This equivalence rests on the existence of a Fermi momentum k_F which is is defined at zero temperature[9,17] and remains T-independent.*

We should, in principle, obtain the number of solitons using the (exact) interacting soliton / phonon theory. Due to the latter's exact equivalence to the BA, this amounts to a "literal" interpretation of the BA formalism, i.e. counting all excitations with $|k| > k_F$ as solitons. The BA approach is technically simpler and it is the one we shall follow. It remains to be seen whether this procedure is consistent with the one used in the soliton counter of Refs. 10 and 11.

We now describe the BA procedure. Given a chain of N atoms and length L, the density of occupied states $\rho(k)$ satisfies in the thermodynamic limit

$$\frac{N}{L} = d = \int_{-\infty}^{\infty} dk\, \rho(k). \tag{9}$$

We will define the number of solitons N_s via

$$\frac{N_s}{L} = \int_{|k|>k_F} dk\, \rho(k). \tag{10}$$

In order to determine the soliton fraction $n_s = N_s/N$ we will further use (i) the relationship[18]

$$\rho_h(k) = -\frac{1}{2\pi} d\left(\frac{\partial \epsilon}{\partial p}\right)_T, \tag{11}$$

valid in the classical limit, and (ii) the closed form solution[8] of an integral equation[6] which determines the quasiparticle energy $\epsilon(k)$ subject to the constraint of given external pressure P, i.e.

$$e^{-\beta\epsilon(k;\beta,p)} = 2\pi g\beta^{3/2}\phi\left(\frac{2\sqrt{\beta}k}{k_F}, \beta(1+p)\right), \tag{12}$$

where $p = P/a$ and

$$\phi(t,\lambda) = \frac{1}{\pi}\frac{\partial}{\partial t}\theta(t,\lambda) \tag{13}$$

$$\theta = arg\, f_+(t,\lambda) \tag{14}$$

$$f_+(t,\lambda) = f_1 + if_2 = \int_0^{\infty} du\, e^{itu-u^2/2}\, u^{\lambda-1}, \tag{15}$$

which is a solution of

$$\frac{d^2 f_+}{dt^2} + t\frac{df_+}{dt} + \lambda f_+ = 0, \tag{16}$$

(parabolic cylinder function). Using Equations (9)-(14) it is straightforward to establish that

$$n_s = 2\int_{2\sqrt{\beta}}^{\infty} dt\left(\frac{\partial\phi(t,\lambda)}{\partial\lambda}\right)_{\lambda=\beta(1+p)} = 1 - \frac{2}{\pi}\left(\frac{\partial\theta(2\sqrt{\beta},\lambda)}{\partial\lambda}\right)_{\lambda=\beta(1+p)}. \tag{17}$$

The expression (17) represents the prediction of the *exact* (BA) statistical mechanics for the fraction of degrees of freedom occupied by solitons. We have not been able to find useful asymptotic expansions for the derivative of θ with respect to the parameter λ. A direct evaluation of the parabolic cylinder functions involved (15) is impractical because of their highly oscillatory character for large values of λ (corresponding to low temperatures). The simplest way to extract useful numbers from the theory is to return to the nonlinear (Ricatti) equation which arises in the context of the closed-form solution[8] and turns out to have a much smoother behavior, suitable for numerical integration.

If $f_+ = Fe^{i\theta}$, and $w = F'/F$, then (9) implies

$$\frac{d\theta}{dt} = \psi \tag{18}$$

$$\frac{d\psi}{dt} = -(t + 2w)\psi \qquad (19)$$

$$\frac{dw}{dt} = -w(w + t) + \psi^2 - \lambda, \qquad (20)$$

with initial conditions

$$\theta(0) = 0 \qquad (21)$$

$$\psi(0) = f_2'(0)/f_1(0) = \sqrt{2}\frac{\Gamma(\frac{\lambda+1}{2})}{\Gamma(\frac{\lambda}{2})} \qquad (22)$$

$$w(0) = 0. \qquad (23)$$

The initial value problem defined by Equations (18-23) can be solved numerically with a high degree of accuracy using the Bulirsch-Stoer method[19]. For a given value of β we have integrated from $t = 0$ to $t = 2\sqrt{\beta}$ for two adjacent values $\lambda = \beta(1 \pm \delta)$ with $\delta = 10^{-4}$. The difference in the resulting θ - values, multiplied by $2/\pi$ is our numerical estimate of the fraction of degrees of freedom occupied by *phonons*. The rest is solitons (cf. Eq.(17)). The results are plotted in Fig.2 and suggest a $T^{1/3}$ dependence.

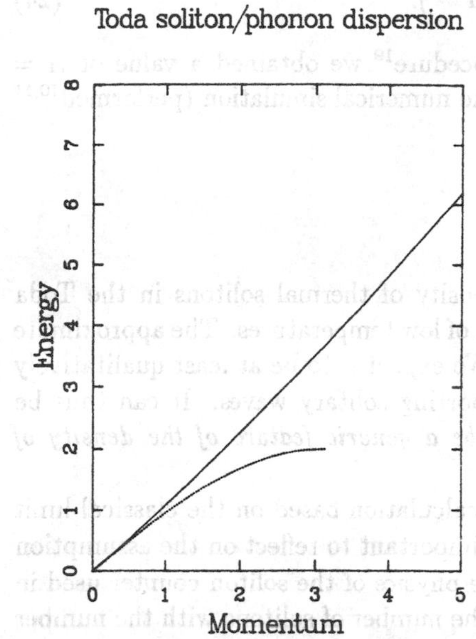

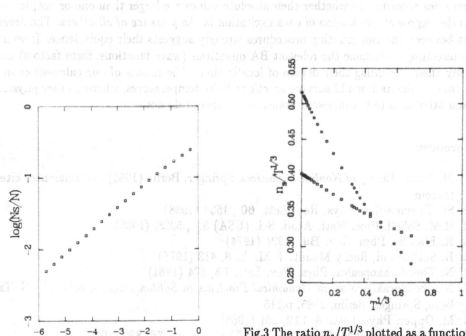

Toda soliton/phonon dispersion

Fig.1 The dispersion curves of solitons (upper curve) and phonons (lower curve).

Fig.2 The fraction of degrees of freedom n_s occupied by solitons as a function of temperature plotted on logarithmic scales.

Fig.3 The ratio $n_s/T^{1/3}$ plotted as a function of $T^{1/3}$. The results extrapolate to a value 0.402 in the limit $T \to 0$. The results of the approximate theory are also exhibited. They extrapolate to a value 0.516

An analysis of the results in terms of such a dependence (Fig.3) provides a value for the prefactor A in the asymptotic dependence

$$\bar{n}_s = A\bar{T}^{1/3} + O(\bar{T}^{2/3}).$$ (24)

Using a standard (Richardson) extrapolation procedure[19] we obtained a value of $A = 0.402$, in excellent agreement with the results of the numerical simulation (performed[10,11] with $T_o = 601\,K$), which suggest $A = 0.394$.

Discussion

We have presented two calculations of the density of thermal solitons in the Toda lattice, both leading to a $T^{1/3}$ power law in the limit of low temperatures. The approximate calculation does not demand exact integrability. We expect it to be at least qualitatively valid for other nonlinear systems capable of supporting solitary waves. It can thus be reasonably argued that *the $T^{1/3}$ power law will be a generic feature of the density of nontopological lattice solitons.*

We conclude with a few remarks on the exact calculation based on the classical limit of the thermodynamics of the Bethe Ansatz. It is important to reflect on the assumption which underlies our calculation and relate it to the physics of the soliton counter used in the numerical simulation[10,11]. We have identified the number of solitons with the number of occupied states whose momenta are higher than k_F. This choice may seem somewhat arbitrary at first sight. It is however forced by the *exact* equivalence[7,9] of the interacting soliton/phonon picture to the BA. On the other hand, the soliton counter, by sorting eigenvalues according to whether their absolute values are larger than one or not, in fact tests the degree of localization of each excitation *in the presence of all others.* The agreement between the two counting procedures strongly suggests their equivalence. It would be interesting to examine the relevant BA quantities (wave functions, form factors) and classify them according their degree of localization. The results of our calculation suggest that localization would survive *exactly* at finite temperatures, whereas other physical soliton attributes (e.g. supersonic transmission speed) do not.

References

1. M. Toda, *Theory of Nonlinear Lattices*, Springer, Berlin (1981) and references cited therein
2. St. Pnevmatikos, Phys. Rev. Lett. **60** , 1534 (1988)
3. H.M. Sobell, Proc. Natl. Acad. Sci. (USA) **82** , 5328, (1985)
4. H. Flashka, Phys. Rev. **B9** , 1924 (1974)
5. B. Sutherland, Rocky Mount. J. Math. **8**, 413 (1978)
6. N. Theodorakopoulos, Phys. Rev. Lett. **53**, 874 (1984)
7. N. Theodorakopoulos, in *Dynamical Problems in Soliton Physics*, edited by S. Takeno, Springer, Berlin, 1985, p.115
8. M. Opper, Phys. Lett. **A 112**, 201 (1985)
9. H. Takayama and M. Ishikawa, Prog. Theor. Phys. **76**, 820 (1986)
10. V. Muto, A.C. Scott and P.L. Christiansen, Phys. Lett. **A 136**, 33 (1989)
11. V. Muto, A.C. Scott and P.L. Christiansen, Physica **D 44**, 75 (1990)
12. N. Theodorakopoulos and F.G. Mertens, Phys. Rev. **B28**, 3512 (1983)

13. N. Theodorakopoulos in *Statics and Dynamics of Nonlinear Systems*, edited by H. Bilz *et al*, Springer, Berlin, (1983), p. 271. A similar account of Toda soliton thermodynamics based on the same physical input (phonon phase shifts due to a single soliton) and also leading to a $T^{1/3}$ dependence of the soliton density has been given by F. Yoshida and T. Sakuma, Phys. Rev. **A25**, 2750 (1982).

14. F.G. Mertens and H. Büttner, Phys. Lett. **A84**, 335 (1981)

15. P. Grüner-Bauer and F.G. Mertens, Z. Phys. **70**, 435 (1988)

16. M. Fowler and N-C. Yu, J. Phys. **A22**, 3095 (1989)

17. N. Theodorakopoulos, in *Proc. 2nd Int. Conf. on Phonon Physics*, edited by J. Kollar *et al*, World Scientific, Singapore, 1985, p.468

18. C.N. Yang and C.P. Yang, J. Math. Phys. **10**, 1115 (1969)

19. W. H. Press *et al*, *Numerical Recipes*, Cambridge (1986)

DESTRUCTION OF SOLITONS IN FRENKEL - KONTOROVA MODELS
WITH ANHARMONIC INTERACTIONS

Andrey Milchev

Institute of Physical Chemistry
Bulgarian Academy of Sciences, 1040 Sofia, Bulgaria

Abstract

The stability of moving topological solitons is investigated in the framework of a Frenkel - Kontorova model with anharmonic (e.g. Morse, Toda, etc.) interactions between the atoms and a lattice misfit between overlayer and substrate. The numerical simulations confirm the analytically derived misfit- and velocity-dependent critical ratio of periodic vs. interatomic potential strength, which determines the existence region of rarefaction kinks. No such limit exists for antikinks. The maximal propagation velocity of solitary excitations increases monotonically from zero with growing misfit and becomes supersonic at positive misfit. The destruction of kinks beyond their stability limits and the resulting formation of cracks are demonstrated numerically in cases of kink collisions with antikinks or with lattice defects. Extension of the model to long-range interactions beyond first neighbors does not change qualitatively these results.

I. Introduction

The simple model of a chain of atoms interacting via nearest-neighbor harmonic forces and placed in a periodic external (substrate) potential, due originally to Frenkel and Kontorova/1/ and designed to describe the structure of the chain near a core of a solid-state dislocation, has become during the last thirty years one of the universal models of one dimensional nonlinear physics. The simplicity of the original model as well as its capability of describing a broad spectrum of phenomena, e.g. charge-density waves, domain boundaries, magnetic structures, etc., in which competing periodicities play a major role, have attracted a great deal of attention from both theoreticians and experimentalists during the last decades.

In a series of papers recently/2-4/ an extension of the model to anharmonic interactions between the atoms of the chain was shown to predict the possibility of crack formation in the overlayer (rupture of the chain) following the collapse of rarefaction kinks after some threshold values of the parameters have been exceeded. It is tempting to suggest this phenomenon as an explanation of the observed formation of cracks from misfit dislocations as reported recently by Franzosi et al./5/. The idea is that over-stretched bonds at the center of the kink may break, notwithstanding the expansive force of the substrate potential, after the tensile strength of anharmonic potential has been surpassed/6-8/. This critical elongation of the bond is reached at the inflection point of the anharmonic potential and it provides a second characteristic length along with the equilibrium atom spacing of the unperturbed chain which will compete with the lattice constant of the substrate.

In this work after briefly defining the model (Section II) we show how this misfit- and velocity- dependent characteristic length may be used to determine the critical effective amplitude of the substrate potential up to which for a given misfit moving rarefaction kinks may still exist (Section III). In Section IV we extend the model so as to allow for distant-neighbor long-range interactions and demonstrate that this leads to small corrections which do not change qualitatively our results. The analytical expression derived for Morse (or Toda) interactions within the continuum limit approach is then verified by means of computer simulation of the discrete chain (Section V A) and found valid also for discrete systems. The behavior of kinks at and beyond these threshold values of the system parameters is studied numerically in cases when kinks interact with other kinks or with lattice inhomogeneities (Section V B). Thus for the first time we indeed witness the formation of cracks after kinks have broken and examine the conditions under which this may occur. In the final section we summarize our results and point out that they should be qualitatively valid for any non-convex interactions.

II. The model

We consider a monatomic chain with nearest-neighbor interactions with a Hamiltonian

$$\mathcal{H} = \sum_n \left\{ \frac{1}{2} M \dot{x}_n^2 + V(x_{n+1} - x_n - b) + \frac{W}{2} \left[1 - \cos(\frac{2\pi}{a} x_n) \right] \right\} \quad (1)$$

where $x_n(t)$ is the coordinate of a particle with mass M and velocity $\dot{x}_n$. The equilibrium distance of the unperturbed chain is denoted by b, while the external (sine) potential has a period a and may be thought of as exerted by a crystalline substrate on the overlayer. The second term in (1) is the anharmonic interaction potential, which can be quite general, but in this work we shall be primarily dealing with Morse interactions:

$$V(x) = E_0(e^{-2\alpha x} - 2e^{-\alpha x}) \quad (2)$$

as more realistic ones, although some results are more easily obtained with the Toda potential

$$V(x) = E_0(e^{-\beta x} + \beta x - 1)/\beta^2 \quad (3)$$

where by changing β between 0 and ∞ one may study the crossover from purely harmonic potential to a "hard core" one. Besides the equilibrium nearest-neighbor distance, the inflection point of the Morse potential (2), $x_{inf} = \ln 2/\alpha$, is the other physically meaningful length which plays, as we shall see, a major role in the behavior of our system. At x_{inf} the maximal tensile strength of the chain is attained (the first derivative of Eq.(2) is maximal) so that its relation to rupture of the chain may be anticipated.

Although the Toda potential, Eq.(3), has its inflection point at $x_{inf} = \infty$ where the second derivative is zero, the restoring force (first derivative of (3)) saturates at its maximal value at $x \cong \beta^{-1}$. Thus, with respect to the tensile strength of the chain, the Toda potential resembles much more the Morse or any other anharmonic potential, than purely convex potentials like the harmonic one.

The application of an external field f, due to mechanical stress or electric field in cases of charged masses, corresponds to a potential

$$V_{ex} = - f \cdot x_n \qquad (4)$$

which makes the substrate potential asymmetric. This limits the values of f below some critical value, f_{cr}, at which the multiple minima vanish.

If the coordinates x_n are replaced by atomic displacements (or phase shifts), $x_n = n \cdot a + (\frac{a}{2\pi})\varphi_n$, measured from the bottoms of the respective potential wells of the substrate, the equations of motion of the system, obtained after differentiating (1), will read

$$M(\frac{a}{2\pi})^2 \ddot{\varphi}_n = - C_M^2 (\alpha')^{-1} \left[e^{-2\alpha'(\varphi_{n+1} - \varphi_n - P)} - e^{-\alpha'(\varphi_{n+1} - \varphi_n - P)} \right.$$

$$\left. - e^{-2\alpha'(\varphi_n - \varphi_{n-1} - P)} + e^{-\alpha'(\varphi_n - \varphi_{n-1} - P)} \right] - \frac{W}{2} \sin(\varphi_n) \qquad (5)$$

where we have introduced the speed of sound, $C_M^2 = 2(\alpha')^2 E_0$, for the Morse potential and $\alpha' = \alpha(\frac{a}{2\pi})$. For brevity in what follows we shall drop the prime at α. The parameter $P = 2\pi(b - a)/a$ allows for the lattice misfit between the unperturbed chain and the periodic (substrate) potential.

III. The continuum limit

In the continuum limit, $\varphi_{n+1} - \varphi_n \cong \nabla\varphi + \frac{1}{2}\nabla^2\varphi$, Eq. (5) yields

$$\varphi_{tt} - [2e^{-2\alpha(\varphi_x - P)} - e^{-\alpha(\varphi_x - P)}]\varphi_{xx} = -\frac{\lambda}{2} \sin\varphi \qquad (6)$$

where the time is measured in dimensionless units and $\lambda = W / C_M^2$.

Looking for solutions of Eq. (6) in the form of travelling waves, $\varphi(x,t) \equiv \varphi(x - v \cdot t) = \varphi(\xi)$, one may convert it into an ordinary differential equation (DE), or equivalently, as a system of two first-order DE which may be represented in autonomous form as:

$$\frac{d\omega}{d\varphi} = \frac{(\lambda/2) \sin\varphi}{[2e^{-2\alpha(\omega - P)} - e^{-\alpha(\omega - P)} - V^2]\omega} \qquad (7)$$

In Eq. (7) the derivative $\omega = d\varphi/d\xi$ describes the strain (displacements differences of neighboring atoms) in the chain and $V = v/C_M$ is the scaled (dimensionless) soliton velocity. It has been shown in /2/ that the root ω_0 of the bracketed expression in (7) adds new special points to the conventional ones, ($\omega = \varphi = 0$) and ($\omega = 0$, $\varphi = \pi$), in the phase portrait of the system. If we solve the resulting quadratic equation $2X^2 - X - V^2 = 0$ which determines the singularity in (7) for $X = \exp[-\alpha(\omega_0 - P)]$, we get

$$\omega_0 = P - \frac{1}{\alpha} \ln \frac{1 + \sqrt{1 + 8V^2}}{4}. \qquad (8)$$

In the case of a static kink, $V = 0$, at zero misfit, $P = 0$, $\omega_0 = x_{inf}$.

Thus ω_0 marks a critical strain whereby the energy bond between two atoms attains its maximal tensile strength. Non-zero velocities and misfits change this value but the physical meaning of the new singularity in the Frenkel - Kontorova model becomes evident.

Eq. (7) may be easily integrated once and one gets for the first integral

$$\alpha^2 \lambda (K - \cos\varphi) = \qquad (9)$$

$$= e^{2\alpha P}[1 - (1 + 2\alpha\omega)e^{-2\alpha\omega}] - 2e^{\alpha P}[1 - (1 + \alpha\omega)e^{-\alpha\omega}] - \alpha^2\omega^2 V^2$$

where K is an integration constant. In the case of a single kink K = 1. A positive value of ω, satisfying Eq. (9), may be found only if the maximum of the λ-dependent left-hand side of (9), which is reached at $\varphi = \pi$, does not exceed the maximum of the right-hand side of (9), which is attained at ω_0. Thus we may claim that the chain retains its integrity, i.e. one still has a rarefaction kink instead of a crack only if λ is below some threshold value $\lambda_{cr} = \lambda_{cr}(P, V)$. The analytical form of this threshold surface is readily obtained if the value of ω_0 from Eq. (8) is inserted in the right-hand-side of Eq. (9):

$$\left(\frac{W}{E_0}\right)_{cr} \leq e^{2\alpha P} - 2e^{\alpha P} - V^2\left(\alpha P - \ln\frac{1 + \sqrt{1 + 8V^2}}{4}\right)^2 + \frac{1 + \sqrt{1 + 8V^2}}{4}\left\{ \right.$$

$$\left. \frac{1 + \sqrt{1 + 8V^2}}{4} - 2\left[\frac{1 + \sqrt{1 + 8V^2}}{4} - 1\right]\left(1 + \alpha P - \ln\frac{1 + \sqrt{1 + 8V^2}}{4}\right)\right\} \qquad (10)$$

For $P = 0$, $V = 0$, Eq. (10) yields $(W/E_0)_{cr} = (2\ln 2 - 1)/4 \cong 0.0965736...$ This means that the maximal effective amplitude of periodic substrate potential is rather small and justifies the use of the continuum approximation in this treatment.

Similarly, for the case of a Toda potential one obtains

$$\left(\frac{W}{E_0}\right)_{cr} \leq \frac{1}{\beta^2}\left\{e^{\beta P} - V^2\left[1 + \ln V^2 - \beta P + \frac{(\ln V^2 - \beta P)^2}{2}\right]\right\} \qquad (11)$$

In Fig. 1 we show the critical surfaces, Eqs. (10), and (11), stressing again that above these surfaces stable topological kinks may not exist.

Another conclusion following from the inspection of these critical surfaces suggests that at negative misfits, P < 0, rarefaction solitons may propagate with *subsonic* velocities only while for P > 0 *supersonic* kinks are also possible. The role of the anharmonic forces as being responsible for the possibility of supersonic velocities of topological solitons has been originally pointed out by Kosevich and Kovalev/9/.

It is interesting to note that this restriction in maximal velocity applies also to *antikinks* for which no limits with respect to the effective ratio W/E_0 exist! Indeed, from Eq. (8) (set for simplicity P = 0) one gets evidently $\omega_0 < 0$ for V > 1. This negative strain, corresponding to a compression in the core of an antikink, denotes a local minimum of the right-hand side of Eq. (9) which must be always non-negative, as the left-hand side is. Thus antikinks cannot propagate faster than kinks and the maximal speed vs. misfit dependence is given simply by the cross-section of the $(W/E_0)_{cr}$ surface, Fig. 1, with the (P, V) - plane.

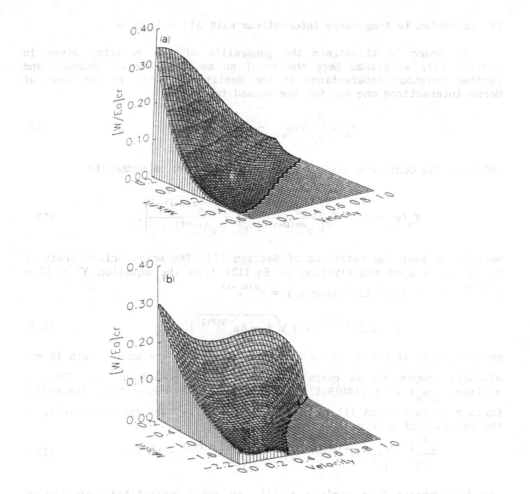

Fig1. Dependence of the critical ratio $(W/E_0)_{cr}$ on incompatibility between substrate and overlayer P and on kinks velocity V for the case of: (a) Morse interactions, Eq. (10), and (b) Toda interactions, Eq. (11). Kinks may exist only for W/E_0 *below* these critical surfaces.

According to Fig. 1, with the misfit becoming more and more negative, the range of permitted velocities is gradually narrowed and for $P < P_{cr} < 0$ (cf. Fig. 1a) no rarefaction solitons whatsoever may exist! As it might be anticipated, $P_{cr} = - \ln2/\alpha \equiv - x_{inf}$, that is, for lattice mismatch, $(a - b)/a$, between substrate and chain larger than the inflection point of the Morse potential even static solitons should not exist.

In the Toda case, Fig. 1b, the shape of the critical surface $(W/E_0)_{cr}$ is qualitatively the same, although $P_{cr} = - \infty$ which, as mentioned in Section II, is in accordance with the inflection point being at infinity.

IV. Extension to long-range interactions with all neighbors

In order to illustrate the generality of the results, given in Section III, we extend here the model so as to allow for second- and further neighbor interactions in the Hamiltonian (1). In the case of Morse interactions one has for the second term in (1):

$$V_\infty = \sum_{l=1}^{\infty} V(x_{n+l} - x_n - b), \qquad (12)$$

which in the continuum limit, $\varphi_{n+1} - \varphi_n \cong 1.\nabla\varphi$, may be summed to

$$V_\infty(\varphi_x = \omega) = E_0 \left[\frac{e^{2\alpha(P+1)}}{e^{2\alpha(\omega+1)} - 1} - \frac{e^{\alpha(P+1)}}{e^{\alpha(\omega+1)} - 1} \right], \qquad (13)$$

whereby we keep the notations of Section III. The equilibrium strain of V_∞ is found upon minimization of Eq.(13) from the equation $Y^2 + [2 - e^{\alpha(P+1)}] + 1 = 0$ for the unknown $Y = e^{\alpha(\omega_0+1)}$:

$$Y = 0.5 e^{\alpha(P+1)} [1 + \sqrt{1 - 4e^{-\alpha(P+1)}}] - 1, \qquad (14)$$

so that, e.g. at $P = 0$, for $\alpha = 3$ $\omega_0 \cong -0.036$ and the whole chain is now slightly compressed as compared to the nearest-neighbor one. The new minimum $V_\infty(\omega_0) \cong -1.11401911...E_0$ is about 11.4% deeper too. Proceeding further as in Section III, one can readily verify that the singularity of the equation of motion is now determined from

$$\frac{Z^2(1 + Z^2)}{(Z^2 - 1)^3} 2e^{2\alpha(P+1)} - \frac{Z(1 + Z)}{(Z - 1)^3} e^{\alpha(P+1)} - V^2 = 0 \qquad (15)$$

for the unknown $Z = \exp[\alpha(\omega + 1)]$. An exact analytical solution of Eq.(15) may be found at least for $V = 0$, but here we shall rather estimate the corrections of the order of $\varepsilon = e^{-\alpha} \cong 0.05$ (for $\alpha = 3$) to the "nearest neighbor" critical strain ω_0. For $V = 0$ we have

$$e^{-\alpha\omega_0} = (1 + 4\varepsilon e^{-\alpha\omega_0} + 5\varepsilon^2 e^{-2\alpha\omega_0} + ...) \frac{e^{-\alpha P}}{2} \qquad (16)$$

which gives practically a shift of $\cong 10\%$ in the value of ω_0. The first integral in the case of all neighbors taken into account is

$$e^{2\alpha(P+1)} \left[\frac{e^{2\alpha(\omega+1)}(1-2\alpha\omega) - e^{4\alpha(\omega+1)}}{[e^{2\alpha(\omega+1)} - 1]^2} + \frac{e^{4\alpha} - e^{2\alpha}}{(e^{2\alpha} - 1)^2} \right] - \alpha^2 V^2 \omega^2 -$$

$$- 2e^{\alpha(P+1)} \left[\frac{e^{\alpha(\omega+1)}(1-\alpha\omega) - e^{2\alpha(\omega+1)}}{[e^{\alpha(\omega+1)} - 1]^2} + \frac{e^{2\alpha} - e^{\alpha}}{(e^{\alpha} - 1)^2} \right] = \alpha^2 \lambda(K+1) \qquad (17)$$

Thus with Eq.(15) and Eq.(17) one could derive the critical surface $(W/E_0)_{cr}$ which, as the estimates show, may be shifted at most 10% with respect to that given in Fig. 1. Evidently this will not change qualitatively the results of Section III, at least for realistic values of the Morse parameter $\alpha = 3 \div 4$.

V. Destruction of kinks and formation of cracks: numeric results

The predicted critical surface for kink destruction $\lambda_{cr}(P, V)$ has been derived within the framework of the continuum approximation whereby discreteness effects of the lattice have been completely ignored. In order to allow for these effects numerical experiments are performed on the discrete system described by Hamiltonian (1). Moreover, using computer simulations we have been able to study the interactions of kinks among themselves as well as with lattice defects in those cases when kinks are existing near their stability limits and thus examine the possible destructive consequences of such interactions.

The numerical scheme for solving the non-linear differential difference equations of motion (5) is a fourth-order Runge-Kutta method of a lattice with typically 400 and 1000 atoms and fixed boundary conditions as described earlier /10,11/. The initial conditions for each run were determined via minimization of the energy (1) with respect to φ_n by the gradient (steepest descend method). As a starting point for convergence we use a configuration of the chain atoms with a kink (antikink) moving with velocity V as given by some analytical function. In this work we use also $\alpha = 3$ and $a = 3$ as fixed parameters in the computations.

A. Existence of kinks and stability

In the course of extensive investigations we found that the critical surface $\lambda_{cr}(P, V)$, as given by Eq. (10), determines with very high accuracy the possibility of finding numerically a kink solution with the same set of parameters. For given values of P and V the minimization procedure converges steadily as long as $\lambda < \lambda_{cr}$ while for $\lambda \geq \lambda_{cr}(P, V)$ one ends up eventually with two pieces of chain separated by an empty well of the sine potential. All $\varphi_n = 0$, or 2π in dependence of whether they are to the left, or to the right of the empty well so that indeed such a configuration may be viewed as an accurate confirmation (better than 10^{-16}) for the predicted non-existence of kink solutions and the inevitable disintegration of the system at the given λ, P and V.

If a kink solution with minimal energy is found to exist, the subsequent molecular-dynamics simulation shows that it propagates as a stable excitation with virtually no emission of radiation. In this respect we have not seen any difference in behavior between sub- and supersonic kinks. Radiative losses and an accompanying slight reduction in speed of the kink are detected only for λ very near to $\lambda_{cr}(P, V)$. Thus, in view of the numeric experiments one may claim that the stability region $\lambda \leq \lambda_{cr}(P, V)$ confines the system to a parameter space for which discreteness effects are of minor importance and the continuum approximation provides a very good description.

B. Breakup of kinks and formation of cracks

Having verified numerically the existence of the breakup threshold $\lambda_{cr}(P, V)$, the question arises as to what happens to kinks which find themselves beyond this safety limit. Evidently, one possible outcome in a situation like that is for the kink to reduce its velocity by means of radiation losses so as to enter the stability region again and survive. The other possibility for the endangered kink is to break, transforming

TABLE I. Summary of the numerical simulation results on cracks formation from rarefaction kinks

Type of action	Break
Collision with impurity	No
Collision with an interface with $\lambda > \lambda_{cr}$	No
Collision with an interface with $P < P_{cr}$	Yes
Kink - kink collisions	No
Kink - antikink collisions	Yes
Acceleration of kinks by external field to velocity $V > V_{cr}$	No*

into a static crack at the site of event. There could be various notions about the way this latter possibility may be realized. In Table 1 we represent some of them and summarize the results of our numerical investigations.

They reveal that cracks are indeed formed in the course of kinks destruction but that this happens in particular cases only. Below these cases are briefly discussed and illustrated.

1. Interaction of a kink with impurity.

Recently/3/ it was shown that the interaction of a kink in the vicinity of its breakup threshold with a local inhomogeneity, represented by a narrow region where the substrate-potential amplitude is decreased or increased, leads to effective attraction and stabilization of the kink in the former case, and to repulsion by the impurity in the latter. Considering the latter case as potentially destructive for the kink, we studied numerically collisions with such inhomogeneities for various sets of parameters, i.e. for a given velocity V we systematically changed the local amplitude W_{imp} with respect to the substrate-potential amplitude W. Depending on the ratio W_{imp}/W kinks behavior changed between transmission

---

* see text for details.

and reflection, and no breakup whatsoever was observed. A typical plot is shown in Fig. 2 where the relative displacements differences, $\varphi_{n+1} - \varphi_n$, in a chain with 400 atoms are shown for different times. The incident supersonic kink, moving to the right, is scattered by the inhomogeneity located at position 200. With P = 0.6, V = 2.0 and $(W/E_0)_{cr}$ = 0.153... from Eq. (10) known, we choose here (W/E_0) = .140 and (W_{imp}/E_0) = 6.25 (Fig. 2a) and (W_{imp}/E_0) = 6.28 (Fig. 2b). Thus the change from transmission to reflection is fixed at a large value of the ratio W_{imp}/W.

(a)

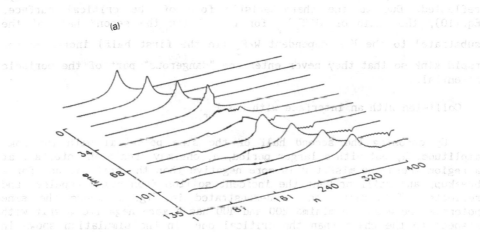

(b)

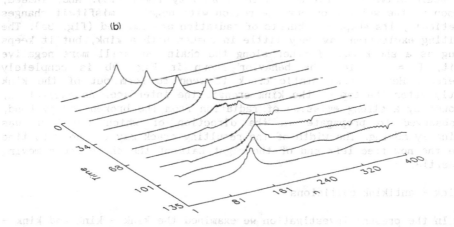

Fig 2. Collision of a supersonic soliton with a single inhomogeneity placed in the middle of the substrate at n = 200. The parameters are P = 0, V = 2.0 and W/E_0 = 0.14 while $(W/E_0)_{cr}$ = 0.153. Transmission is observed up to W_{imp}/E_0 = 6.24 – (Fig. 2a) and reflection – from W_{imp}/E_0 = 6.28 on – (Fig. 2b).

For subsonic kinks this change occurs at much smaller ratios: for P = 0, V = 0.3 and (W/E_0) = 0.06 this is observed at (W_{imp}/E_0) = 0.14. The form of the kink is temporarily distorted and radiation losses manifest the inelasticity of the collision, but the kink later recovers and attains its stable form.

2. Collision with an interface with $\lambda > \lambda_{cr}$.

We simulate the case in which the incident kink collides with an interface between two semi-infinite parts of the substrate, characterized by the same lattice constant a, but with different amplitude W/E_0. It was expected that if W/E_0 in the second half of the system is chosen higher than $(W/E_0)_{cr}$ of a *static* kink, the incident soliton may still enter the second half and, despite reducing speed to zero, break. However, as in the previous case, also here we never observe a breakup of the kink no matter how high its initial velocity is. Instead, the kink is always reflected. Due to the characteristic form of the critical surface, Eq.(10), the ratio of $(W/E_0)_{cr}$ for V = 0 (in the second half of the substrate) to the V - dependent W/E_0 (in the first half) increases for rapid kink so that they never enter the "dangerous" part of the periodic potential.

3. Collision with an interface with $P < P_{cr}$.

By choosing the second half of the sine potential with the same amplitude, W, but with a larger period, a, one may create an interface at a region where the misfit P is more negative than the critical one for a breakup, and still prevent the incident soliton from being repulsed and reflected. This situation is demonstrated in Fig. 3 where the sine potential between the minima 200 and 400 has a more negative misfit with respect to the chain than the critical one. In the simulation shown in Fig. 3 we have V = 0.3, W/E_0 = 0.062 (the critical value is 0.064) and P = 0 in the left half of the substrate while any negative P in the right half would drive the kink beyond the stability limit (10). And, indeed, as soon as the soliton enters the region with negative misfit it changes dramatically its shape and bursts of radiation are emitted (fig. 3a). The resulting excitation has very little in common with a kink, but it keeps moving as a shock wave further along the chain. At still more negative misfit, P = - 0.14, the behavior shown in Fig. 3b is completely different. Here a real static crack is seen to form out of the kink shortly after the top of the kink crosses the interface. An excitation, followed by a strong emission of radiation from the broken energy bond, is observed to propagate in the direction of which the kink was originally moving. The width of the resulting crack may change with time since the now free left end of the right half of the chain keeps moving by inertia.

4. Kink - antikink collisions

In the present investigation we examined the kink - kink and kink - antikink collisions of kinks existing in the vicinity of the breakup surface $(W/E_0)_{cr}$. In all cases of kink - kink collisions the kinks go smoothly through each other causing no visible distortions in shape. As a possible explanation one may assume that the repulsive interaction

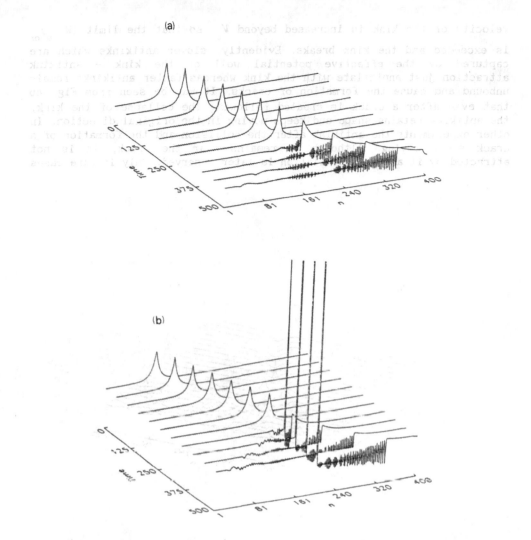

Fig 3. Propagation of a kink through the interface between two substrates with different misfit P as regards the overlayer. P = 0 for 1 < n < 200. For 201 < n < 400 P = - 0.01 (Fig. 3a) and P = - 0.14 - (Fig. 3b). The other parameters are W/E_0 = 0.062 and V = 0.3. The height of the crack exceeds the framework of the drawing.

between the kinks reduces to some extent their speed thus increasing effectively $(W/E_0)_{cr}$ and rendering the kinks more stable.

The more interesting case of kink - antikink collisions is represented in Fig. 4 where a static kink is approached by an antikink moving with speed V = - 0.1 (Fig. 4a) and V = -.3 (Fig. 4b). In both cases we have P = 0 and W/E_0 = 0.095 while $(W/E_0)_{cr}$ = 0.09657...One might speculate here that because of the kink - antikink attraction the velocity of the kink is increased beyond V_{cr} so that the limit $(W/E_0)_{cr}$ is exceeded and the kink breaks. Evidently, slower antikinks which are captured by the effective potential well of the kink – antikink attraction just annihilate with the kink whereas faster antikinks remain unbound and cause the formation of cracks. It may be seen from Fig. 4b that even after a crack is created following the collapse of the kink, the antikink retains shape and keeps moving in the original direction. In other experiments the antikink after the collision and the formation of a crack stops moving. Although it remains near the crack, it is not attracted by it and no annihilation is later observed. Only in rare cases

(a)

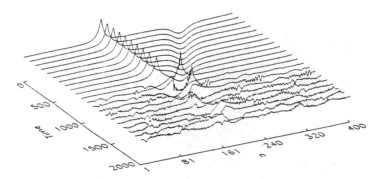

(b)

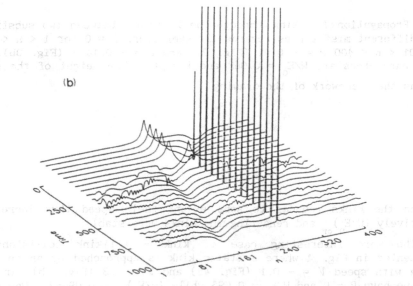

Fig4. Annihilation (a) and formation of a crack (b) after a kink – antikink collision. P = 0, W/E_0 = 0.095 and the kink is static while the antikink moves with V = - 0.1 (Fig. 4a) and with V = - 0.3 (fig. 4b).

when a crack is created within the outskirts of the now static antikink a subsequent recombination after a finite period of time is observed.

Our experience with kink – antikink collisions suggests also that static, or slowly ($V \leq 0.17$) moving kinks break thereby easily while with growing own velocity kinks become much more stable. Kink – antikink collisions then result in an inelastic impact with large energy radiation, but subsequently the kinks recover shape corresponding to the reduced velocity.

For a given kink velocity, V_k, at the respective $(W/E_0)_{cr}$, we found the minimal antikink velocity, V_{ak}, for which a transition from annihilation to crack formation is observed: (i) $V_k = 0.13$, $V_{ak} = -0.28$ at $(W/E_0)_{cr} = 0.089$; (ii) $V_k = 0.14$, $V_{ak} = -0.31$ at $(W/E_0)_{cr} = 0.088$; (iii) $V_k = 0.15$, $V_{ak} = -0.81$ at $(W/E_0)_{cr} = 0.086$ and (iv) $V_k = 0.16$, $V_{ak} = -0.83$ at $(W/E_0)_{cr} = 0.085$. Evidently for $V_k > 0.14$ kinks gain much larger stability – an interpretation of this finding, however, would require more comprehensive investigations.

5. Acceleration of kinks by an external field

We considered the acceleration of a kink by an external field as another method to drive the kink beyond its stability limits and initiate a breakup. In fig. 5 we show the behavior of such kinks in cases of applied external field, f, with varying strength. As expected, the acceleration of the kink causes bursts of energy radiation, but generally no cracks are formed. At small fields, $f = -0.001$, the kink preserves its shape (Fig. 5a) while with increasing field intensity, $f = -0.002 \div -0.004$ (Fig. 5b ÷ 5c) the kink is gradually transformed into an excitation of the type observed for $P \leq P_{cr}$ (see above Case 3). In Fig. 5c even the formation of cracks may be seen but this happens only after

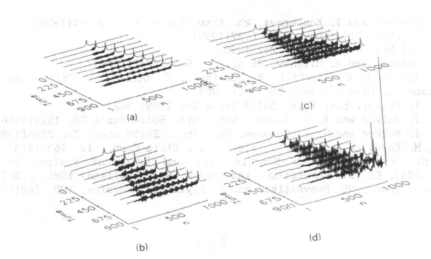

Fig. 5. Acceleration of a kink in external field with different intensity f: (a) $f = -0.001$; (b) $f = -0.002$; (c) $f = -0.003$ and (d) $f = -0.004$.

the front of the excitation has reached the fixed boundaries of the system (which in reality might be represented by a heavy impurity atom). However, in simulations with the same set of parameters and *periodic* boundary conditions the excitation front propagates without any formation of cracks.

V. Concluding remarks

Summarizing the results of the present investigation we should like to point out that the qualitative conclusions should be viewed as independent from the concrete type of anharmonic forces (Morse, Toda) used in the analytical treatment. These conclusions remain valid even if long-range interactions with distant-neighbor atoms are accounted for in an extended Frenkel - Kontorova model.

The critical limits for kinks existence show that topological kinks may be
(i) only subsonic for negative misfits,
(ii) both sub- and supersonic - for positive misfits,
and that,
(iii) discreteness effects are negligible and the continuum approximation - very accurate for any kink in a model with realistic (anharmonic) interatomic forces as long as the kink does not exist immediately at the threshold surface λ_{cr} (P, V).

Considering the cases of kinks destruction which have been investigated in the course of our computer experiments, we should like to point out those where a crack formation is indeed observed as a proof of the theoretical predictions. Two certain sources for cracks formation appear to be kinks running into regions with increased lattice constant as well as kink - antikink collisions. A real experiment in this aspect would be both interesting and useful. Contrary to expectations/4/, however, a single outstanding inhomogeneity is seemingly not able to cause destruction after a kink has collided with it. Investigations of collisions with different types of defects as well as a more quantitative picture of the cases when breakup is observed are beyond the scope of this report but a subject of future investigations.

References

1. J. Frenkel and T. Kontorova, Zh. Eksp. Theor. Fiz. **8**, 89(1938)
2. A. Milchev, Phys. Rev. **B42**, 6727(1990)
3. A. Milchev, Physica **D41**, 262(1990)
4. B. Malomed and A. Milchev, Phys. Rev. **B41**, 4240(1990)
5. P. Franzosi, G. Salviati, M. Scaffardi, F. Genova, S. Pellegrino and A. Stano, J. Cryst. Growth **88**, 138(1988)
6. A. I. Melker, Sov. Phys. Solid State **24**, 1809(1982)
7. A. I. Melker and A. V. Ivanov, Sov. Phys. Solid State **28**, 1912(1986)
8. A. I. Melker and S. V. Govorov, Sov. Phys. Solid State **30**, 2066(1988)
9. A. M. Kosevich and A. S. Kovalev, Solid State Comm., **12**, 763(1973)
10. St. Pnevmatikos, in *Singularities and Dynamical Systems* (Math. Stud. **103**), Ed. St. Pnevmatikos (Amsterdam: North Holland) 1985, p. 397
11. M. Peyrard, St. Pnevmatikos and N. Flytzanis, Physica, **19D**, 268(1986)

PROTON POLARIZABILITY OF HYDROGEN BONDED SYSTEMS DUE TO COLLECTIVE PROTON

MOTION - WITH A REMARK TO THE PROTON PATHWAYS IN BACTERIORHODOPSIN

Georg Zundel and Bogumił Brzezinski[x)]

Institute of Physical Chemistry
University of Munich
Theresienstr. 41
D-8000 München 2
Germany

Dedicated in memory of Stephanos Pnevmatikos

ABSTRACT

Hydrogen bonds with double minimum proton potential show so-called
proton polarizabilities which are about two orders of magnitude larger
than usual polarizabilities of electron systems. Such hydrogen bonds are
indicated by continua in the infrared spectra. It is shown experimentally
as well as theoretically that symmetrical or largely symmetrical hydrogen-
bonded chains show particularly large proton polarizabilities due to collective proton motion. Protons can be conducted within psec. along these
chains. The proton transfer processes in the active center of the bacteriorhodopsin molecule as well as the proton conduction process from the
active center to the outside of the purple membrane are discussed on the
basis of these results.

INTRODUCTION

Hydrogen bonds $B^+H\cdots B \rightleftharpoons B\cdots H^+B$ or $AH\cdots A^- \rightleftharpoons A^-\cdots HA$ bonds, respectively, i.e. bonds with which the donor and the acceptor is the same type
of group are structurally symmetrical. They are called homoconjugated
bonds. Hence within these bonds double minimum proton potentials or potentials with broad flat well are present, being symmetrical if they are considered without environment. These bonds show so-called proton polarizabilities caused by shifts of the proton within these bonds. It is particularly important that these proton polarizabilities are about 2 orders of
magnitude larger than the usual polarizabilities caused by distortion of
electron systems as shown by theoretical treatments[1-4]. These hydrogen
bonds cause intense continua in the infrared spectra[5]. Fig. 1 shows such a
continuum which arises if a strong acid is added to pure water. The protons are present and fluctuate in the hydrogen bond of the $H_5O_2^+$ groups.
The continua arise due to the strong interactions of these hydrogen bonds
with their environments caused by their large proton polarizability[4-5].

x) Permanent address: Faculty of Chemistry, A. Mickiewicz University, ul.
   Grunwaldzka 6, 60-780, Poland

Proton Transfer in Hydrogen-Bonded Systems
Edited by T. Bountis, Plenum Press, New York, 1992

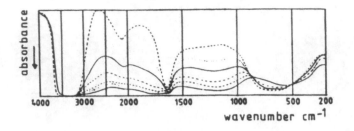

Fig. 1. IR spectra: Aqueous solutions of HCl (at 28° C). Layer thickness 13.6 μ. In the order of sequence of increasing absorbance of the background 0.0, 1.21, 2.43, 4.85, 7.28, 8.49 and 10.95 mol/dm^3

The fluctuation frequency of the proton within all homoconjugated hydrogen bonds is larger than 10^{13} sec^{-1}.

Heteroconjugated $AH\cdots B \rightleftharpoons A^-\cdots H^+B$ hydrogen bonds may show such large proton polarizabilities, too. With these bonds the double minima are, however, usually only created by their environments[4,7]. With these bonds the fluctuation frequencies of the protons are, however, at least 2 orders of magnitude smaller than with the homoconjugated ones.

But not only single hydrogen bonds may show large proton polarizability. Particularly large proton polarizabilities are shown caused by collective proton motion in hydrogen-bonded systems[8] - as shown in the following.

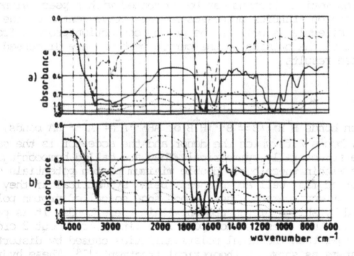

Fig. 2. a) IR spectra of $(L-Lys)_n + KH_2PO_4$ systems: (-··-), pure $(Lys)_n$; (——), Lys:KH_2PO_4 = 3:1; (---), Lys:KH_2PO_4 = 1:1; (···), Lys:KH_2PO_4 = 1:2.

b) IR spectra of $(L-Glu)_n + KH_2PO_4$ systems: (-··-), pure $(Glu)_n$; (——), Glu:KH_2PO_4 = 2:1; (---), Glu:KH_2PO_4 = 1:1; (···)

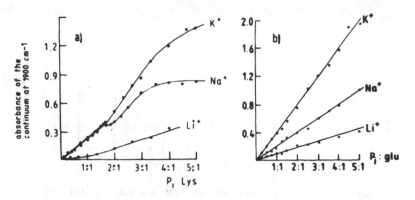

Fig. 3. Absorbance of the continuum at 1900 cm^{-1} as a function of the phosphate (P$_i$) : residue ratio; a) (L-Lys)$_n$ + KH$_2$PO$_4$ systems, b) (L-Glu)$_n$ + KH$_2$PO$_4$ systems.

PROTON POLARIZABILITY DUE TO COLLECTIVE PROTON MOTION IN POLY-α-AMINOACID + PHOSPHATE SYSTEMS

Fig.2a shows polylysine + KH$_2$PO$_4$ systems. With increasing phosphate content a very intense continuum arises. The same is true with polyglutamic acid + KH$_2$PO$_4$ systems - as shown in Fig. 2b. In Fig. 3 are given the intensities of these continua as a function of the phosphate to side chain ratio. Fig. 3a shows that with the polylysine systems, the intensity of the continuum increases up to three phosphates per lysine residue if Na$^+$ ions or up to five if K$^+$ ions are present, respectively. Fig. 3b shows that in the case of the polyglutamic acid systems the intensity of the continuum increases up to a phosphate : glutamic acid residue ratio of 5:1, the highest phosphate content which could be measured. Similar results are obtained with polyhistidine + KH$_2$PO$_4$ and with polytyrosine + hydrogen phosphate systems. All these results, together with the original references, are summarized in ref. (9), Tab. II.

These results demonstrate that the side chains form with the phosphates, hydrogen-bonded chains with large proton polarizabilities which are caused by collective proton motion within these chains. Probably these chains are structurally symmetrical. Then with the polylysine + phosphate systems in the case of the Na$^+$ system 7 and with the K$^+$ system 11 hydrogen bonds are built up, respectively, whereas in the polyglutamic acid + dihydrogen phosphate systems extended hydrogen-bonded networks are formed.

INTRAMOLECULAR HYDROGEN-BONDED CHAINS AS MODELS TO STUDY THE PROTON POLARIZABILITY OF HYDROGEN-BONDED CHAINS

The first system which we studied was a structurally symmetrical system (the monosalt of 1,2,3-benzenetricarboxylic acid, Fig. 4) with two hy-

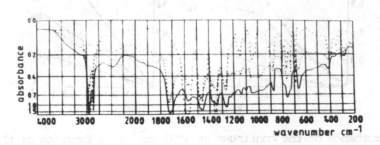

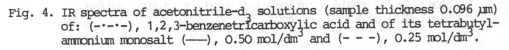

Fig. 4. IR spectra of acetonitrile-d$_3$ solutions (sample thickness 0.096 μm) of: (-·-·-), 1,2,3-benzenetricarboxylic acid and of its tetrabutyl-ammonium monosalt (——), 0.50 mol/dm^3 and (- - -), 0.25 mol/dm^3.

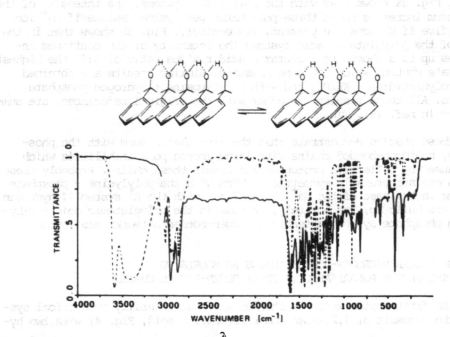

Fig. 5. FTIR spectra of 0.1 mol/dm^3 CH$_2$Cl$_2$ solutions of: (---), 1,11,12,13,14-pentahydroxymethylpentacene; and (——), of its tetrabutyl-ammonium monosalt

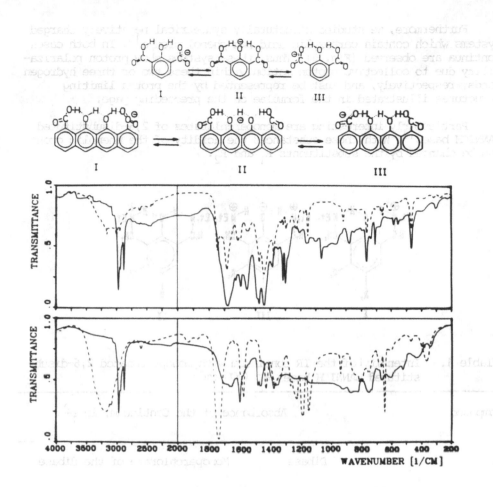

Fig. 6. IR spectra of acetonitrile-d₃ solutions (---), of the substance and (——), of the tetrabutylammonium monosalt: a) 2-nydroxysisophtalic acid, b) 11,12-dihydroxy-1, 10-naphtacene dicarboxylic acid.

drogen bonds formed between two carboxylic acid and one carboxylate group[10]. The intense continuum observed for the monosalt demonstrates that this system formed by two intramolecular hydrogen bonds shows large proton polarizability due to collective proton motion and must be represented by three proton limiting structures.

In the case of the monosalt of 1,11,12,13,14-pentahydroxypentacene[11] also a continuum is observed extending down to 350 cm⁻¹. It demonstrates that the hydrogen-bonded chain shows large proton polarizability due to collective proton motion and must be represented by five proton limiting structures. With this molecule the intensity of the continuum is small since the hydrogen-bonded atoms are connected by π-electron systems. Thus, the dipole fluctuation in the hydrogen-bonded chain is compensated to a large extent by the fluctuation of the electrons which follows the motion of the protons. Therefore, we synthesized 1,11,12,13,14-pentahydroxymethylpentacene, in which this mesomeric effect is interrupted by CH₂ groups (Fig. 5)[12]. The intensity of the continuum is much larger since with this system the electrons can only follow the proton motion by an inductive effect. The formulae of these compounds show two of the five limiting structures by which this system must be represented (see preceeding page).

Furthermore, we studied structurally symmetrical negatively charged systems which contain carboxylic acid and phenol groups[13]. In both cases continua are observed (Fig. 6). Thus, these systems show proton polarizability due to collective proton motion within these two or three hydrogen bonds, respectively, and must be represented by the proton limiting structures illustrated in the formulae on the preceeding page.

Particularly interesting are monoperchlorates of 2,6-disubstituted MANNICH bases[14]. With these substances the acidity of the phenolic group can be changed by the substituents R_1 and R_2.

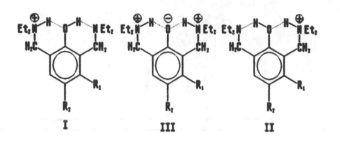

Table 1. Intensity of the IR continuum with monoprotonated 2,6-disubstituted MANNICH bases at 1900 cm^{-1}

Compound		Absorbance of the Continuum, $\ln \frac{I_o}{I}$	
R_1	R_2	Dibase	Monoperchlorate of the dibase
H	OBu	0.000	0.000
H	F	0.010	0.213
H	Ph	0.043	0.264
H	Cl	0.051	0.284
Cl	Cl	0.064	0.261
H	$COOCH_3$	0.084	0.178
H	$COOC_2H_5$	0.097	0.088
H	NO_2	0.113	0.037
NO_2	NO_2	0.199	bands at 3283 and 3120 cm^{-1}

In this Tab. the acidity of the phenolic group increases with the substitutents from top to bottom. In the case of the monoperchlorates with increasing acidity of the phenolic group the absorbance of the continuum increases. If the acidity increases still further the intensity decreases again and finally instead of the continuum, NH^+ stretching vibration bands are observed. Thus, the proton polarizability due to collective proton motion in these two hydrogen bonds first increases and if the acidity becomes still larger it decreases again and vanishes finally. How this behaviour can be understood? With increasing acidity of the phenolic group the collective proton fluctuation between the two proton limiting structures (I) $N^+H\cdots OH\cdots N$ $\rightleftharpoons N\cdots HO\cdots H^+N$ (II) of the system is favored, and thus, the proton polarizability increases. With further increasing acidity a third proton limiting structure, the structure $N^+H\cdots O^-\cdots H^+N$ (III) obtains weight. The system must now be represented by the three proton limiting structures shown above. If the acidity increases still further the weights of the limiting structures I and II decrease and thus, the proton polarizability of the system becomes smaller. Finally, only limiting structure III remains. The protons are now localized at the nitrogens and instead of a continuum, bands are observed.

Similar results are obtained[15] with the monoperchlorates of the respective di-N-oxides. With the system with the most acidic phenolic group, however, only one broad OH^+ stretching vibration band is observed since the O atoms of the NO groups are more acidic than the amine groups.

THEORETICAL TREATMENTS

We calculated by an ab initio SCF procedure[8] the energy and dipole moment surfaces of the systems formic acid-water-formate and formic acid-water-water-formate.

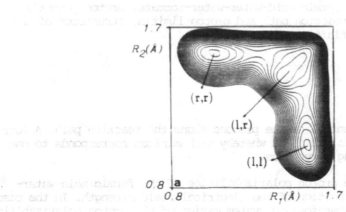

Fig. 7a demonstrates that the energy surface of the first system shows three minima and that the collective transfer of the protons does not occur simultaneously but step by step. This result becomes particularly obvious if one considers the structural formulae in Fig. 7b. This Fig. shows the potential of the protons along the reaction path. Herewith, each minimum corresponds to one of the three proton limiting structures. The formic acid-water-water-formate system has a four dimensional energy surface.

Fig. 7a The system formic acid-water-formate (definition of R_1 and R_2 see ref. 8). Energy surface

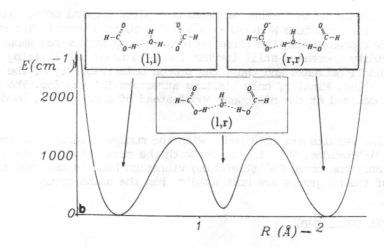

Fig. 7b The system formic acid-water-formate (definition of R_1 and R_2 see ref. 8). Proton potential of this system along the reaction path and proton limiting structures.

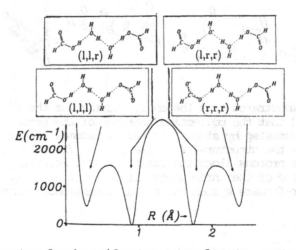

Fig. 8. The system formic acid-water-water-formate, proton potential along the reaction path and proton limiting structures of the system (definition of R see ref. 8)

Fig. 8 shows the potential of the protons along the reaction path. A four minima proton potential is found whereby each minimum corresponds to one of the proton limiting structures.

Fig. 9 shows the proton polarizabilities of the formic acid-water-formate system as a function of the electrical field strength. In the case of the formic acid-water-formate system maxima of the proton polarizability are observed with electrical fields of $\pm$ 0.6x10^7 V/cm (only the maximum for positive fields is shown in the Fig.. At this field strength the proton po-

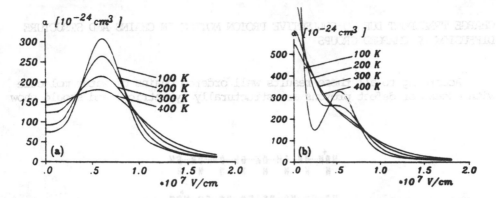

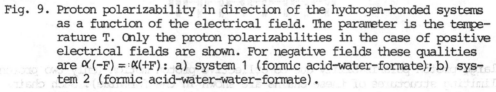

Fig. 9. Proton polarizability in direction of the hydrogen-bonded systems as a function of the electrical field. The parameter is the temperature T. Only the proton polarizabilities in the case of positive electrical fields are shown. For negative fields these qualities are $\alpha(-F) = \alpha(+F)$: a) system 1 (formic acid-water-formate); b) system 2 (formic acid-water-water-formate).

larizability amounts to $(150-200) \times 10^{-24}$ cm^3. Thus, the proton polarizability of this system is more than 2 orders of magnitude larger than polarizabilities due to distortion of the electron systems. Fig. 9b shows the proton polarizability of the formic acid-water-water-formate system. At higher temperatures only one maximum of the polarizability is observed. It is found if no electrical field is present at the hydrogen-bonded system. If no electrical field is present the proton polarizability amounts to $(500-900) \times 10^{-24}$ cm^3. Thus going from the formic acid-water-formate to the formic acid water water formate system the proton polarizability - due to collective proton motion - increases, i.e., the proton polarizability increases with increasing length of the hydrogen-bonded chains.

No longer chains can be calculated in this way ab initio. Therefore, we proceeded our calculations with model proton potentials. Fig. 10 shows the proton polarizability as a function of the number of minima. With increasing number of minima it becomes 3-4 orders of magnitude larger than usual polarizabilities due to distortion of electron systems.

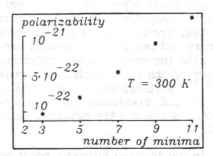

Fig. 10. Proton polarizability of hydrogen-bonded chains as a function of the number of the minima of the proton potential. Results of calculations with model potentials.

CHARGE TRANSPORT DUE TO COLLECTIVE PROTON MOTION IN CHAINS AND STRUCTURE
DIFFUSION OF CHARGED GROUPS

According to all these results well ordered chains of water molecule
with excess or defect protons are structurally symmetrical and should show

$$HOH\cdots OH\cdots OH\cdots OH\cdots OH\cdots OH\cdots OH\cdots OH$$

$$HO\cdots HO\cdots HO\cdots HO\cdots HO\cdots HO\cdots HO\cdots HOH$$

large proton polarizability due to collective proton motion (only two proton
limiting structures of these chains are shown in the formulae). Such chains
should be very effective proton pathways, whereby the transport of the charge
occurs in less than one ps.

In contrast to this type of charge transport, a much slower transport
is the transport due to structure diffusion of the charged groups. For instance, in aqueous acid solutions the excess proton fluctuates in the hydrogen bond of $H_5O_2^+$ with a frequency larger than 10^{13} sec.$^{-1}$. As described in
detail in[5] and[16], p. 89, the excess charge is only shifted within the
hydrate structure network, if the excess proton changes its role with one of
the H of the two water molecules of the $H_5O_2^+$. With this event many relaxation processes in the hydrate structure network occur and therefore the rate
of this process is much less. Careri and coworkers[17] [further refs. there,
see also Careri's contribution to these proceedings] studied the proton
conductivity in two and three dimensional networks of hydrogen bonds. He
found that this conductivity is of decisive importance for the function of
biological systems. The rate of this process is, however, much less than
that of charge shifts due to collective proton motion in hydrogen-bonded
chains with large proton polarizability. Hence, probably the mechanism of
these charge shifts may be similar to that of the proton conductivity in
aqueous acid solutions, i.e. a structure diffusion of the charged groups.

SINGLE HYDROGEN BONDS WITH LARGE PROTON POLARIZABILITY
AND HYDROGEN-BONDED CHAINS WITH THIS PROPERTY

Studying IR continua with model systems we found that a large number
of homoconjugated hydrogen bonds (Cys-Cys)$^-$, (Lys-Lys)$^+$, (Tyr-Tyr)$^-$,
(His-His)$^+$, (Glu-Glu)$^-$, and (Asp-Asp)$^-$ as well as heteroconjugated hydrogen bonds Tyr-Arg, Cys-Lys, Tyr-Lys, Glu-His, and Asp-His show large proton polarizability [summary and original references are given in ref. 9,
Tab. I]. It is of particular interest that all hydrogen bonds which may
form in the active center of the bacteriorhodopsin molecule may show large
proton polarizability[18,19]. Furthermore, a large number of hydrogen bonds
formed between side chains and phosphates may show large proton polarizability. All these hydrogen bonds and the original references are summarized in ref. 9, Tab. II.

The proton motion in neighboring hydrogen bonds with large proton polarizability is coupled via proton dispersion forces[5]. Thus, hydrogen-
bonded chains which are built up from single hydrogen bonds with large

proton polarizability should also show large proton polarizability due to collective proton motion.

THE PROTON PATHWAYS IN BACTERIORHODOPSIN (BR)

Gerwert et al.[20,21] [see also these proceedings and ref. 22] have demonstrated by time resolved FTIR spectroscopy that, when the intermediate M_{412} is formed, the retinal Schiff base is deprotonated and the Asp 85 is protonated simultaneously. Afterwards, the retinal Schiff base is reprotonated by Asp 96. $OH \cdots N \rightleftharpoons O^- \cdots H^+ N$ bonds between Asp residues and the Schiff base show large proton polarizability[18]. Thus, changes of local fields or changes of specific interactions may easily perform this charge shift. These changes may arise due to secondary and tertiary structural changes occurring during the photocycle[23-25]. According to ref. 25 the distance between Asp 96 and the N atom of the Schiff base amounts to about 10 Å. The resolution of the three dimensional map in ref. 25, vertically to the membrane is, however, only 10 Å, therefore a direct connection of Asp 96 and the N atoms of the Schiff base is not excluded. It was shown that H_2O molecules are present near the active centre[24,26,27]. Hence, Asp 96 and the N atom of the Schiff base may be bridged by water molecules[23,26,27]. These molecules are highly ordered due to entropy reasons since they are in a hydrophobic environment[25]. In this way a hydrogen-bonded chain may form a proton pathway, which shows still proton polarizability, since it was shown in ref. 28 that in the case of a multiminima proton potential the proton polarizability is not very strongly influenced if single potential wells are lifted or lowered.

Heberle and Dencher[27] have shown that the proton arises at the outer surface at the same time when M_{412} is formed. Thus, the proton conduction process from the active centre to the outside has to be very fast. A process with such a large rate of the proton transport can only be performed by a hydrogen-bonded chain with large proton polarizability due to collective proton motion.

According to Henderson et al.[25] the external channel contains much more hydrophilic amino acid residues. Therefore, it is not probable that besides these residues a structurally symmetrical protonated water channel with large proton polarizability - as discussed above - is formed. Earlier we discussed on the basis of a CPK model of BR such a chain formed by one Asp for Tyr and one Glu residue[29]. Other chanals have been postulated in refs. 30 and 31. On the basis of the structure of Henderson et al.[25] such a chanal could be formed wherby this chanal, however, enter in the bulk water face (see, however, above remark on the resolution vertically to the membrane. Of course, all hydrogen bonds with large proton polarizability formed between side chains can be included in such a chanal (see above) and also Tyr can be substituted by a water molecule[28]. Indeed, very recently in the FTIR difference spectra of L_{550} - BR_{570} an IR continuum was observed in the region 2800 - 800 cm^{-1} which indicate in the intermediate L_{550} a hydrogen-bonded chain with large proton polarizability. In the difference M_{412} - BR_{570} this continuum has been vanished and two broad bands in the region 2800 - 1600 cm^{-1} are found, indicating an asymmetrical strongly hydrogen-bonded structure[32].

ACKNOWLEDGEMENT

Our thanks are due to Diplomchemiker C. Nadolny for the technical assistance on the preparation of this manuscript.

REFERENCES

1. E. G. Weidemann and G. Zundel, Field-dependent Mechanism of Anomalous Proton Conductivity and the Polarizability of Hydrogen Bonds with Tunneling Protons, Z. Naturforschung 25a:627 (1970).

2. R. Janoschek, E. G. Weidemann, H. Pfeiffer, and G. Zundel, Extremely High Polarizability of Hydrogen Bonds, J. Amer. Chem. Soc. 94:2387 (1972).

3. R. Janoschek, E. G. Weidemann, and G. Zundel, Calculated Frequencies and Intensities Associated with Coupling of the Proton Motion with the Hydrogen Bond Stretching Vibration in a Double Minimum Potential Surface, J. Chem. Soc. Faraday Trans. II 69:505 (1973).

4. G. Zundel and M. Eckert, IR Continua of Hydrogen Bonds and Hydrogen-bonded Systems, Calculated Proton Polarizabilities and Line Spectra, J. Mol. Struct. 200:73 (1989).

5. G. Zundel, Easily Polarizable Hydrogen Bonds - their Interactions with the Environment - IR Continuum and Anomalous Large Proton Conductivity, in: "The Hydrogen Bond - Recent Developments in Theory and Experiments," P. Schuster, G. Zundel, and C. Sandorfy, eds., Vol. II, North Holland Publ. Co. (1976).

6. M. Eckert and G. Zundel, Proton Polarizability, Dipole Moment and Proton Transitions of an $AH \cdots B \rightleftharpoons A^- \cdots H^+B$ Proton Transfer Hydrogen Bond as a Function of an External Electrical Field - an ab initio SCF Treatment, J. Phys. Chem. 91:5170 (1987).

7. R. Krämer and G. Zundel, Influence of Specific Interaction Effects on the Proton Transfer Equilibrium Intermolecular Hydrogen Bonds of Carboxylic Acids and Amines, JCS Faraday Trans. II 82:301 (1990).

8. M. Eckert and G. Zundel, Energy Surfaces and Proton Polarizability of Hydrogen-Bonded Chains - an ab initio Treatment with Respect to the Charge Conduction in Biological Systems, J. Phys. Chem. 92:7016 (1988), see also 93:5324 (1989).

9. G. Zundel, Hydrogen-bonded Systems as Proton Wires Formed by Side Chains of Proteins and by Side Chains and Phosphates in: "Transport through Membranes: Carriers, Channels and Pumps.," A. Pullman, J. Jortner and B. Pullman eds., Kluver Acad. Publ., Dortrecht (1988).

10. B. Brzezinski, G. Zundel, and R. Krämer, Proton Polarizability Caused by Collective Proton Motion in a System with two Intramolecular Hydrogen Bonds, Chem. Phys. Letters 124:395 (1986).

11. B. Brzezinski, G. Zundel, and R. Krämer, An Intramolecular Chain of Four Hydrogen Bonds with Proton Polarizability due to Collective Proton Motion, Chem. Phys. Letters 157:512 (1989).

12. B. Brzezinski and G. Zundel, An Intramolecular Chain of Four Hydrogen Bonds in the 1,11,12,13,14-Pentahydroxymethylpentacene Tetrabutylammonium Salt, Chem. Phys. Letters 178:138 (1991).

13. B. Brzezinski, G. Zundel, and R. Krämer, Proton Polarizability Caused by Collective Proton Motion in Intramolecular Chains Formed by Two and Three Hydrogen Bonds - Implications for the Charge Conduction in Bacteriorhodopsin, J. Phys. Chem. 91:3077 (1987).

14. B. Brzezinski, H. Maciejewska, G. Zundel, and R. Krämer, Collective Proton Motion and Proton Polarizability of Hydrogen-bonded Systems in Disubstituted Non-protonated and Protonated MANNICH Bases, J. Phys. Chem. 94:528 (1990).

15. B. Brzezinski, H. Maciejewska, and G. Zundel, Proton Polarizability due to collective Proton Motion in Intramolecular Hydrogen-bonded Systems in Monoperchlorates of 2,6-Disubstituted MANNICH Bases Di-N-Oxides, J. Phys. Chem. 94:6983 (1990).

16. G. Zundel and J. Fritsch, Interaction and Structure of Ionic Solvates - Infrared Results, in: Chemical Physics of Solvation, Vol. II, Ch. 2, R. R. Dogonadze, E. Kálmán, A. A. Kornyshev, and J. Ulstrup, eds., Elsevier, Amsterdam (1986).

17. G. Careri, in: "Proc. NATO Advanced Study Institute," Cargèse (1990). H. E. Stanley and N. Ostrowski, eds., Kluver Academic Press, Dortrecht (1990).

18. H. Merz and G. Zundel, Thermodynamics of Proton Transfer in Carboxylic Acid-Retinal Schiff Base Hydrogen Bonds with Large Proton Polarizability, Biochem. Biophys. Res. Comm. 138:819 (1986).

19. H. Merz, U. Tangermann, and G. Zundel, Thermodynamics of Proton Transfer in Phenol - Acetate Hydrogen Bonds with Large Proton Polarizability and the Conversion of Light Energy into Chemical Energy in Bacteriorhodopsin, J. Phys. Chem. 90:6535 (1986).

20. K. Gerwert, B. Hess, J. Soppa, and D. Oesterhelt, Role of Aspartate-96 in Proton Translocation by Bacteriorhodopsin, Proc. Natl. Acad. Sci. USA 86:4943 (1989).

21. K. Gerwert, G. Souviguier, and B. Hess, Simultaneous Monitoring of Light-induced Changes in Protein Side-group Protonation, Chromophore Isomerization, and Backbone Motion of Bacteriorhodopsin by Time-resolved Fourier-transform infrared Spectroscopy, Proc. Natl. Acad. Sci. USA 87:9774 (1990).

22. O. Bouschê, M. Braiman, Y.-W. He, T. Marti, H. G. Khorana, and K. J. Rothschild, Vibrational Spectroscopy of Bacteriorhodopsin Mutants, J. Biol. Chem. 266:11063 (1991).

23. M. H. J. Koch, N. A. Dencher, D. Oesterhelt, H.-J. Plöhn, G. Rapp, and G. Büldt, Time-resolved X-ray Diffraction Study of Structural Changes Associated with the Photocycle of Bacteriorhodopsin, Embo Journal 10:521 (1991).

24. J. Heberle and N. A. Dencher, Bacteriorhodopsin in Ice. Accelerated Proton Transfer from the Purple Membrane Surface, FEBS Letters 277:277 (1990).

25. R. Henderson, J. M. Baldwin, T. A. Ceska, F. Zemlin, E. Beckmann, and K. H. Downing, Model for the Structure of Bacteriorhodopsin Based on High-resolution Electron Cryo-microscopy, J. Mol. Biol. 213:899 (1990).

26. P. Hildebrandt and M. Stockburger, Role of Water in Bacteriorhodopsin's Chromophore: Resonance Raman Study, Biochem. 23:5539 (1984).

27. G. Papadopulos, N. A. Dencher, G. Zaccai, and G. Büldt, Water Molecules and Exchangeable Hydrogen Ions at the Active Centre of Bacteriorhodopsin Localized by Neutron Diffraction, J. Mol. Biol. 214:15 (1990).

28. M. Eckert and G. Zundel, Motion of one Excess Proton between Various Acceptors - Theoretical Treatment of the Proton Polarizability of Such Systems, J. Mol. Struct. Theoret. Chem. 181:141 (1988).

29. G. Zundel, Proton Transfer in and Proton Polarizability of Hydrogen Bonds - IR and Theoretical Studies Regarding Mechanisms in Biological Systems, J. Mol. Struct. 177:43 (1988).

30. W. Stoeckenius, in: Membrane Transduction Mechanisms", R. A. Cone, J. E. Dowling eds., New York, Raven Press (1979).

31. V. V. Krasnogolovets, N. A. Protsenko, P. M. Tomchuk, and V. S. Guriev, The Mechanism of Bacteriorhodopsin Functioning, I. The Light-induced Proton Throw-over by Retinal, Internat. J. Quantum Chem. 33:327 (1988).

32. J. Olejnik, B. Brzezinski, and G. Zundel, A Proton Pathway with Large Proton Polarizability in Bacteriorhodopsin - Fourier-Transform Difference Spectra of Photoproducts of Bacteriorhodopsin and of its Pentademethylretinal Analogue, in preparation (1991).

PERCOLATION AND DISSIPATIVE QUANTUM TUNNELING OF PROTONS IN HYDRATED PROTEIN POWDERS

G. Careri

Dipartimento di Fisica, Universita' di Roma "La Sapienza"
Piazzale Aldo Moro, 2 - 00185, Rome Italy

INTRODUCTION

Previous work from this laboratory has shown that lysozyme powders exhibit dielectric behavior due to proton conductivity assisted by water molecules adsorbed on surface (Careri et al., 1985), and that this behavior can be described in the frame of percolation theory (Careri et al., 1986; Careri et al., 1988). This statistical-physical model, has been shown to be applicable to a wide range of processes where spatially random events and topological disorder are of intrinsic importance. A typical physical application of the percolation theory[1,2] is to the electrical conductivity of a network of conducting and non-conducting elements. One of the most appealing aspects of the percolation process is the presence of a sharp transition, where long-range connectivity among the elements of a system suddenly appear at a critical concentration of the carriers. A similar 2-dimensional protonic percolation has been detected in powdered samples of purple membrane of *Halobacterium Halobium* (Rupley et al., 1988). In both cases the emergence of biological function, respectively enzyme catalysis and photoresponse, has been found to coincide with the critical hydration for protonic percolation h_c. More recently the above room temperature studies have been extended to samples of viable biological systems to confirm the close connection between protonic percolation threshold and the onset of biological function in nearly anhydrous biosystems (see Table 1).

TABLE 1. ONSET OF BIOLOGICAL FUNCTION AT PROTON CONDUCTIVITY THRESHOLD *

Nearly Anhydrous System	Dimensionality in scaling law	Hydration threshold coincides with
Lysozyme-saccaride	2	enzymatic-activity (1)
Purple membrane	2	photoresponse (2)
Artemia cysts	3	pre-metabolism (3)
Maize components	2	germination (4)

(*)   G. Careri, in "Symmetry in Nature", Scuola Normale Superiore, Pisa, p. 213 (1989).
(1)   G. Careri, A. Giansanti and J.A. Rupley - Phys. Rev. A 37 2703 (1988).
(2)   J.A. Rupley, L. Siemankowski, G. Careri and F. Bruni - Proc. Nat. Acad. Sci. U.S.A. 85, 9022 (1988).
(3)   F. Bruni, G. Careri and J.S. Clegg - Biophys. J. 55, 331, (1989).
(4)   F. Bruni, G. Careri and A.C. Leopold - Phys. Rev. A 40 2803 (1989).

In the typical example of a network of conducting and non conducting elements, percolation theory predicts the critical concentration P_C of the conducting elements for the onset of the percolative process, and the critical exponent t for the conductivity σ dependence on P above this threshold

$$\sigma = \sigma_C + k(P-P_C)^t \tag{1}$$

In eq. (1) the kinetic coefficient k depends on the specific process in question, while P_C and t are universal quantities which are only dependent from the dimensionality D of the system . In Table 2, I have collected the appropriate data in the litterature together with our ones.

TABLE 2. CRITICAL EXPONENTS OF PERCOLATION CONDUCTIVITY*

2D systems	
• glass spheres and silver coated spheres (1)	1.25 ± 0.10
• lysozyme powder low hydration (2)	1.29 ± 0.05
• purple membrane fragments (3)	1.23 ± 0.05
• corn embryos pellets (4)	1.23 ± 0.05
- theory, by finite-size scaling (5)	1.26 ± 0.05
- theory, by transfer matrix (6)	1.28 ± 0.03

3D systems	
• amorphous cermet film (7)	1.9 ± 0.2
• amorphous carbon and teflon powder (8)	1.85 ± 0.2
• silver coated glass and teflon powder (8)	2.0 ± 0.2
• acetanilide microcrystalline (9)	1.72 ± 0.05
• Artemia cysts (10)	1.65 ± 0.05
- theory, by finite-size scaling (5)	1.87 ± 0.04
- theory, by series expansion (11)	1.95 ± 0.03

(*)   G. Careri, in "Symmetry in Nature", Scuola Normale Superiore, Pisa, p. 213 (1989).
(1)   J.P. Clerc, G. Giraud, S. Alexander and G. Guyon - Phys. Rev. B 22, 2489 (1980).
(2)   G. Careri, A. Giansanti and J.A. Rupley - Phys. Rev. A 37 2703 (1988).
(3)   J.A. Rupley, L. Siemankowski, G. Careri and F. Bruni - Proc. Nat. Acad. Sci. U.S.A. 85 , 9022 (1988).
(4)   F. Bruni, G. Careri and A.C. Leopold - A40 2803 (1989).
(5)   M. Sahimi, B.D. Hughes, L. E. Schriven and H.I. Davies -J. Phys. C: Solid State Phys. 16 L 521 (1983).
(6)   B. Derrida and J. Vannimenns - J. Phys. A: Math. Gen. 13 L 147 (1982).
(7)   B. Abels, H.L. Pinch and J.I. Gittleman - Phys. rev. Lett. 35, 247 (1975).
(8)   Y. Song, T.W. Noh, S.I. Lee and J. Gaines - Phys. Rev. B 33 904 (1986)
(9)   G. Careri and E. Compatangelo - (unpublished result from this laboratory).
(10) F. Bruni, G. Careri and J.S. Clegg - Biophys. J. 55, 331 (1989)
(11) R. Frish and A.B. Harris - Phys. Rev. B 416 (1977).

It is very satisfactory to see that the scaling law for conductivity is followed with such accuracy in quite different organic and inorganic materials, for both electronic and protonic carriers. The importance of Table 2 is that it offers complete evidence for the validity of the percolation model in several anhydrous biosystems. Moreover this model provides a clear molecular-level picture that can convey a novel insight of a process. For instance in our nearly-anhydrous biosystems, the conductivity reflects motion of protons along threads of hydrogen-bonded water molecules, with long-range proton displacement in the extended network which is allowed only above the percolation threshold.

Here I report some recent results on the low temperature protonic conductivity of slightly hydrated biomaterials, to show the occurence of proton quantum tunneling in hydrogen bonded water molecules adsorbed on these materials (Careri et al., 1990; Careri et al., 1991).

RESULTS AND DISCUSSION

The dielectric tecnique has been already described (Careri et al., 1985) as well as the procedure to evaluate the d.c. conductivity σ of the sample (Rupley et al., 1988). In this work the insulated-electrode capacitor was reduced to a two layer composite capacitor, one layer being the 1.8 mm teflon sheets and the other one the 4.5 mm powdered sample at constant water content h. This capacitor was cooled to 170 K by cryogenic apparatus and data from 10 KHz to 1 MHz have been recorded while raising the temperature at a rate of about 1 K min^{-1}. A typical run lasted about 6 hours and included about 300 conductivity vs temperature data.

In hydrated lysozyme powders the conductivity is found to increase with increasing temperatures , and at Tg2 ~ 200 K it displays a slight break associated with protein glass transitions (Careri and Consolini, 1991). In the high temperature region the Arrhenius law is accurately followed, with an activation energy H slightly increasing with hydration level. Our values of H are close to 7 Kcal/mole, the activation energy detected by NMR for water-reorientation correlation time (Andrew et al., 1983), suggesting that the mobility of the adsorbed water molecules must be the major controlling factor for proton transport, if the proton number density is assumed to be temperature independent. Thus at temperatures above about 260 K the rate process is controlled by a thermally activated hopping of charged defects over an energy barrier which is temperature independent. The isotopic factor is close to $2^{1/2}$, expected for a classical rate process and previously found at room temperature.

In the following we shall consider the temperature region where tunneling may prevail. A general theory of quantum tunnel out of a metastable state interacting with an environment at temperature T has been produced (Grabert, Weiss and Hanggi, 1984) (GWH) with the finding that for damping of arbitrary strenght, the tunneling decay rate always matches smoothly with the Arrhenius factor at a crossover temperature and that heat enhances the tunneling probability at T=0 K by a factor exp [A(T)]. For undamped system A(T) is exponentially small, whereas for a dissipative system A(T) grows algebrically with temperature. Of particular interest here is the case of tunneling centers in solids, where A(T) increases proportional to T^n at low temperature, with n=4 or 6. In fig. 1 we have plotted the log (conductivity) data versus T^6, and we find that in a temperature range of about 40K the conductivity data $\ln\sigma(h,T)$ can be fitted by straight lines originating at Tg2 and σ_2. This linear plot is accurately followed up to a crossover temperature T ~ 271 K, where it merges with the lower temperature side of the Arrhenius law as required by GWH theory. We have fitted our data with different values of n, but only n=4 gave results comparable with n=6 shown above, as predicted by GWH theory. Although the linear dependence of ln σ on T^6 requested by GWH is certainly fullfilled, this theory requires that proton tunneling start at T=0 K. We believe that these last apparent contraddictions between GWH theory and our data can be easily overcome by suggesting that charged defects are free to tunnel with an appropriate effective mass across an energy barrier only above the protein glass transition temperature, when the adsorbed water clusters behaves as a supercooled fluid.

The above reported work on lysozyme powders (Careri and Consolini, 1991) has been extended to intact viable biosystems already studied at room temperature, in order to detect the likely presence of proton tunneling (Careri et al., 1990). Dry seeds, pollen, and anhydrobiotic microscopic animals like *Artemia*, were chosen as model systems primarily because their hydration properties are well known, and their availability in large quantities. Hand-dissected corn (*Zea mays* L.) embryo and endosperm have been ground to produce pellets whose diameter d was 0.5<d<1 mm. These last data are in very good agreement with the behaviour found in lysozyme powders, considering the intrinsic non reproducibility of the glass transition and the higher complexity of the samples here investigated. This suggests that

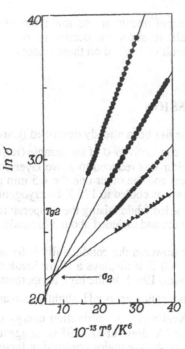

Fig. 1   Natural logaritm of part the conductivity data plotted vs the sixth power of the absolute temperature T^6. Lysozyme H_2O-hydrated samples at hydration level h=0.31 (pentagons), 0.22 (circles), 0.17 (squares), 0.13 (triangles). Solid lines are best fit through data. From these plots Tg 2 = 203 ± 5 K and σ_2 = 8.8 ± 4x 10-9 mho m-1. Adapted from Ref. 6.

the same protonic tunneling process must take place in water clusters adsorbed in all samples so far investigated, since adsorbed water clusters are known to be present both in lysozyme powders and in a great variety of biological tissues.

In conclusion, our tecnique can offer a direct way to investigate proton transfer in water adsorbed inside samples of biological materials and can reveal significant deviations from Arrhenius behaviour to be attributed to proton tunneling. Moreover our data show that theory can be used as a guide to describe dissipative quantum tunneling in a molecular process.

REFERENCES

1 -   E.R. Andrew, D.J. Bryant and T.Z. Rizvi, Chem. Phys. Lett. 95, 463 (1983).
2 -   G. Careri, M. Geraci, A. Giansanti and J.A. Rupley, Proc. Natl. Acad. Sci. U.S.A. 82, 5342 (1985).
3 -   G. Careri, A. Giansanti and J.A. Rupley, Proc. Natl. Acad. Sci. U.S.A. 83, 6810 (1986).
4 -   G. Careri, A. Giansanti and J.A. Rupley, Phys. Rev. A 37, 2703 (1988).
5 -   G. Careri, G. Consolini and F. Bruni, Biophys. Chemistry, 37 165 (1990)
6 -   G. Careri, in "Correlations and Connectivity" (H.E. Stanley and Ostrowsky eds.), Kluwer, Dordrecht 1990, p. 262
7 -   G. Careri and G. Consolini, Ber. Bunsengesel. Phys. Chem. 95, 376 (1991).
8 -   H.V. Grabert, U. Weiss and P. Hanggi, Phys. Rev. Lett. 52, 2193 (1984); Z. Phys. B, 56, 171 (1984).
9 -   J.A. Rupley, L. Siemankowski, G. Careri and F. Bruni, Proc. Natl. Acad. Sci. U.S.A. 85, 9022 (1988).

LIGHT-TRIGGERED OPENING AND CLOSING OF AN HYDROPHOBIC GATE CONTROLS VECTORIAL PROTON TRANSFER ACROSS BACTERIORHODOPSIN

Norbert A. Dencher[1], Georg Büldt[2], Joachim Heberle[1], Hans-Dieter Höltje[3] and Monika Höltje[3]

[1]Hahn-Meitner-Institute, BENSC-N1, Glienicker Str. 100, W-1000 Berlin 39, FRG; [2]Dept. of Physics/Biophysics, Arnimallee 14 and [3]Dept. of Pharmacy, Königin-Luise-Straße 2+4, Free University, W-1000 Berlin 33, FRG

INTRODUCTION

Bacteriorhodopsin (BR) in the purple membrane (PM) of *Halobacterium halobium* is considered as a prototypic membrane protein. This integral membrane protein is the only protein species of the purple membrane and functions as a light-energized proton pump[1]. The protein consists of a single polypeptide chain of 248 amino acids (M_r 26486 Da) traversing the lipid bilayer (about 8 lipids per BR) in seven α-helical segments (Fig. 1; see also Fig. 1 of ref. 2, this book). Photons are absorbed by the chromophore retinal, which is covalently linked via a protonated Schiff's base to lysine-216 of the protein moiety. The antenna retinal is responsible for the characteristic purple colour of BR with its absorption maximum at 568 nm in the ground state. Upon photon absorption, BR undergoes a cyclic photoreaction with an overall half-time of about 10 ms via a series of at least five intermediates of different colour with rise times of femto- to milliseconds[3,4]. During the photochemical cycle one[5,6,7] proton is vectorially translocated across BR, causing the formation of an electrochemical proton gradient. (Further information is reviewed elsewhere[2,4,8]). This light-generated proton electrochemical potential across the energy-transducing membranes is utilized by the halobacteria as a driving force for ATP synthesis, active transport processes, and rotation of flagella. It is quite obvious that bacteriorhodopsin is a very attractive system for the investigation of molecular steps in light-driven vectoral H^+-translocation. Bacteriorhodopsin is one of the most promising candidates to elucidate the link between the structure of a membrane transport protein, its dynamics, and the function.

LOCALIZATION OF THE ACTIVE CENTER AND OF THE PROTON PATHWAY IN BACTERIORHODOPSIN

Since in the purple membrane, BR is arranged as clusters of three molecules in a well ordered two-dimensional hexagonal lattice, its structure can be studied by electron microscopy[9], X-ray diffraction, and neutron diffraction. In combination with functional studies, these techniques allow both identification and localization of the two most important structural domains of this proton pump, i.e., the active center and the proton pathway.

Retinal Location in Bacteriorhodopsin

Retinal is a multifunctional component in BR. Bound in a one-to-one ratio to the protein moiety, this chromophore is the antenna for the absorption of photons. In addition, retinal directly participates in vectorial proton pumping as an active element. Upon photon absorption, the all-*trans* to 13-*cis* isomerization of its polyene chain triggers structural changes in the protein adjacent to the ring and the Schiff's base end of the chromophore[10,11,12] (see below). This accompanies the deprotonation of the Schiff's base nitrogen[13] and release of one proton per M-intermediate to the external medium[5,6,7]. Furthermore, the presence of retinal regulates the passive proton pathway, which connects the active center in BR with both surfaces of the membrane[14,15,16]. It is obvious that knowledge of the location of retinal in BR is not only required for the understanding of its interaction with the protein moiety as well as for the elucidation of the molecular mechanism of proton pumping, but can give a first hint for the location of the active center and the proton pathway.

By X-ray diffraction we have determined the location of the heavy-atom label Br or HgCl of three different retinal analogues in the plane of the PM and in this way the orientation of the chromophore in BR[17]. The three analogues, i.e., 9-bromoretinal and 13-bromoretinal labelled in the polyene chain as well as the ring labelled HgCl-retinal, were incorporated into BR either biosynthetically using a retinal-deficient mutant strain of *Halobacterium halobium* or with photochemically bleached bacterioopsin. All BR samples regenerated with retinal analogues were functional active as a proton pump. The diffraction data show that the cyclohexene ring of retinal is situated in the corner formed by helix 4E and 5D, the 13-methyl group adjacent to helix 6C, and therefore the Schiff's base nitrogen about midway between helix 6C and 2G (Fig. 1). The 9-bromo-label is found slightly off the line connecting the two other labels, directed towards helix 3F, suggesting torsion of the polyene chain or slight incline of the ring. The position and orientation of retinal obtained by our experiments are in agreement with data from neutron diffraction[18,19,20] and high-resolution electron microscopy[9]. By neutron diffraction, the transmembrane location of retinal was recently determined, indicating that the Schiff's base is located near the middle of the membrane[21].

The Proton Conducting Pathway Across Bacteriorhodopsin

The position of retinal and of the Schiff's base not only determines the active center within the protein moiety of BR. It also gives a first hint for the location of the passive proton conducting pathway, which connects the active center in BR with both surfaces of the membrane (Fig. 1), since retinal was shown to

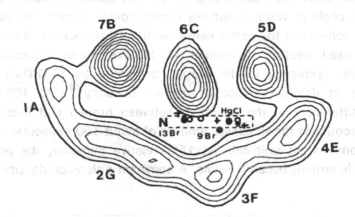

Figure 1. Part of the unit cell showing one bacteriorhodopsin molecule with the determined label positions, indicating the orientation and position of retinal (dashed line) in the projected structure and the calculated position of the Schiff's base nitrogen (indicated by N). The heavy-atom positions are marked by filled circles. Open circles denote the positions of three deuterium labelled retinals[19]. The star depicts the location of the centre of mass for perdeuterated retinal[18]. The three crosses mark the centre of the retinal ring and the positions for C-9 and C-13 as given in the atomic model of Henderson et al.[9]. (From ref. 17.)

directly participate in this H^+-pathway[14,15,16]. This fact was proven by comparing the rate of transmembrane proton/hydroxide ion diffusion through the protein in the presence and absence of the chromophore. Fast pH-jumps were imposed across reconstituted lipid vesicles containing native BR, chromophore-free bacterioopsin, or regenerated BR. Alterations in the proton concentration in the vesicle's internal aqueous bulk phase were monitored as fluorescence changes of the entrapped pH-probe pyranine. Always, the H^+/OH^- flux through bacterioopsin was considerably faster (up to ninefold) than through native BR and through regenerated BR[16]. This convincingly demonstrates that *retinal controls the proton conducting pathway*. In the dark, the chromophore blocks this transmembrane path.

BR is strongly affected in its functional properties by the degree of hydration. Thermally and photochemically energized transitions between the ground state species all-*trans* BR and 13-*cis* BR (light-dark adaptation), as well as the molecular processes involved in vectorial proton translocation across this transmembrane protein rely on the presence of water[22,23]. Water molecules and exchangeable hydrogens might be important elements of the active centre in BR, formed by the chromophore retinal, the Schiff's base linking it to lysine-216, and the amino acids composing the binding pocket. In addition, a hydrogen bonded chain of water molecules might provide a conductance pathway for protons connecting the active centre with both surfaces of the membrane. We have used neutron diffraction to localize water molecules and/or exchangeable hydrogens in the purple membrane by H_2O/D_2O exchange experiments at different values of relative humidity[24]. At 100% relative humidity, differences in the hydration between protein and lipid areas are observed accounting for an excess amount of about 100 molecules of water in the lipid domains per unit cell. At 15% relative humidity, the positions of exchangeable protons became visible. A dominant difference density peak

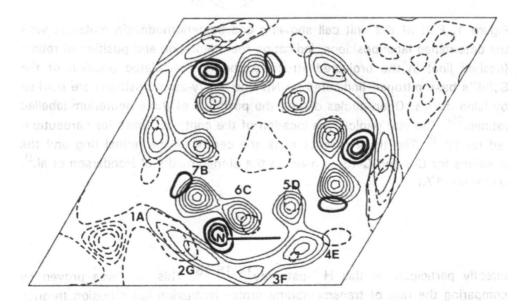

Figure 2. Localization of water molecules and of exchangeable hydrogens in the plane of the purple membrane at 15% relative humidity. The most prominent peaks of the two-dimensional difference Fourier map are shown, calculated from difference intensities between an oriented film of PM in D_2O and H_2O at 15% relative humidity (boldfaced contour lines: 90%, 75% and 50% between zero density level and the maximum positive level). The in-plane position and orientation of the retinal is depicted. N indicates the calculated position of the Schiff's base nitrogen. (From ref. 24.)

corresponding to 11 ± 2 exchangeable protons was detected in the central part of the projected structure of bacteriorhodopsin at the Schiff's base end of the chromophore (Fig. 2). A difference density map obtained from data on purple membrane films at 15% relative humidity in D_2O and the same sample after complete drying in vacuum revealed that about 8 of these protons belong to 4 water molecules. This is *direct evidence for tightly bound water molecules close to the chromophore binding site of bacteriorhodopsin*, which could participate in the active steps of H^+-translocation as well as in the proton pathway across this membrane protein[24].

EXAMINATION OF CONFORMATIONAL CHANGES DURING THE BACTERIO-RHODOPSIN PHOTOCYCLE

Although to date no well diffracting 3-dimensional crystals exist, the structure of BR is known to moderate resolution (3.5 - 7.8 Å)[9]. However, even if atomic resolution will be obtained in the future, the molecular mechanism of vectorial proton translocation might still be hidden. Membrane protein conformational changes are expected to be involved in transport mechanisms. Therefore, only examination of structural changes accompanying the respective functional events, will lead to the elucidation of the transport mechanism at the atomic level.

Light-induced Changes of the Protein

Neutron diffraction experiments have for the first time established *significant reversible structural changes in bacteriorhodopsin*, without loss of crystalline order, during the light-induced transition from the BR_{568} ground state to the M_{412} intermediate of the photocycle[10]. These changes were observed on native purple membranes, in which the decay of M_{412} to BR_{568}, which under physiological conditions occurs in about 10 ms, was artificially retarded by the presence of 2M guanidine hydrochloride at alkaline pH. The M_{412} intermediate was then accumulated by illumination at 6°C and stabilized at -180°C[11]. The difference density map (M_{412} - BR_{568}) shown in Figure 3 displays strong peaks at helix G and B and between helix D and E, caused by a shift of the projected density in the neighborhood of the cyclohexe ring and at the Schiff's base end of the chromophore retinal during M_{412} formation. These intensity changes in the resolution range 60-7 Å are indicative of *alterations in the tertiary structure*, such as a small shift or a 1-2° tilt of the helices G, B, and E. Also positional changes of four or five amino acids over distances of 3 - 5 Å would be in line with these data[11].

In view of the conflicting evidence regarding the extent of the conformational changes previously presented in the literature, it was necessary to study the changes directly under physiological conditions. Furthermore, to examine the structure-function relationship of BR at the molecular level, the kinetics of the light-induced conformational changes have to be correlated with spectroscopic states of the photocycle. In order to achieve this, we have performed time-resolved X-ray diffraction experiments on native and mutated BR[12]. Taking advantage of the high brilliance of synchrotron radiation sources, the light-induced structural changes in BR could be monitored with a time-resolution of 15 ms under physiological conditions. The data confirm the validity of the neutron scattering results and indicate that the observed changes are inhe-

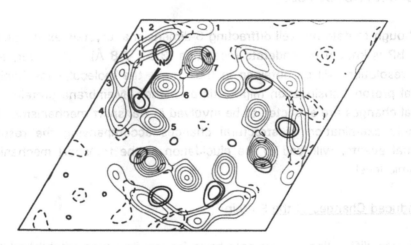

Figure 3. Superimposed on the projected structure of BR in the M_{412} state at -180°C (thin contour lines), the highest positive (bould contour lines) and negative (dashed contour lines) levels of the two-dimensional difference Fourier map between the density of the M_{412} intermediate and the BR_{568} ground state are shown. The bold bar indicates the in-plane orientation of the chromophore retinal and N marks the position of the Schiff's base nitrogen at the end of the polyene chain. (From ref. 11.)

rent in the photocycle of BR. The difference electron density maps obtained show pronounced changes in the vicinity of helix G and smaller ones at helix B and E, as also found by neutron diffraction (Fig. 3) . This was observed for all of the samples examined under various conditions applied, indicating the importance of structural alterations of the protein in the vicinity of the Schiff's base for the pumping mechanism. The X-ray synchrotron data suggest that after the *light-induced structural changes generated during the BR_{568} to*

M_{412} transition (which could not be resolved in time) *BR relaxes to its original conformation during the N_{550} to BR_{568} transition*[25].

It is worth mentioning that the transition from all-*trans* BR_{568} to 13-*cis* BR_{548} (light-dark adaptation) is not accompanied by significant structural changes in the protein moiety of BR[26]. Why is the all-*trans* to 13-*cis* isomerization of the chromophore retinal accompanied by localized structural changes in the protein only for the transition from all-*trans* BR_{568} (light-adapted ground state) to the M-intermediate, but not for the all-*trans* BR_{568} to 13-*cis* BR_{548} (light-dark adaptation) reaction? The answer seems to be obvious if one considers the fact that the latter, but not the former process involves an additional isomerization around the C15=N Schiff's base bond, i.e., the chromophore in all-*trans* BR_{568} has a 13-*trans*, 15-*anti* (=*trans*) configuration and in the photocycle intermediate M 13-*cis*, 15-*anti*, but is 13-*cis*, 15-*syn* (=*cis*) in 13-*cis* BR_{548} (see refs. in 26). The 15-*syn* bond in the 13-*cis* isomer of BR_{548} allows the chromophore to be accommodated in an approximately linear binding pocket of about the same dimensions as required for the all-*trans*, 15-*anti* isomer. Therefore, no pronounced change in the protein conformation will be induced by this double isomerization. On the other hand, isomerization solely around the C13=C14 bond occurring during M formation has to lead to a large steric force on the protein with the observed structural changes in the vicinity of the chromophore binding pocket.

Light-induced Changes of the Chromophore Retinal

During the photocycle of BR, retinal undergoes a transient isomerization from all-*trans* in the BR_{568} ground state to 13-*cis* in the M_{412} intermediate. Since it was not yet clear, if and how this isomerization is related to the observed structural changes in the protein moiety of BR, we have employed neutron diffraction to visualize light-triggered positional alterations of the chromophore[20]. One perprotonated and two specifically deuterated retinals were used to show possible movement of the ring (D11-retinal) and of the Schiff's base portion (D5-retinal). Structural data were recorded at 90 K on PM in the light-adapted ground state and in the trapped M state as described[11]. According to the data, the cyclohexene ring does not change its position during the BR_{568} to M_{412} transition (less than 0.4 Å, as compared to the error of 1 Å inherent in determination of the label position)[20]. In contrast, the D5 label positon (close to the Schiff's base) shifts by 1.5 Å towards the ring position[20]. This *alteration of the D5 label position between BR_{568} and M_{412} suggests a tilt of the retinal*, which seems to be fixed with its ring in the protein pocket, increasing the angle between the membrane plane and the polyene chain. This positional change of the chromophore might trigger the alterations in the tertiary structure as well as the active proton translocation across bacteriorhodopsin.

EXAMINATION OF PROTON TRANSFER REACTIONS DURING THE BACTE-
RIORHODOPSIN PHOTOCYCLE

Bacteriorhodopsin, in contrast to the mobile monomeric visual pigment rhodopsin of vertebrates, is arranged in a two-dimensional hexagonal lattice of protein trimers which are separated from neighbouring trimer clusters by one shell of lipids. Interpretation of measured functional data is hampered by this extraordinary organization of BR in the PM. To construct a working model for the BR proton-transport mechanism, the basic functional unit of this pump has to be defined; the functional unit could be the monomer, the trimer, or even the lattice. By the application of time-resolved laser spectroscopy in combination with optical pH-indicators that allows determination of the kinetics and stoichiometry of light-induced proton release from purple membrane sheets, vesicles, and cells in single-turnover experiments, *the BR monomer has been proven to be the essential transport unit*[7]. The mobile monomeric bacterio-rhodopsin is able to pump protons[27] and to generate a transmembrane elec-trochemical potential difference with an efficiency very similar to that of the aggregates[7].

One promising approach to unravel the proton pumping mechanism relies on the temporal correlation of light-induced structural changes in bacterio-rhodopsin with functional steps (spectroscopic intermediates, proton release/reuptake) of the reaction cycle. Since protons interact strongly with the membrane surface[28] and are retained there for several hundreds of micro-seconds before being released into the aqueous bulk phase[2,25,29], optical pH-indicators residing in the aqueous bulk phase distant to the membrane will not monitor the real kinetics of proton pumping by BR. With pH-indicator dyes, the kinetics of changes in the H^+-concentration occurring in microseconds or faster can only be determined by probes attached to the extracellular surface of BR[2]. The advantages and features of measurements with covalently linked pH-indicators are described in detail in ref. 2 (this book). Fluorescein cova-lently bound to the extracellular surface of the protein via lysine-129 detects the pumped proton about 13-times faster than the indicator pyranine residing in the aqueous bulk phase. Pyranine does not record appearance of the proton at the extracellular side of BR, but the transfer of protons from the surface to the bulk, which is also reflected in the first deprotonation phase of fluorescein. *The proton appears at the extracellular surface of BR concomitant with the rise of the M_{412} intermediate* and both indicators monitor proton reuptake during the N_{550} to BR_{568} transition.

STRUCTURE-FUNCTION RELATIONSHIP OF THE PROTON PUMPING MECHANISM

At present, we are still far away from understanding the details of each step in light-energized proton pumping by BR. However, based on the data presented above, the important structural investigaton by Henderson et al.[9], and the experimental work of others cited in refs. 9 and 30, some key elements and reaction steps inherent in vectorial transmembrane proton transfer will be emphasized. It should be noted that the ideas presented in the following are not original, but are under discussion in the scientific community during the past two decades. Due to the recent vast increase in structural and functional knowledge about BR, we can now select and highlight those suggestions which are supported by firm experimental facts, i.e., the involvement of conformational alterations of both the chromophore and the protein, and the participation of water molecules as well as specific amino acids such as Asp85 and Asp96.

Figure 4 depicts a computer graphics model of the functionally important part of BR, constructed from the coordinates provided by Henderson et al.[9]. Only the functional relevant amino acid side chains, polar hydrogens involved in hydrogen bonding, water molecules, and the chromophore are shown. These water molecules together with exchangeable hydrogens are suggested to be necessary elements of the active centre in BR, formed by the chromophore retinal, the protonated Schiff's base linkage to lysine-216 of the protein moiety, and the amino acids composing the binding pocket. In addition, hydrogen-bonded chains of water molecules provide the conductance pathway for protons connecting the active centre with both surfaces of the membrane. The structure was calculated with the program SYBYL (Tripos Assoc. Inc., St. Louis; general details are described elsewhere[31]) without, on purpose, any optimization of the geometry as well as changes in the conformation of side chains. Only Arg82 was slightly displaced to relieve steric contacts and to allow formation of an hydrogen bond with Glu204. (A refined structure derived from molecular dynamics calculations is currently in progress. The energy-minimized structure for the retinal binding pocket of BR recently published[32] is very similar to the structure illustrated in Figure 4, but lacks water molecules.)

Only three aspects emerging from Figure 4 will be discussed here.

(i) Participation of water molecules in proton transfer. A very relevant result for the structure-function relationship of BR is the localization of 4 ± 1 water molecules and of some exchangeable hydrogens at the retinal Schiff's base end in the projected map at 15 % rel. humidity (Fig. 2). These water molecules are strongly associated with BR, since they are still present in part even at 0% rel. humidity and can only be removed by the application of vacuum[10,24]. From resonance Raman spectroscopy, tightly bound water molecules at the chromophore site have been suggested[33], which should serve as proton donors/acceptors for the Schiff's base nitrogen and stabilize the ion-pair structure between the protonated positively charged Schiff's base nitrogen and its negative counterion(s) in the protein. It is very probable that at least one of the four water molecules localized by our neutron diffraction

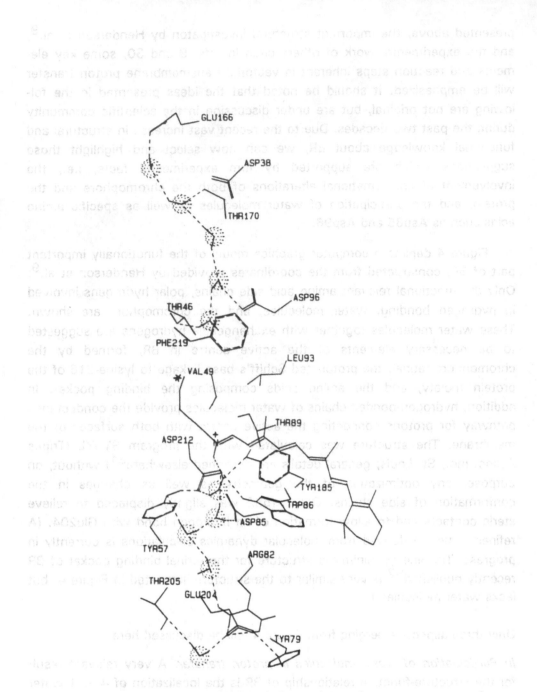

Figure 4. Computer graphics model of the pathway for protons across bacteriorhodopsin, emphasizing two chains of hydrogen-bonded water molecules. Only side chains and explicit polar hydrogens involved in hydrogen-bonding are plotted. The position of water molecules derived from the program GRIN/GRID (Molecular Discovery Ltd., Oxford, UK) are marked by dotted circles. Lys216 is indicated by *. The bottom of the structure represents the extracellular surface of BR towards which the proton is ejected.

study is in contact with the Schiff's base. In fact, one of the ten predicted water molecules in our modelled structure was positioned in the calculation in the direct vicinity of the Schiff's base nitrogen, between the side chains of Asp212 and Asp85 (Fig. 4). The Schiff's base nitrogen reversibly deprotonates during the photocycle[13] and is in an environment concomitantly undergoing structural alterations (Fig. 3). Therefore, one or several water molecules at this site could be directly involved in the processes of light-energized vectorial proton transport. The value of 4 water molecules localized in the projected structure of the Schiff's base (Fig. 2) was obtained for relatively dry PM samples, i.e., at 15 % relative humidity. Obviously, the number of water molecules at 100 % relative humidity, where BR is active as proton pump, has to be known. Because of lack of measuring time with neutrons, the necessary measurements at 40 - 95 % relative humidity cannot be performed during the next 1-2 years. Therefore, we rely on the value of 10 water molecules positioned by the program GRIN/GRID across the BR structure in Figure 4. These water molecules and the exchangeable hydrogens might be components of the proton conductance pathway between the active centre and both surfaces of BR. A hydrogen bonded chain formed from amino acid side chains and some bound water has been previously proposed as a "proton wire"[34,35]. (Various mechanisms by which protons are transferred across and along biological structures will be presented in contributions to this conference by G. Careri, by D.W. Deamer, by M. Gutman, by J.F. Nagle, by S. Scheiner, and by G. Zundel and will therefore not discussed here.)

(ii) Transient encounter of the Schiff's base nitrogen and Asp85. Upon photon absorption by the chromophore, the Schiff's base nitrogen deprotonates, amino acid aspartate-85 becomes protonated, and a proton appears at the extracellular surface of BR, on the time scale of the L to M transition ($\tau = 60$ μs at 22 °C; compare ref. 2, this book). It is generally believed that Asp85 is the primary acceptor of the Schiff's base proton. However, in the *ground-state* structure of BR (Fig. 4) the proton of the Schiff's base is forming a hydrogen-bond with Asp212. Asp85 is too far away, even a change in the conformation of its side chain would not allow formation of a hydrogen-bond to the Schiff's base. We therefore propose that during the BR_{568} to M_{412} transition, Asp85 and the Schiff's base transiently get closer, leading to the formation of a hydrogen-bond and subsequent transfer of the proton to Asp85. This transient approach is induced either by the observed movement of the Schiff's base linkage of the chromophore[20] or by the detected alterations in the protein moiety (Fig. 3). The importance of adequate distances between the Schiff's base proton and the primary proton acceptor, as well as between the primary and the secondary proton acceptor, is demonstrated by the observation that replacement of Asp85 by Glu in the Asp85Glu mutant BR (resulting in an elongation of the side chain by only one C-C bond and maybe in a slight pK-change) leads to an 50 times faster formation of M_{412} (deprotonation of the

Schiff's base). On the other hand, instead of following this acceleration, the appearance of the proton at the extracellular BR surface is strongly delayed as compared to wild-type BR[25] (see also ref. 2 in this book for further details). Movement of the Schiff's base linkage of the chromophore[20] and/or alterations in the tertiary structure of the protein could be the driving force for the active step(s) of proton transfer. It has been suggested[36] that alteration of the H-bond geometry can result in transfer of a proton by reversing the order of pKs of two residues. Therefore, conformational changes that affect the H-bond geometry would lead to pK shifts, which in turn induce active proton translocation.

(iii) A hydrophobic gate controls vectorial proton transfer. It is evident from Figure 4 that the proton conducting pathway is interrupted for about seven angstroms (compare also ref. 9). Between Thr46 and Thr89, just above the Schiff's base, the hydrophobic amino acids Phe 219, Leu93, and Val49 are gathered, leaving no space for water molecules. Here, the hydrogen-bonding system is interrupted. This explains the experimental finding that the proton pathway through BR, represented by the network of hydrogen bonds formed from the amino acid side chains and the water molecules indicated in Figure 4, is completely blocked in the ground state of BR[16]. Obviously, it is essential that the proton pathway is closed as long as BR is not active. This property prevents uncontrolled proton backflow via this specific pathway, which would otherwise lead to undesired collapse of any electrochemical proton gradient across the membrane. Only during the short period of conformational change (Fig. 3) resulting in an "opening" of this hydrophobic gate, i.e., in a transient populating with 2-3 water molecules, can protons surmount the barrier in the proton conducting pathway. (Formation of a transient aqueous channel during parts of the pumping cycle is also considered by J. Nagle, this book.) Furthermore, this only transiently passable hydrophobic gate can explain and is required for the vectoriallity of this proton pump. In the first part of the pumping cycle, the Schiff's base proton protonates Asp85, and via a "domino effect" a proton is ejected nearly simultaneously from the extracellular surface of BR. In the second part of the cycle, via the transiently formed network of hydrogen bonds in this gate, the Schiff's base is reprotonated from Asp96, which in turn is later reprotonated by uptake of a cytoplasmic proton. During the N_{550} to BR_{568} transition three important molecular events occur: (i) the conformational relaxation of the protein, (ii) reisomerization of the chromophore retinal and (iii) uptake of a cytoplasmic proton by the protein[12].

ACKNOWLEDGEMENTS. We thank C. Bark for her excellent technical assistance. This work was supported by the Deutsche Forschungsgemeinschaft (SFB 312 B4/D2) and by the German Federal Minister for Research and Technology (BMFT, contract number 03-BU2FUB-3). Figures 1-3 are reproduced by copyright permission of Pergamon Press (Photochemistry and Photobiology), Academic Press (J. Molecular Biology), and the National Academy of Sciences (Proc. Natl. Acad. Sci. USA).

REFERENCES

1. Oesterhelt, D. and Stoeckenius, W. (1973) Functions of a new photoreceptor membrane. *Proc Natl Acad Sci USA* **70**: 2853-2857.

2. Heberle, J. and Dencher, N.A., Proton transfer in the light-harvesting protein bacteriorhodopsin: An investigation with optical pH-indicators. *This book.*

3. Dencher, N.A. and Wilms, M (1975) Flash photometric experiments on the photochemcal cycle of bacteriorhodopsin. *Biophys Struct Mechanism* **1**: 259-271.

4. Oesterhelt, D. and Tittor, J. (1989) Two pumps, one principle: Light-driven ion transport in Halobacteria. *TIBS* **14**: 57-61.

5. Drachev, L.A., Kaulen, A.D. and Skulachev, V.P. (1984) Correlation of photochemical cycle, H^+ release and uptake, and electric events in bacteriorhodopsin. *FEBS Lett.* **178**: 331-335.

6. Grzesiek, S. and Dencher, N.A. (1986) Time-course and stoichiometry of light-induced proton release and uptake during the photocycle of bacteriorhodopsin. *FEBS Lett.* **208**: 337-342.

7. Grzesiek, S. and Dencher, N.A. (1988) Monomeric and aggregated bacteriorhodopsin: Single-turnover proton transport stoichiometry and photochemistry. *Proc. Natl. Acad. Sci. USA* **85**: 9509-9513.

8. Dencher, N.A. (1983) The five retinal-protein pigments of halobacteria: Bacteriorhodopsin, Halorhodopsin, P 565, P 370, and slow-cycling rhodopsin. *Photochem Photobiol* **38**: 753-767.

9. Henderson, R., J.M. Baldwin, T.A. Ceska, F. Zemlin, E. Beckmann and K.H. Downing (1990) Model for the structure of bacteriorhodopsin based on high-resolution electron cryo-microscopy. *J. Mol. Biol.* **213**, 899-929.

10. Dencher, N.A., D. Dresselhaus, G. Maret, G. Papadopoulos, G. Zaccai and G. Büldt (1988) Light-induced structural changes in bacteriorhodopsin and topography of water molecules in the purple membrane studied by neutron diffraction and magnetic birefringence. *Proceedings of the Yamada Conference XXI*, 109-115.

11. Dencher, N.A., D. Dresselhaus, G. Zaccai and G. Büldt (1989) Structural changes in bacteriorhodopsin during proton translocation revealed by neutron diffraction. *Proc. Natl. Acad. Sci. USA* **86**: 7876-7879.

12. Koch, M.H.J., N.A. Dencher, D. Oesterhelt, H.-J. Plöhn, G. Rapp and G. Büldt (1991) Time-resolved X-ray diffraction study of structural changes associated with the photocycle of bacteriorhodopsin. *EMBO J.* **10**: 521-526.

13.Lewis, A., J. Spoonhower, R.A. Bogomolni, R.H. Lozier and W. Stoeckenius (1974) Tunable laser resonance Raman spectroscopy of bacteriorhodopsin. *Proc. Natl. Acad. Sci. USA* **71**: 4462-4466.

14.Konishi, T. and L. Packer (1978) A proton channel in bacteriorhodopsin. *FEBS L.* **89**: 333-336.

15.Dencher, N.A., P.A. Burghaus and S. Grzesiek (1986) Determination of the net proton-hydroxide ion permeability across vesicular lipid bilayers and membrane proteins by optical probes. *Meth. Enzymol.* **127**: 746-760.

16.Burghaus, P.A. and N.A. Dencher (1989) The chromophore retinal hinders passive proton/hydroxide ion translocation through bacteriorhodopsin. *Arch. Biochem. Biophys.* **275**: 395-409.

17.Büldt, G., K. Konno, K. Nakanishi, H.-J. Plöhn, B.N. Rao and N.A. Dencher (1991) Heavy-atom labelled retinal analogues located in bacterio-rhodopsin by X-ray diffraction. *Photochem. Photobiol.* **54**: in press.

18.Jubb, J.S., D.L. Worcester, H.L. Crespi and G. Zaccai (1984) Retinal location in purple membrane of *Halobacterium halobium*: a neutron diffraction study of membranes labelled in vivo with deuterated retinal. *EMBO J.* **3**: 1455-1461.

19.Heyn, M.P., J. Westerhausen, I. Wallat and F. Seiff (1988) High-sensitivity neutron diffraction of membranes: Location of the Schiff base end of the chromophore of bacteriorhodopsin. *Proc. Natl. Acad. Sci. USA* **85**: 2146-2150.

20.Hauß, T., Heyn, M.P., Büldt, G. and Dencher, N.A. (1991) Movement of retinal in bacteriorhodopsin during the BR_{570} to M_{411} transition: A neutron diffraction study. *Jahrestagung der Deutschen Gesellschaft für Biophysik, Homburg*, abstract P32.

21.Hauß, T., Grzesiek, S., Otto, H., Westerhausen, J. and Heyn, M.P. (1990) Transmembrane location of bacteriorhodopsin by neutron diffraction. *Biochemistry* **29**: 4904-4913.

22.Korenstein, R. and Hess, B. (1977) Hydration effects on the photocycle of bacteriorhodopsin in thin layers of purple membrane. *Nature* **270**: 184-186.

23.Varo, G. and Keszthelyi, L. (1983) Photoelectric signals from dried oriented purple membranes of *Halobacterium halobium*. *Biophys. J.* **43**: 47-51.

24.Papadopoulos, G., N.A. Dencher, G. Zaccai and G. Büldt (1990) Water molecules and exchangeable hydrogen ions at the active centre of bacteriorhodopsin localized by neutron diffraction. *J. Mol. Biol.* **214**, 15-19.

25.Dencher, N.A., Heberle, J., Bark, C., Koch, M.H.J., Rapp, G., Oesterhelt, D., Bartels, K. and Büldt, G. (1991) Proton translocation and conformational changes during the bacteriorhodopsin photocycle: Time-resolved studies with membrane-bound optical probes and X-ray diffraction. *Photochem. Photobiol.* **54**: in press.

26. Dencher, N.A., G. Papadopoulos, D. Dresselhaus and G. Büldt (1990) Light- and dark-adapted bacteriorhodopsin, a time-resolved neutron diffraction study. *Biochim. Biophys. Acta* **1026**: 51-56.

27. Dencher, N.A. and Heyn M.P. (1979) Bacteriorhodopsin monomers pump protons. *FEBS Lett* **108**: 307-310.

28. Grzesiek, S. and N.A. Dencher (1986) Dependency of pH-relaxation across vesicular membranes on the buffering power of bulk solutions and lipids. *Biophys. J.* **50**: 265-276.

29. Heberle, J. and N.A. Dencher (1990) Bacteriorhodopsin in ice: Accelerated proton transfer from the purple membrane surface. *FEBS Lett.* **277**: 277-280.

30. Tittor, J. (1991) A new view of an old pump: bacteriorhodopsin. *Current Opinion in Structural Biology* 1: 534-538.

31. Höltje, M. and Höltje, H.-D. (1991) Molecular modelling study on the negative inotropic potencies of 1,4-dihydropyridines. *Pharm. Pharmacol. Lett.* 1: 19-13.

32. Mathies, R.A., Lin, S.W., Ames, J.B. and Pollard, W.T. (1991) From femtoseconds to biology. *Annu. Rev. Biophys. Biophys. Chem.* **20**: 491-518.

33. Hildebrandt, P. and Stockburger, M. (1984). Role of water in bacteriorhodopsin's chromophore: resonance Raman study. *Biochemistry* **23**: 5539-5548.

34. Nagle, J. F. and Tristram-Nagle, S. (1983). Hydrogen bonded chain mechanism for proton conduction and proton pumping. *J. Membr. Biol.* **74**: 1-14.

35. Deamer, D. W. & Nichols, J. W. (1989). Proton flux mechanisms in model and biological membranes. *J. Membrane Biol.* **107**: 91-103.

36. Scheiner, S. and Hillenbrand, E.A. (1985) Modification of pK values caused by change in H-bond geometry. *Proc. Natl. Acad. Sci. USA* **82**: 2741-2745.

PROTON TRANSFER IN THE LIGHT-HARVESTING PROTEIN BACTERIO-RHODOPSIN: AN INVESTIGATION WITH OPTICAL pH-INDICATORS

Joachim Heberle and Norbert A. Dencher

Hahn-Meitner-Institut, BENSC
Glienicker Str. 100
W-1000 Berlin 39, FRG

INTRODUCTION

Bacteriorhodopsin (BR) is an integral membrane protein in Halobacteria which uses the sunlight to generate a proton gradient across the cell membrane. Proton gradients play a fundamental role in energy transduction. They are found in photosynthesis as well as in the respiratory chain. According to Mitchell's chemiosmotic theory, a proton gradient can be utilized by a proton consumer (H^+-ATP-synthase) to produce ATP (adenosine-triphosphate), the universal storage device for the energy required in every metabolic process of life.

H^+-generator and H^+-consumer of natural cells are interconnected by a hydrogen bonded network. This network consists of the ubiquitous water molecules and the protonable groups of the protein and of the lipids on the membrane surface. A central question in the chemiosmotic theory is: How are pumped protons transferred to the H^+-consumer? Do they migrate along the membrane surface or do they pass the bulk water phase? These two different mechanisms are known as localized and delocalized chemiosmotic hypothesis, respectively. Still, this question is vehemently discussed *(Dilley, this book)*. Long-range connectivity has been proven to exist along natural membranes *(Careri et al., 1986; Rupley et al., 1988)*. The method presented in this article might give insight into the molecular mechanism underlying proton transfer in biological systems.

But BR is not an abstract H^+-generator. Although the tertiary structure is not resolved down to the atomic level, BR is one of the best characterized proton-pumping proteins *(Dencher et al., this book)*. A complete set of biophysical studies now makes it possible to elucidate structural elements important for the H^+-pathway within the protein. Time-resolved FTIR-spectroscopy *(Souvignier, this book)* is a very powerful tool to study such H^+-transfer reactions. With modern genetic techniques the role of specific amino acids in the H^+-transport can be investigated.

Theoretical considerations let *Nagle & Morowitz (1978)* first propose a 'hydrogen bonded chain' (HBC) model to account for the H^+-transport inside the protein. Large H^+-polarizability detected by IR-spectroscopy *(Merz & Zundel, 1981)* supplements this model. Water molecules *(Papadopoulos et al., 1990)* can act as bridges where the distance between the protonable amino acid side chains is too long for an effective H⁺-transfer. *Nagle (1987)* extended this argument to the 'transient hydrogen bonded chain' (tHBC). Especially in the case of BR, the tHBC model has to be demanded. On the basis of the structural data by *Henderson et al. (1990)* part of the proposed H⁺-pathway consists of hydrophobic residues. Only a transient conformational change of the protein *(Koch et al., 1991)* will allow proton passage. Although 20 years have passed already *(Oesterhelt & Stoeckenius, 1971)* with a lot of hard work on BR, a precise knowledge of how and why H⁺ are translocated through BR is still lacking.

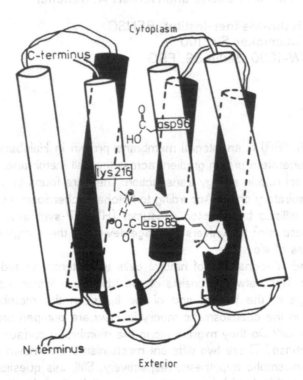

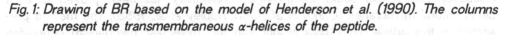

Fig. 1: Drawing of BR based on the model of Henderson et al. (1990). The columns represent the transmembraneous α-helices of the peptide.

In order to understand the complexity of this system, a brief introduction of BR will follow.
BR comprises of a polypeptide with 248 amino acids which spans the membrane as α-helices seven times (Fig. 1). Retinal as a prosthetic group forms a Schiff base (SB) linkage with Lys 216 of the peptide. In the purple membrane (PM), a well-defined region of the plasma membrane of halobacteria, BR is the only protein.

Upon light-absorption BR undergoes a photocycle (Fig. 2) with various intermediates. During this photocycle the SB is deprotonated and one H^+ is pumped from the protein interior into the extracellular medium. Subsequently, a H^+ is taken up from the cytoplasm to reprotonate the SB. From Fig. 2 it is obvious that the M intermediate is the key intermediate for H^+-translocation. M is the only intermediate in the photocycle with a deprotonated SB.

From the beginning of the work on BR, several investigators used pH indicators to monitor the time-course of H^+-release and H^+-reuptake. But they always observed

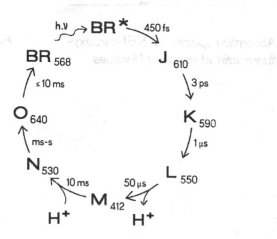

Fig. 2: *The photocycle of BR. Indices correspond to absorption maxima of the respective intermediate. Adopted from Oesterhelt & Tittor (1989)*

a time-delay between the rise of the M intermediate and the H^+-release detected by a pH indicator residing in the surrounding water phase. As will be shown in this article, only pH indicators covalently bound to the protein are able to overcome this discrepancy. With the pH-indicator technique presented here, it is possible to monitor time-resolved and site-specific H^+-transfer reactions in biological systems.

MONITORING PROTON TRANSFER WITH pH INDICATORS

Fig. 3A shows the pH dependent absorption spectra of the pH indicator 5(6)-carboxy-fluorescein (CF). Fig. 3B shows the titration curve. With decreasing pH of the solution the absorbance at the maximum of 489 nm also decreases. Absorption spectra at different pH values for another pH indicator, pyranine (8-hydroxy-1,3,6-pyrene-trisulfonate), are depicted in Fig. 4A. The titration curve at 457 nm (Fig. 4B) is similar to that of CF: A decrease in absorption with decreasing pH. Thus, H^+-release by BR will be accompanied by a transient decrease in absorption of CF and pyranine, respectively.

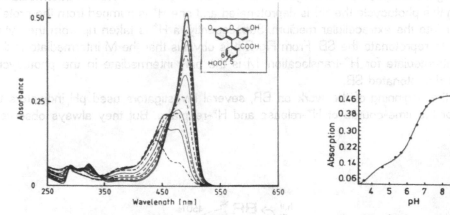

Fig. 3A: Absorption spectra of 5(6)-carboxy-
fluorescein at various pH-values

Fig. 3B: Corresponding
titration curve

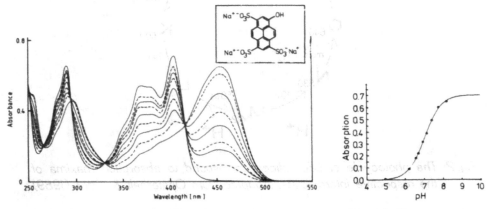

Fig. 4A: Absorption spectra of pyranine
at various pH-values

Fig. 4B: Corresponding
titration curve

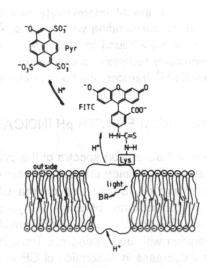

Fig. 5: Sketch of fluorescein (FITC) covalently bound to the extracellular
surface of bacteriorhodopsin (BR) via a thiourea linkage.

With a reactive analogue of CF, fluorescein-5(6)-isothiocyanate (FITC), it is possible to link the pH indicator fluorescein covalently to BR. FITC binds under the applied conditions to Lys 129 at the extracellular side of BR (Fig. 5). This is a prerequisite for measuring H⁺-transfer in BR because BR actively transports a proton from the interior of the protein into the extracellular medium.

The time-resolved absorption changes of BR and of the pH indicators pyranine and covalently bound fluorescein are measured by laser flash photolysis with a time-resolution of 50 ns (*Heberle & Dencher, 1990*). In Fig. 6 the respective absorption changes are plotted on a logarithmic time-scale covering 8 decades. The upper trace (M) represents rise and decay of the photocycle intermediate M which plays a major

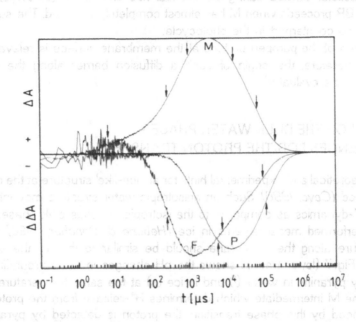

Fig. 6: Kinetic comparison of the M intermediate (M) with the response of the pH-indicators fluorescein (F, covalently attached to BR) and pyranine (P, located in the aqueous bulk phase) in water; T = 3 °C, 150 mM KCl, c_{BR} = 13 μM, λ_M = 412 nm, c_F = 13 μM, λ_F = 489 nm, c_P = 40 μM, λ_P = 457 nm;
The time constants for the kinetic processes are indicated by arrows:
$\tau(M_1)$ = 6.8 μs, $\tau(M_2)$ = 311 μs, $\tau(M_3)$ = 1.2 ms, $\tau(M_4)$ = 20 ms, $\tau(M_5)$ = 82 ms, $\tau(M_6)$ = 247 ms;
$\tau(F_1)$ = 287 μs, $\tau(F_2)$ = 9.2 ms, $\tau(F_3)$ = 111 ms; $\tau(P_1)$ = 6.5 ms, $\tau(P_2)$ = 135 ms.

role in the proton pumping process of BR. The lower traces are absorbance changes of fluorescein (F) bound to the surface of BR, and pyranine (P) which resides in the aqueous bulk phase due to its high hydrophilicity. Both of the pH-indicators respond to the light-energized proton pumping activity of BR with a transient decrease in absorbance. This corresponds to pH-decrease in the vicinity of the respective indicator. However, the protonation reaction of surface bound fluorescein is faster by one order of magnitude at least than the protonation of pyranine. The relaxation to the original pH-value proceeds for pyranine in a monoexponential manner whereas for fluorescein this process is biexponential.

Fig. 5 illustrates the molecular processes underlying the responses of the respective pH-indicators. Upon light excitation a proton is pumped from the interior to the extra-cellular membrane surface of BR ($\tau = 287 \mu s$ at 3 °C). Thus, the surface-pH is transiently lowered which is detected by an absorption decrease of surface bound fluorescein. The pumped proton dwells for 6.5-9.2 ms at the membrane surface before it is released into the surrounding aqueous water phase. This surface/bulk H^+-transfer is monitored by fluorescein with an increase in absorption and by pyranine with a decrease in absorbance. The H^+-reuptake necessary to complete proton- and photocycle, is detected by the relaxation of the absorbance of both pH-indicators to the initial value ($\tau = 111$ ms and 135 ms, respectively).

Comparison with the kinetics of the M intermediate reveals that the proton is pumped to the extracellular surface during the the fast rise time of M ($\tau = 311 \mu s$). The H^+-reuptake by BR proceeds when M has almost completely decayed. The surface/bulk transfer has no counterpart in the photocycle.

The retardation of the pumped protons at the membrane surface is relevant for bioenergetics. Therefore, the origin of such a diffusion barrier along the membrane surface has to be evaluated.

VARIATION OF THE BULK WATER PHASE - CONSEQUENCES FOR THE PROTON TRANSFER

There are theoretical and experimental hints for an 'ice-like' structure at the membrane/water interface (Cevc, 1990). Such an anisotropic water structure may influence the observed H^+-dynamics as compared to the isotropic aqueous bulk phase.

We have performed measurements in ice (Heberle & Dencher, 1990) where the water structure along the membrane should be similar to that of the surrounding bulk phase. Fig. 7 (lower traces) shows the pH-changes in the surrounding medium monitored by pyranine in water (l) and in ice (s) at the same temperature. Although the rise of the M intermediate which determines H^+-release from the protein interior, is not influenced by this phase transition, the proton is detected by pyranine much faster in ice than in liquid water. This demonstrates that in ice the surface/bulk transfer is no longer rate-limiting for the pH-change in the surrounding medium. Obviously, equalization of the water structure at the membrane/water interface and the bulk phase by freezing abolished the diffusion barrier for protons which accounts in liquid water for the dwell time of the pumped protons at the membrane surface. Moreover, pyranine detects now the pumped proton concomitantly with the surface bound fluorescein as well as with the rise of the M intermediate.

The H^+-reuptake kinetic is not changed by the water/ice transition. This reaction proceeds in a time domain where H^+-diffusion is not rate-limiting for the detection time of pyranine. The rate for H^+-reuptake is solely determined by the protein.

In order to gain further insight into the molecular mechanism of proton transfer in biological matter, we investigated the influence of H_2O/D_2O-exchange on the pH-indicator response (Fig. 8). The upper trace demonstrates that the kinetics of the M intermediate are slowed down. The rise of M corresponding to the deprotonation of the protonated retinal Schiff Base (PRSB), exhibits an isotope effect $\tau(D_2O)/\tau(H_2O) = 4$ whereas the decay kinetics are less affected with an isotope effect of 1.6. The appearance of the pumped proton at the membrane surface of BR detected by fluorescein (Fig. 8, middle trace) shows the same isotope effect as the rise of M. From this ex-

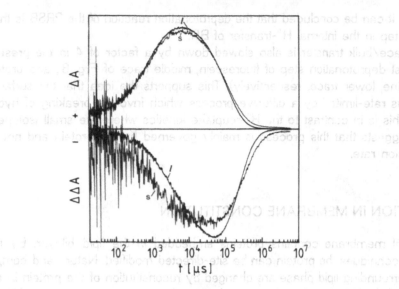

Fig. 7: Effect of the water/ice transition on the kinetics of the pH-indicator pyranine and the M intermediate. T = -6°C, other conditions as in Fig. 6; smooth trace (l): in water; noisy trace (s): in ice.

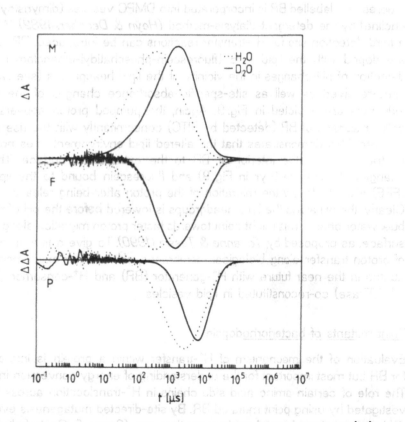

Fig. 8: Effect of isotopic exchange on the kinetics of the M intermediate (M) and the proton transfer reactions detected by pyranine (P) and surface-bound fluorescein (F). Solid lines: BR in D_2O; dotted lines: BR in H_2O; T = 22°C, other conditions as in Fig. 6.

periment it can be concluded that the deprotonation reaction of the PRSB is the rate-limiting step in the internal H$^+$-transfer of BR.

The surface/bulk transfer is also slowed down by a factor of 4 in the presence of D$_2$O (first deprotonation step of fluorescein, middle trace of Fig.8, and protonation of pyranine, lower trace, respectively). This supports the idea that the surface/bulk transfer is rate-limited by a diffusive process which involves breaking of hydrogen-bonds. This is in contrast to the H$^+$-reuptake kinetics where the small isotope effect of 1.6 suggests that this process is mainly governed by the protein and not by the H$^+$-diffusion rate.

VARIATION IN MEMBRANE CONSTITUTION

A natural membrane contains proteins embedded in the lipid bilayer. By modern genetic techniques the protein can be site-directed modified. Nature and composition of the surrounding lipid phase are changed by reconstitution of the protein in artificial lipid vesicles. With the pH-indicator technique the consequences on the H$^+$-transfer reactions of BR can be evaluated.

Bacteriorhodopsin reconstituted in DMPC vesicles:

Fluorescein-labelled BR is incorporated into DMPC vesicles (dimyristoyl-phosphatidyl-choline) by the detergent-dialysis-method *(Heyn & Dencher, 1982)*. With this system a third detection site for H$^+$-transfer reactions can be introduced. BR/DMPC vesicles are doped with the lipid FPE (fluorescein-phosphatidyl-ethanolamine) allowing the detection of pH-changes in the vicinity of the lipid headgroups (see sketch in Fig.9). Time-resolved as well as site-specific absorbance changes of the employed pH-indicators are depicted in Fig.9. Again, the pumped proton appears at the extra-cellular surface of BR (detected by FITC) concomitantly with the rise of the M inter-mediate. This demonstrates that the altered lipid environment does not influence the H$^+$-transfer from the interior of BR to the extracellular surface. The absorbance changes of pyranine (Pyr in Fig.9) and fluorescein bound to the lipid headgroups (FPE) allow to follow the migration of the proton after being released to the surface. Clearly, the pH along the lipid headgroups is lowered **before** the pH of the surrounding bulk water phase. This might point towards faster proton migration along the membrane surface, as proposed by *Tocanne & Tessie (1990)*. To give a final answer to the rate of proton transfer along biological membranes, pH-indicator experiments will be conducted in the near future with H$^+$-generator (BR) and H$^+$-consumer (for example, a H$^+$-ATPase) co-reconstituted in lipid vesicles.

Point mutants of bacteriorhodopsin:

Evaluation of the mechanism of H$^+$-transfer within a protein is interesting not only for BR but most important for the understanding of energy conversion in bioenergetics. The role of certain amino acid side chains in H$^+$-translocation across BR can be in-vestigated by using point mutated BR. By site-directed mutagenesis every amino acid of the protein can be replaced by another one *(Soppa & Oesterhelt, 1989)*.

We have investigated two point mutants of BR *(Dencher et al., 1991)*. In the Asp 85→Glu-mutant, an aspartate which is located in the H$^+$-ejection pathway (see Fig.1), is replaced by glutamate. This prolongs the amino acid side chain just by

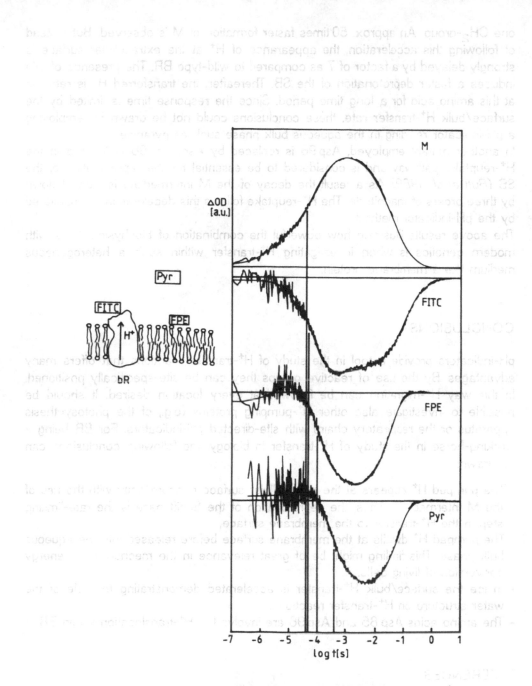

Fig. 9: Site-directed detection of the kinetics of the proton transfer
reactions along BR/DMPC vesicles.
M: rise and decay of the M intermediate.
FITC: Fluorescein covalently linked to BR.
FPE: Fluorescein covalently linked to the lipid headgroups.
Pyr: Pyranine residing in the bulk water phase.

one CH_2-group. An approx. 50 times faster formation of M is observed. But instead of following this acceleration, the appearance of H^+ at the extracellular surface is strongly delayed by a factor of 7 as compared to wild-type BR. The presence of Glu induces a faster deprotonation of the SB. Thereafter, the transferred H is retained at this amino acid for a long time period. Since the response time is limited by the surface/bulk H^+-transfer rate, these conclusions could not be drawn by employing a pH-indicator residing in the aqueous bulk phase such as pyranine.

In another mutant employed, Asp 96 is replaced by Asn. Asp 96 is located in the H^+-reuptake pathway and is considered to be essential for the reprotonation of the SB *(Butt et al., 1989)*. As a result the decay of the M intermediate is slowed down by three orders of magnitude. The H^+-reuptake follows this deceleration as monitored by the pH-indicator method.

The above results illustrate how powerful the combination of biophysical tools with modern genetics is when investigating H^+-transfer within such a heterogeneous medium like a membrane protein.

CONCLUSIONS

pH-indicators provide a tool in the study of H^+-transfer reactions that offers many advantages. By the use of reactive groups they can be site-specifically positioned. In this way H^+-migration can be followed at every location desired. It should be possible to investigate also other H^+-pumping proteins (e.g., of the photosynthesis apparatus or the respiratory chain) with site-directed pH-indicators. For BR being a working-horse in the study of H^+-transfer in biology, the following conclusions can be drawn:

- The pumped H^+ appears at the extracellular surface concomitantly with the rise of the M intermediate. Thus, the deprotonation of the Schiff base is the rate-limiting step in the H^+-transfer to the membrane surface.
- The pumped H^+ dwells at the membrane surface before released into the aqueous bulk phase. This finding might be of great relevance in the mechanism of energy conversion of living cells.
- In ice the surface/bulk H^+-transfer is accelerated demonstrating the role of the water structure on H^+-transfer reactions.
- The amino acids Asp 85 and Asp 96 are involved in H^+-translocation within BR.

REFERENCES

H.-J. Butt, K. Fendler, E. Bamberg, J. Tittor & D. Oesterhelt, EMBO J. **8**, 1657 (1989)

G. Careri, A. Giansanti & J.A. Rupley, Proc. Natl. Acad. Sci. USA **83**, 6810 (1986)

G. Cevc, Biochim. Biophys. Acta **1031**, 311 (1990)

N.A. Dencher, G. Büldt, J. Heberle, H.-D. Höltje, M. Höltje, this book

N.A. Dencher, J. Heberle, C. Bark, M.H.J. Koch, G. Rapp, D. Oesterhelt, K. Bartels, G. Büldt, Photochem. Photobiol. 54, (1991) in the press

R.E. Dilley, this book

J. Heberle & N.A. Dencher, FEBS Lett. 277, 277 (1990)

R. Henderson, J.M. Baldwin, T.A. Ceska, F. Zemlin & E. Beckmann, J. Mol. Biol. 213, 899 (1990)

M.P. Heyn & N.A. Dencher, Methods Enzymol. 88, 31 (1982)

M.H.J. Koch, N.A. Dencher, D. Oesterhelt, H.-J. Plöhn, G. Rapp, & G. Büldt, EMBO J. 10, 521 (1991)

H. Merz & G. Zundel, Biochem. Biophys. Res. Commun. 101, 540 (1981)

J.F. Nagle & H.J. Morowitz, Proc. Natl. Acad. Sci. USA 75, 298 (1978)

J.F. Nagle, J. Bioenerg. Biomembr. 19, 413 (1987)

D. Oesterhelt & W. Stoeckenius, Nature New Biol. 233, 149 (1971)

D. Oesterhelt & J. Tittor, Trends Biochem. Sci. 14, 57 (1989)

G. Papadopoulos, N.A. Dencher, G. Zaccai, & G. Büldt, J. Mol. Biol. 214, 15 (1990)

J.A. Rupley, L. Siemankowski, G. Careri & F. Bruni, Proc. Natl. Acad. Sci. USA 85, 9022 (1988)

J. Soppa & D. Oesterhelt, J. Biol. Chem. 264, 13043 (1989)

G. Souvignier, this book

J.-F. Tocanne & J. Tessie, Biochim. Biophys. Acta. 1031, 111 (1990)

DIFFUSION OF PROTON IN THE MICROSCOPIC SPACE OF THE PhoE CHANNEL

Tsfadıa Yossi and Gutman Menachem

Laser Laboratory for Fast Reactions in Biology
Department of Biochemistry
The George S. Wise Faculty of Life Sciences
Tel-Aviv University, Ramat,Aviv (Israel)

INTRODUCTION

Most biochemical reactions take part in microscopic subcellular
cavities such as enzyme-binding sites, or between membranal structures
like mitochondria. The dimensions of these intracellular spaces can be
characterized in time domain, which is the time duration needed for a free
diffusing particle to propagate across it. An example for a very fast
diffusion process is the proton transfer, the most common reaction in the
biosphere. In this study we shall employ a free (hydrated) proton to probe
a well defined biological structure.

There are some advantages in using proton as a gauge particle. First,
it has a well known chemical reactivity and diffusion mechanism. The
second advantage is experimental one, by using the laser induced proton
pulse (1) one can easily generate a proton pulse in situ with a high degree
of sinchronicity, and by employing sensitive indicators, submicro molar
quantities are easily detectable. Finally, due to its small dimensions, it
serves as a good model for electrostatic calculations. If we can limit the
monitoring to a very short time frame, after a proton has been released in
a definend site, we can use the current theories on the rates of proton
transfer and its diffusion to obtain specific physical information about
the environment where the reaction takes place. Thus, parameters like
activity of water (2), viscosity of the solvent and electrostatic
interactions can be measured (3,4).

One type of the most fascinating microenvironments in biology are the
ion channels in membranes. These protein structures are found in all
living cells, where their role is to allow specific charged materials to
get in and out of the cell through the lipid barrier. In this study we
shall investigate the physical properties of the aqueous micropore of the
PhoE channel.

PhoE is an anion specific porin of E. coli with a well recognized
structure and known amino acid sequence (5,6). At the extracellular side
of the outer membrane, there is a large aqueous vestibule ~35A long with an
elliptical opening ~27A along the major axis and ~18A along the minor axis.
The vestibule has an extended narrower channel that is ~10A long, having an
average inner diameter of ~10A, (scheme 1.).

Proton Transfer in Hydrogen-Bonded Systems
Edited by T. Bountis, Plenum Press, New York, 1992

The technique we used in this study is a time-resolved fluorescence measurments, which allows us to follow, over a 20 nsec period, the propagation of the proton which was generated inside the aqueous cavity of the ion channel. The probing reaction is the reversible interaction of proton with excited pyranine anion (8-hydroxypyrene 1,3,6 trisulfonate) inserted in the PhoE channel. Upon photo-excitation of the dye, its pK is shifted from pK=7.7, of the ground state, to pK*=1.4, causing fast dissociation of the hydroxy proton with time constant of ~100 psec (see scheme 2). Due to the high time resolution of our observation we can monitor the excited dye reprotonation by a proton still confined whithin the Coulomb cage of the excited anion, a process called geminate recombination.

Geminate recombination is a reaction common to all reversible dissociation processes. It takes place when the two products are formed at

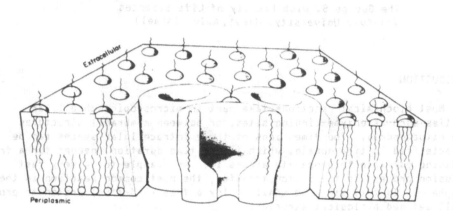

Scheme 1. SCHEMATIC MODEL OF PHOE PORIN TRIMMERS DERIVED FROM THE STRUCTURAL INFORMATION taken from Jap B.K. (1989), ref. 6.

an ultimate proximity to each other; thus, there is a certain probability that the pair will recombine to reform the parent molecule. In a case where the two products attract each other electrostatically or are slowly diffusing, the probability of recombination increases.

The mechanism of geminate recombination in bulk water has recently been analyzed by Pines et. al. (7), Agmon et. al. (8) and Agmon and Szabo (9). The analysis is based on the reconstruction of the observed signals using the differential form of the Debye-Smoluchowsky equation for the time-dependent translational diffusion of a charged particle in a Coulombic potential field.

MATERIALS AND METHODS

The PhoE protein was a generous gift from Prof. Jan Tommassen from State University of Utrecht, The Netherlands. The protein was incorporated into Brij 58 micelles and equilibrated with pyranine by dialysis against the dye in 1.4% Brij. Since the pore of the protein has a net charge of +3, it binds the pyranine (Z=-3) with high affinity. The dissociation constant of the dye-protein complex is K=0.127μM.

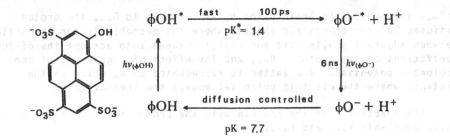

$$\phi OH^* \xrightarrow[\text{pK}^* = 1.4]{\text{fast} \quad 100\,\text{ps}} \phi O^{-*} + H^+$$

$hv_{(\phi OH)}$ $6\,\text{ns} \mid hv_{(\phi O^-)}$

$$\phi OH \xleftarrow{\text{diffusion controlled}} \phi O^- + H^+$$

pK = 7.7

Scheme 2. STRUCTURE OF 8-HYDROXYPYRENE 1,3,6 TRISULFONATE which is a proton emitter and schematic presentation of dynamic steps invoked in photodissociation of the dye.

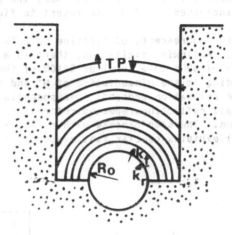

Scheme 3. SCHEMATIC PRESENTATION OF THE REACTION SPACE FOR PROTON-EXCITED PYRANINE ANION RECOMBINATION IN THE PHoE CHANNEL.

R_o - The radius of the reaction sphere.

k_f - The probability constant for separation of the proton
     from the excited anion to the water.

k_r - The probability constant for reprotonation back to OH*.
     The protons diffuse in concentric shells with $TP_{i+1/i}$ - the
     probability to move forward to the next shell and $TP_{i/i+1}$ to return
     one shell backwards.

Time-resolved fluorometry was measured with 10-psec resolution as described by Pines et al. (7).

The analysis of the results used a computer program of N. Agmon which was adapted to a spatial configuration shown in scheme 3. Since the PhoE is in a micelles, our model assumes that the narrower channel is blocked by the lipidic phase of the micelle.

The diffusing proton propagates in hemi-spherical 3-dimensional space up to the protein walls of the channel (scheme. 3), then the diffusion becomes 1-dimension along the cylinder axis. The inner boundary R_o represents the pyranine. The constants k_f and k_r are the rates of proton production and consumption respectively, at $r=R_o$. At a distance

r=R_{max} the proton is lost to the bulk. From R_0 to R_{max} the proton
diffuses between concentric shells, where the probability for transition
between adjacent shells r(i) and r(i+1)   takes into account the diffusion
coefficient of the proton - D_{H+}, and the effect of a gradient of the
Coulombic potential. The latter is represented by R_c which is the
distance where the electric potential equals the thermal energy.

The reactivity of the protein with the proton is quantitised by the
rates constants k_{ass} and k_{diss}.

RESULTS AND DISCUSSION

The binding of the dye to the protein changes its fluorescence
emission spectrum. The dye, when dissolved in water, emits mostly from its
anionic state ($\phi0^{*-}, \lambda$ = 515nm). Only ~5% of the ϕOH^* molecules have a
chance to decay radiativly (λ = 440nm) before they exercise proton
dissociation (see scheme 2). In the pore of PhoE, the probability to decay
before dissociation increases to 21% (see insert to figure 2).

The time-resolved fluorescence of pyranine in bulk water is shown in
fig. 1 (dotted line). Following excitation, there is a rapid exponential
decay of the emission of ϕOH^* to a level of 5-10% of the initial
amplitude, corresponding to the dissociation of H^+ and formation of
$\phi0^{-*}$. After that the relaxation proceeds nonexponentially in a shape of
a very low shallow tail reflecting repeated reformation of ϕOH^* by
geminate recombination. The same process with pyranine bound to the PhoE
channel (upper curve) exhibits similar fast exponential decay, but the

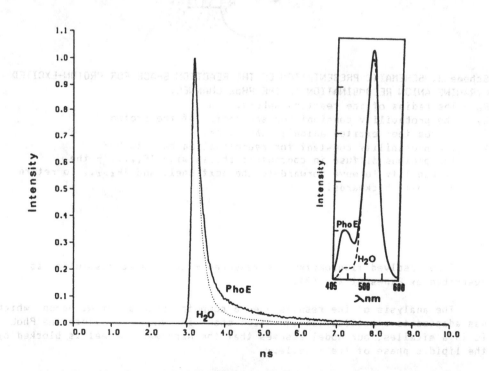

Fig. 1 TIME RESOLVED FLUORESCENCE OF PYRANINE AT THE WAVELENGTH OF MAXIMUM
ϕOH^* EMISSION. <u>Curve A</u> was measured with pyranine dissolved in water
(pH=5.5). <u>Curve B</u> was measured with pyranine-PhoE complex 1:1 (pH=5.5).
Inset: Steady-state fluorescence spectra of pyranine in water and in
PhoE channel. Emission of ϕOH^* is at 440 nm, emission of $\phi0^{-*}$ at 510 nm.

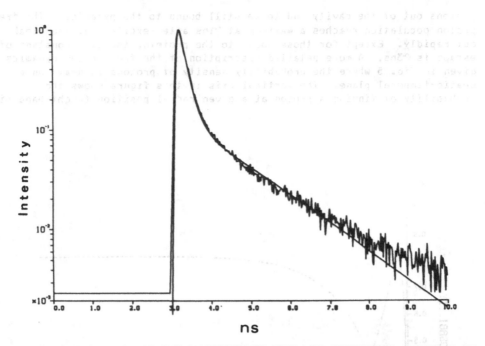

Fig. 2 RECONSTRUCTION OF THE FLUORESCENCE DECAY OF PROTONATED PYRANINE IN
PhoE CHANELL. The curve is drawn on a logarithmic Y axis. The continuous
line superpositioned over the experimental one, is a numerical
reconstruction using Agmon's computer program with the parameters listed in
table 1.

fluorescence tail is significantly larger indicating enhanced geminate
recombination.

The theoretical treatment we used can also reconstruct the spatio-
temporal distribution of the proton. Fig. 3. depicts the dynamics of the
free proton population, those bounded to the protein, the accumulation of
The theoretical reconstruction of the experimental signal of pyranine-
PhoE complex is shown in fig. 2, where for better visualization the curve
is drawn on a semi-logarithmic scale. The parameters required for the
reconstructed dynamics are given in table 1.

TABLE 1

PARAMETERS FOR FITTING THE FLUORESCENCE DECAY OF PYRANINE BOUND TO
PhoE CHANNEL.

Parameter	water	channel of PhoE porin
D (cm²/sec)	9.3×10^{-5}	5.5×10^{-5}
ϵ	78	24
k_+ (sec^{-1})	7×10^{9}	5.5×10^{9}
k_- (sec^{-1})	7×10^{9}	4.3×10^{9}
r_o (A)	5.50	6.0
$k_{...}$	Not applicable	7×10^{-5}

protons out of the cavity and those still bound to the pyranine. The free proton population reaches a maximum at ~1ns after excitation, then leaks out rapidly. Except for those bound to the protein, the time constant of escape is ~3ns. A more detailed description of the free proton dynamics is given in fig. 5 where the probability density of protons is drawn on a spatio-temporal plane. The vertical axis in this figure shows the probability of finding a proton at a given radial position (right-hand side

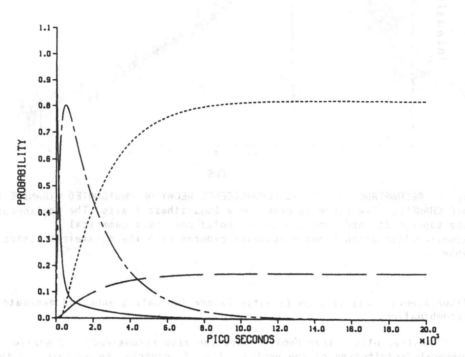

Fig. 3 THEORETICAL RECONSTRUCTION OF THE DYNAMICS INVOLVED WITH PROTON DISSOCIATION FROM PYRANINE IN THE PHOE CHANNEL. The values are consistent with the reconstruction of the experimental curve.
(———) The dynamics of ϕOH^*. This line is identical with that given in fig. 2.
(— – —) The free proton population within the channel.
(—— ——) Protons bound to the protonated groups of the protein within the pore.
(------) Protons which escaped to the bulk.

abcissa) and time (left hand side abcissa). During the observation time the proton population is depleted without major change of the intracavity distribution profile. As seen in this presentation, the protons can reach the opening of the channel whithin ~2ns, yet during the whole time their distribution is higher near the origin, where they are attracted by the strong electrostatic field of the dye. The figure demonstrates that, inspite of strong electrostatic attraction andspatial confinement, the pore cannot maintain a proton for a time frame larger than a few nanoseconds.

Finally, we wish to focus the attention on two principal aspects; the dielectric constant and the diffusion coefficient.

I. The dielectric constant near a boundary ($\epsilon_1 = \epsilon_2$) is a function of the polarity of both matrices and its value varies with the distance from the boundary ($\epsilon_{(r)}$) and its shape. In our model we represent the whole space with a constant value for the dielectric constant that is the space average of all $\epsilon_{(r)}$. The value that we get for the reconstruction dynamics $\epsilon_{eff}=24$, is significantly lower than the dielectric constant of bulk water ($\epsilon=78$). It reflects intensified electric

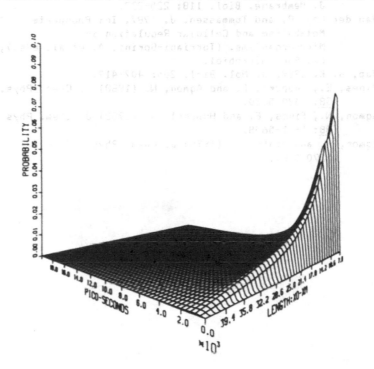

Fig. 4 SPATIO-TEMPORAL DESCRIPTION OF THE DISTRIBUTION OF FREE PROTONS IN THE PHOE ION CHANNEL. The figure denotes on the ordinate the probability of finding a proton at given distance (r_i) and time as defined by the right and left abcisseae. The curves were calculated using the parameters listed in table 1.

field inside the pore. The low dielectric constant points to the fact that even a relativly large space that contains ~500 water molecules is not large enough to behave like bulk water. Thus, we can conclude that smaller cavities will also deviate from bulk water properties.

II. Proton diffusion mechanism depends on the randomicity of hydrogen-bonds of the water. The decreased diffusion coefficient of the proton (see table 1) indicates that water molecules within the pore are more ordered. The ordering is also indicated by the decrease rate of proton dissociation (k_f), an observation common to systems where the rotational diffusion of water is diminished (2).

ACKNOWLEDGEMENTS: This research is supported by the United States-Israel Binational Science Foundation, grant no. 870035 and the US Navy, ONR, grant no. N00014-89-J1662.
     The authors are grateful to Prof. J. Tommassen for the generous gift of the Pho-E protein.

References

1. Gutman, M. 1984, Methods Biochem. Anal. 30: 1-103.
2. Huppert, D., Kolodney, E., Gutman, M. and Nachliel,
        E. 1982, J. Am. Chem. Soc. 104: 6949-6953.
3. Gutman, M., Nachliel, E. and Moshiach, S. 1989,
        Biochemistry. 28: 2936-2941.
4. Rochel, S., Nachliel, E., Huppert, D. and Gutman, M. 1990,
        J. Membrane. Biol. 118: 225-232.
5. Van der Ley, P. and Tommassen, J. 1987, In: Phosphate
        Metabolism and Cellular Regulation in
        Microorganisms. (Torriani-Gorini, A. et al. eds.),
        Am. Soc. Microbiol.
6. Jap, B. K. 1989, J. Mol. Biol. 205: 407-419.
7. Pines, E., Huppert, D. and Agmon, N. (1988) J. Chem. Phys.
        88: 5620-5630.
8. Agmon, N., Pines, E. and Huppert, D. (1988) J. Chem. Phys.
        88: 5631-5638.
9. Agmon, N. and Szabo, A. (1990) J. Chem. Phys. 92:
        5270-5284.

GLASS TRANSITIONS IN BIOLOGICAL SYSTEMS

Polycarpos Pissis

National Technical University of Athens
Department of Physics
Zografou Campus, 15773 Athens, Greece

ABSTRACT

This work deals with detailed investigations of the dynamics of the dielectric relaxations in hydrated lysozyme powders and in plant tissue in the temperature region where transitions have been reported to occur. We use the method of thermally stimulated depolarization currents (TSDC), which is very sensitive to transitions and has been widely used in the study of glass transitions in synthetic polymeric systems. The relaxations are due to proton transport, side chains reorientation and space charge polarization. Two features of the relaxations, namely the dependence of their dynamics on water content and the dependence of their activation energies on temperature, reveal, in analogy to synthetic polymeric systems, the existence of two glass transitions, probably due to the hydration water and the matrix structure.

INTRODUCTION

Recent theoretical and experimental studies suggest the occurence of glass or glass-like transitions in a variety of biological systems. A difficulty often faced in the experiments is that the classical method of detecting glass transitions in synthetic amorphous polymers and in glass forming liquids and of studying its dynamics, differential scanning calorimetry (DSC), is by far less sensitive in biological systems, with the result that there are only a few reports[1-3] on DSC studies of glass transitions in biological systems. In most cases the occurence of glass or glass-like transitions in biological systems is deduced from the observation that the mobility of different components of the system increases with increasing temperature more or less steeply, typically at around 200K. A variety of experimental techniques have been used to study these mobility changes with temperature, such as inelastic neutron scattering[4], Mössbauer spectroscopy[5], X-ray diffraction[6], Rayleigh scattering of Mössbauer radiation[7] and nuclear magnetic resonance[8]. Other methods used include ac dielectric techniques[9] and infrared spectroscopy[10].

It is not clear from these studies as to whether even a hydrated protein, the system best studied, undergoes only one glass transition as a whole system or we are dealing with two transitions, one due to the hydration water and the other due to the protein structure itself. The water

Proton Transfer in Hydrogen-Bonded Systems
Edited by T. Bountis, Plenum Press, New York, 1992

molecules form a network of hydrogen bonds and a coupling between water and the protein is provided by the water of hydration. However, as the protein structure ifself is stabilized by the formation of hydrogen bonds, mobility changes observed at about 200K may be due to the melting of solid water as well as to rearrangements of the protein structure hydrogen bonds. We think that measurements with systematically varying the water content in the system should help to clarify this point.

This work deals with detailed investigations of the dynamics of the dielectric relaxations in hydrated lysozyme powders and in plant tissue in the temperature region where glass transitions have been reported to occur. We use the method of thermally stimulated depolarization currents (TSDC), which is very sensitive to transitions and has been widely used in the study of glass transitions in synthetic polymeric systems[11]. We make use of experience accumulated from dielectric studies of the dynamics of the glass transition in synthetic polymers:dependence of glass transition temperature on water content[12], dependence of activation energy on temperature[13], range of values of activation energies and entropy factors[10], validity of compensation laws[14].

EXPERIMENTAL DETAILS

The thermally stimulated depolarization current (TSDC) method consists of studying the thermally activated release of stored dielectric polarization. The method is as follows[11]:The sample is polarized by an applied electric field E_p at a temperature T_p. This polarization is subsequently frozen in by cooling the sample to a temperature T_0 sufficiently low to prevent depolarization by thermal energy. The field is then switched off and the sample is warmed at a constant rate b, while the depolarization current, as the dipoles relax, is detected by an electrometer. Thus for each polarization mechanism an inherent current peak can be detected. A major feature of the TSDC method, which makes it attractive for our purposes, is that it offers the possibility to experimentally resolve relaxation processes arising from sets of dipoles with slightly different relaxation times. In the case of a single relaxation process obeying the Arrhenius equation, $\tau(T) = \tau_0 \exp(W/kT)$, the depolarization current density $J(T)$ is given by the equation

$$J(T) = (P_0/\tau_0) \exp(-W/kT) \exp\left[-(1/b\tau_0) \int_{T_0}^{T} \exp(-W/kT') \, dT'\right], \quad (1)$$

where τ is the relaxation time, W the activation energy of the relaxation, τ_0 the pre-exponential factor, T the absolute temperature, k Boltzmann's constant and P_0 the initial polarization. The analysis of the shape of this curve makes it possible to obtain the activation energy, W, the pre-exponential factor, τ_0, and the contribution, $\Delta\varepsilon$, of a peak to the static permittivity[11].

Experimental details concerning the preparation of the samples and the adjustment of the water content have been given elsewhere[15,16]. Following common practice, the water content h is defined as the ratio of weight of the absorbed water to the weight of dry sample (dry basis) for lysozyme and as the difference between the weight of the sample during the dielectric measurement and the dry weight divided by the weight during the measurement (wet basis) for plant tissue.

A standard experimental apparatus for TSDC measurements was used[17]. Typical experimental errors were ±0.005 for the water content h, ±1K for the temperature of current maximum (peak temperature) T_M, ±0.02 eV for the activation energy W, a factor of about two for the pre-exponential factor

τ_0 in the Arrhenius equation and ±13% for the contribution $\Delta\varepsilon$ of a peak to the static permitivity.

RESULTS

Figure 1 shows a TSDC plot measured on a lysozyme sample with a water content of 0.174. Two separate and independent dispersions can be distinguished: a broad low-temperature (LT) peak and a rather complex high-temperature (HT) band. Both dispersions have been studied in detail by using several experimental techniques offered by the TSDC method[15,18]. The LT peak is due to the dielectric relaxation of loosely bound and free water molecules[18]. Four different relaxation mechanisms contribute to the HT band: the vibrational motions of the polypeptide backbone of lysozyme, the localised, short-range hopping of protons in their transport through the bulk sample, the percolative proton transfer along threads of hydrogen-bonded water molecules on the surface of single macromolecules and the drift of ions towards, and their resulting interaction with, the metal electrodes[15,19]. In the following we focus our attention on the HT band and study its relation to glass or glass-like transitions.

We investigated in detail the dynamics of the different relaxation processes contributing to the complex TSDC thermogram of Fig.1 by using the partial heating and the thermal sampling (TS) techniques[11,15,18]. By the PH technique the sample was polarized in the usual way and then partially depolarized by partial heating, separated by rapid cooling, up to a series of cut temperatures that spanned the whole temperature region of the dispersions. The activation energy W of the different contributions was then calculated by the initial rise method[11]. By the TS technique the sample was continuously cooled down at a constant rate and the polarizing field was switched on at T_p and switched off at T_d, i.e., the relaxation processes within a narrow temperature range were sampled. The temperature window T_p-T_d was 5K. The activation energy W of the thermal sampling responses was calculated by the method of fitting Eq.(1) to the experimental peaks. Thus, the detailed spectra of activation energies W were determined. In Fig.2 we show the interesting regions of the W spectra at different water contents. The dependence of W on T_m is, in general, approximately linear at low temperatures and shows significant and characteristic deviations from this linear dependence in the temperature region of the HT band. The arrows in Fig.2 indicate the peak temperature of the HT band, which increases with decreasing water content. Plots similar to those in Fig.2 have been used to study the dynamics of the glass transition in synthetic polymers[13], where T_g is known from DSC measurements. $W=W(T_m)$ is expected to be linear in the temperature region of relaxation processes, which are noncooperative, and to show deviations from this linearity, of the type shown in Fig.2, in the temperature region of cooperative relaxation processes (for which the Eyring activated state of entropy is nonzero), such as the glass transition. The glass transition temperature T_g, determined by calorimetry, was found to coincide with the temperature of maximum activation energy W[13]. Bearing that in mind, our results in Fig.2 suggest that the HT band measured on lysozyme is related to glass (or glass-like) transitions in the samples. It is not clear at this stage whether the whole sample undergoes only one glass transition, similar to an amorphous homopolymer, or more than one transition occur in the temperature region of the HT band, similar to copolymers and polymer blends.

Figure 3 shows the TSDC thermogram measured on a leaf of Eucalyptus globulus with a water content of 0.64. The low-temperature (LT) peak is attributed to the reorientation of loosely bound water molecules[16]. The intermediate-temperature (IT) peak is either due to the reorientation of tightly bound water molecules or to the relaxation of some botanical materials rather than water[20]. Finally, the high-temperature (HT) band is

209

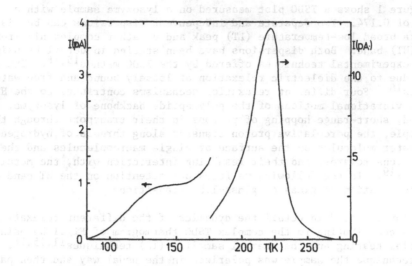

Fig.1. TSDC plot measured on a lysozyme sample with a water content of 0.174 gr H_2O/gr dry material.

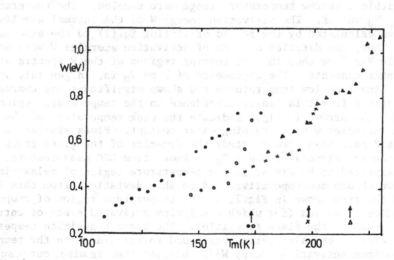

Fig.2 Activation energy W vs. temperature T_m of the TS response or PH cut temperature for lysozyme with four different water contents: 0.61 (● PH), 0.35 (○ TS), 0.25 (x TS), 0.18 (▲ PH). The arrows indicate T_m of the HT band.

attributed to water-assisted space charge polarization connected with dc ionic conductivity[20]. In Fig.4 we show a plot of activation energy W against peak temperature T_m of the thermal sampling responses isolated in the complex thermogram of Fig.3. The most interesting result is the clear evidence for two regions of deviations from the linear dependence $W=W(T_m)$, corresponding to the IT peak at about 180K and the HT band at about 200K, respectively, and indicating the existence of two glass transitions.

The pre-exponential factor τ_0 in the Arrhenius equation was calculated for each thermal sampling response using the following equation:

$$\tau_0 = (kT_m^2 /bW) \exp (-W/kT_m), \qquad (2)$$

where T_m is the peak temperature of the TS response. Figure 5 shows a plot of $\ln\tau_0$ as a function of W for hydrated lysozyme samples at two different water contents. We are looking for linear relationships, known as the compensation effect[14,16]

$$\ln\tau_0 = \ln\tau_0' - (W/kT_c), \qquad (3)$$

where τ_0' a constant and T_c the so-called compensation temperature. The compensation effect has been revealed in many processes that require activation energy in order to proceed. The physical meaning of the compensation temperature is still a subject of investigation and controversy[14]. In synthetic polymers it has often been interpreted in terms of glass transition temperatures. Because of different relaxation mechanisms contributing to the complex thermogram of Fig.1 we don't expect a single linear relationship to include all the TS responses at a fixed water content. The straight lines in Fig.5 include mainly the TS responses which deviate from the linear relationship $W=W(T_m)$ of Fig.2 and give compensation temperatures $T_c=(196\pm15)$K(h=0.354) and (221 ± 15)K(h=0.25), i.e. slightly larger than the corresponding HT band temperatures. The results obtained with plant leaves (Fig.6) are more impressive. They suggest the existence of two compensation laws, one for the TS responses of the LT peak with $T_c=181\pm 6$K and the other for those TS responses of the HT band, which deviate from the linear relationship $W=W(T_m)$ of Fig.4, with $T_c=210\pm 15$K.

In synthetic polymers the glass transition temperature T_g measured by a variety of experimental techniques has been found to decrease with increasing water content[12], while the magnitude of the TSDC thermogram associated with the glass transition increases with water content[21]. Similar results are shown in Figs.7 and 8 for the HT bands measured on lysozyme and plant tissue samples. The effective band temperature in these figures is defined as the geometric centre of gravity of the band. The depolarization charge of a TSDC band is a measure of the number of relaxing units contributing to the band. The results are consistent with the HT bands being associated with glass transitions which shift to lower temperatures with increasing water content.

DISCUSSION

Our results suggest that the high-temperature (HT) TSDC bands in hydrated lysozyme powders and in plant tissue at about 200K in fully hydrated systems and at higher temperatures for systems partially hydrated are related with glass transitions. From the methodological point of view our approach to the topic resembles that of Iben et al.[10], in the sense that we are looking at fundamental properties and characteristics of the glass transition utilizing experience accumulated from detailed investigations of the dynamics of the glass transition in synthetic polymers. In the report by Iben et al.[10] these fundamental glasslike properties were the

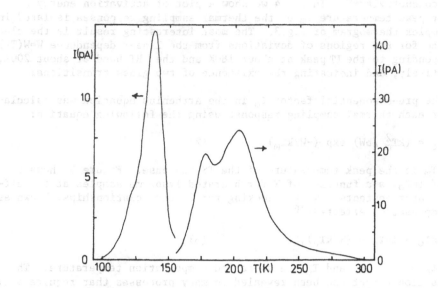

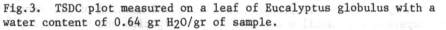

Fig.3. TSDC plot measured on a leaf of Eucalyptus globulus with a water content of 0.64 gr H_2O/gr of sample.

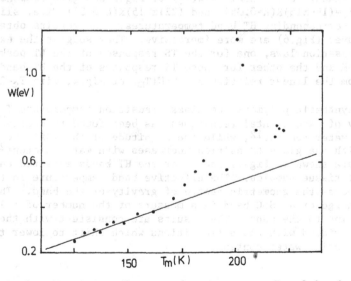

Fig.4. Activation energy W vs. peak temperature T_m of the thermal sampling responses for a leaf of Eucalyptus globulus with a water content of 0.64 gr H_2O/gr of sample. The line is a guide for the eyes.

metastability at low temperatures and the nonexponential time and the non-Arrhenius temperature dependence of the protein relaxations near 200K. In this work the fundamental glasslike properties and attributes studied are the dependence of glass transition temperature on water content (Figs. 7 and 8), the dependence of the activation energy of the thermal sampling responses and the partial heating initial rises on temperature (Figs. 2 and 4), the high values of activation energy combined with unrealistic, small values of the pre-exponential factor τ_0 close to the glass transition temperatures (Figs. 2,4-6), and the validity of the compensation law (Figs. 5 and 6). In the following we discuss these properties in some more detail.

1) The glass transition temperature in synthetic amorphous polymers and in glass forming liquids measured by DSC and by a variety of other less direct experimental techniques has been found to shift significantly to lower temperatures with increasing water content of the samples[12,21]. Several empirical and theoretical expressions based on the concepts of plasticizing action of water and of polymer blends have been used to describe this shift[12]. The magnitude of the TSDC peak associated with the glass transition in thermoplastic polyurethanes has been found to increase with increasing water content h[21]. Similar results obtained within this work (Figs. 7 and 8) suggest that the HT TSDC bands measured on hydrated lysozyme powders and on plant tissue are associated with glasslike transitions. In contrast to many synthetic polymers and glass forming liquids, several different relaxation mechanisms contribute to the HT TSDC bands measured within this work, especially in the case of hydrated lysozyme powders. It is not known a priori how each of these relaxation mechanisms is related to and influenced by the glass transition and what is the extent of its contribution to the HT TSDC band. Therefore we avoid, at this stage, fitting any expression to the experimental points.

2) The activation energy W of the thermal sampling responses, isolated in the complex TSDC thermogram of poly (aryl ether ketone ketone), plotted against the polarization temperature was found to be linear in the temperature region of non-cooperative relaxation processes, such as β and γ, and to show deviations from this linearity in the temperature region of the glass transition[13]. Moreover, the glass transition temperature T_g, determined by DSC, was found to coincide with the temperature of maximum activation energy. The W(T) plot was suggested as a sensitive method of detecting glass transitions in systems where DSC fails[13]. We suggest that this method is improved if the peak temperature T_m of the TS responses replaces the polarization temperature and if, in addition to the TS analysis, partial heating (PH) analysis is also used and the activation energy W calculated by the initial rise method[11] is plotted against the cut temperature T_m. The results in Fig. 2 clearly indicate glasslike transitions in hydrated lysozyme powders in the temperature region of the HT TSDC bands, which shift to lower temperatures with increasing water content. It is not, in general, clear whether the whole sample undergoes only one glass transition, or two glass transitions occur. However, the sample with the lowest water content, h=0.18, clearly exhibits two glass transitions, at about 200 and 220 K. It is noteworthy that dielectric measurements in hydrated lysozyme powders have been interpreted in terms of two glass transitions at about 180 and 200 K, independently of water content, the first due to the network of hydrogen-bonded water molecules and the second due to the protein structure[9]. Our measurements might suggest that at higher water contents the whole system undergoes only one glass transition, similar to polymer blends with good mixing of the phases. Our results on plant tissue are more conclusive, clearly showing two glass transitions at about 180 and 200 K (Fig. 4), similar to measurements on corn embryo[9].

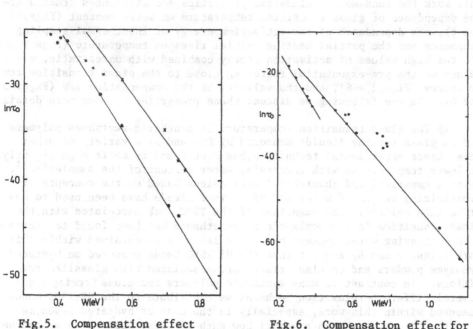

Fig.5. Compensation effect in hydrated lysozyme samples at two different water contents: 0.35(●) and 0.25 (x).

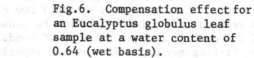

Fig.6. Compensation effect for an Eucalyptus globulus leaf sample at a water content of 0.64 (wet basis).

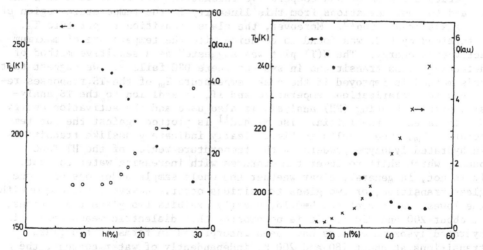

Fig.7. Effective band temperature T_b (●) and depolarization charge Q(o) of the high-temperature TSDC band in lysozyme vs. water content h.

Fig.8. Effective band temperature T_b (●) and depolarization charge Q(x) of the high-temperature TSDC band in Eucalyptus globulus leaves vs. h.

214

The temperature dependence of the relaxation close to the glass transition is non-Arrhenius. However, it follows aproximately the Arrhenius relation over small temperature intervalls, so that the thermal sampling responses and the initial rises obtained by the partial heating analysis can be evaluated in the usual way[11]. Following this evaluation we obtained values for the activation energy W which are high close to Tg, relatively to the temperature region of the relaxation processes, higher than 1 eV for plant tissue (Figs.2 and 4). The corresponding values of the preexponential factor τ_0 are then very small, typically 10^{-25}s (Figs.5 and 6). These results are very similar to results obtained with synthetic polymers and provide evidence for the glasslike character of the relaxations studied.

The significance of the compensation effect (Figs.5 and 6) and the physical meaning of the compensation temperature T_c are subjects of controversy[14]. The most striking result we have obtained in this connection is the clear evidence that the compensation law is valid for the dielectric relaxation of water molecules in plant tissue and that the compensation temperature for this relaxation, $T_c=181\pm6$K, coincides with the peak temperature of the intermediate temperature peak (Fig.3). Bearing in mind that this IT peak exhibits glasslike properties (Fig.4), strong evidence is obtained that the glass transition at about 180 K is due to the water of hydration. The second glass transition at about 200 K is then attributed, in a less clear way, to the underlying matrix structure.

REFERENCES

1. W. Doster, A. Bachleitner, R. Dunau, M. Hiebl, and E. Luscher, Thermal properties of water in myoglobin crystals and solutions at subzero temperatures, Biophys.J. 50 : 213 (1986).
2. F. Parak and H. Hartmann, Structural disorder in myoglobin at low temperatures, in "Tunneling", J.Jortner and B. Pullman, eds., Reidel, Dordrecht (1986).
3. R.J. Williams and A.C. Leopold, The glassy state in corn embryos, Plant Physiol. 89 : 977 (1989).
4. W. Doster, S. Cusack, and W. Petry, Dynamic instability of liquidlike motions in a globular protein observed by inelastic neutron scattering, Phys.Rev.Lett. 65 : 1080 (1990).
5. F. Parak, M. Fischer, and G.U. Nienhaus, The similarity in the dynamics of myoglobin and glycevol as seen from Mossbauer spectroscopy on ^{57}Fe, J.Mol.Liq. 42 : 145 (1989).
6. H. Frauenfelder, F. Parak and R.D. Young, Conformational substates in proteins, Ann.Rev.Biophys.Biophys.Chem. 17 : 451 (1988).
7. I.V. Kurinov, Yu.F. Krupyanskii, A.R. Pachenko, I.P. Suzdalev, I.V. Uporou, K.V. Shaitan, A.B. Rubin, and V.I. Goldanskii, Intramolecular dynamics of hydrated DNA studied by Royleigh scattering of Mossbauer radiation, Hyp.Int. 58 : 2355 (1990).
8. E.R. Andrew, D.J. Bryant, and T.Z. Rizvi, The role of water in the dynamics of proton relaxation of proteins, Chem.Phys.Lett. 95 : 463 (1983).
9. G. Carreri, G. Consolini and F. Bruni, Proton tunneling in hydrated biological tissues near 200 K, Biophys.Chem. 37 : 165 (1990).
10. I.E.T. Iben, D. Braunstein, W. Doster, H. Frauenfelder, M.K. Hong, J.B. Johnson, S. Luck, P. Ormos, A. Schulte, P.J. Steinbach, A.H. Xie, and R.D. Young, Glassy behaviour of a protein, Phys.Rev.Lett. 62 : 1916 (1989).
11. J.van Turnhout, Thermally stimulated discharge of electrets, in "Topics in Applied Physics, Vol. 33, Electrets", G.M. Sessler, ed., Springer, Berlin (1980).
12. L.S.A. Smith and V. Schmitz, The effect of water on the glass transition temperature of poly (methyl methacrylate), Polymer 29 : 1871 (1988).

13. B.B. Sauer, P. Avakian, H.W. Starkweather, Jr., and B.S. Hsiao, Thermally stimulated current and dielectric studies of poly (aryl ether ketone ketone), Macromolecules 23 : 5119 (1990)

14. J.P. Crine, A new analysis of the results of thermally stimulated measurements in polymers, J.Appl.Phys. 66 : 1308 (1989).

15. P. Pissis and A. Anagnostopoulou-Konsta, Protonic percolation on hydrated lysozyme powders studied by the method of thermally stimulated depolarization currents, J.Phys.D 23 : 932 (1990).

16. P. Pissis, The dielectric relaxation of water in plant tissue, J. Exp. Botany 41 : 677 (1990).

17. P. Pissis, D. Diamanti and G. Boudouris, Depolarization thermocurrents in frozen aqueous solutions of glucose, J.Phys.D 16 : 1311 (1983).

18. P. Pissis, Dielectric studies of protein hydration, J.Mol.Liq. 41 : 271 (1989).

19. R. Pethig, Dielectric studies of proton transport in proteins, Ferroelectrics 86 : 31 (1988).

20. P. Pissis, A. Anagnostopoulou-Konsta, and L. Apekis, A dielectric study of the state of water in plant stems, J. Exp.Botany 38 : 1528 (1987).

21. P. Pissis, A. Anagnostopoulou-Konsta, L. Apekis, D. Daoukaki-Diamanti and C. Christodoulides, Dielectric effects of water in water-containing systems, J. Non-Crystalline Solids 131-133 : 1174 (1991).

INTERACTION BETWEEN ACETIC ACID AND METHYLAMINE IN WATER

MOLECULAR DYNAMICS AND AB-INITIO MO STUDIES

Janez Mavri and Dušan Hadži

Boris Kidrič Institute of Chemistry

Hajdrihova 19, P.O.B. 61115, Ljubljana, Slovenia

1. INTRODUCTION

Importance for biological structures and mechanisms of the proton transfer (PT) between carboxylic and amine groups has stimulated numerous MO treatments of model systems which usually consist of the simplest representatives, formic acid and methylamine or ammonia [1][2][3]. Calculations of the proton potential in vacuo invariably show the energetic preference of the neutral, hydrogen bonded complex. The actual shape of the potential strongly depends on the level of theory. Inclusion of the reaction field of the polar solvent strongly influences the proton potential in that the ionic form, resulting from PT from the carboxylic to the amine group, becomes energetically preferred [2]. Even in this type of calculation results that are at least in qualitative agreement with experience [4] can be obtained only by using sufficiently flexible basis sets [2]. Extensive geometry optimization also is very important.

The ultimate goal of the study of PT between carboxylic acids and amines is the simulation at molecular level of the reaction in aqueous solution. This is a very high set goal in view of the quantum effects involved in PT and the strong coupling between solute and solvent[5][6]. A more modest goal is the simulation of the solution structure and thermodynamics of association of a carboxylic acid-amine complex with and without PT, i.e. of the neutral and ionic forms of a model system.

In this paper we report the results of molecular dynamics (MD) simulations of the aqueous solution of the acetic acid-methylamine (AMA) complex using thermodynamic integration (TI) done in order to examine the interionic potential of mean force (pmf). In one set of the simulation runs we treated AMA in its ionic form to obtain information on association, i.e. which of the two possible types, the contact ion pair (CIP) or the solvent separated ion pair (SSIP), are more stable. In the other simulation we considered the free energy of the neutral versus the ionic form of the complex.

From the trajectories created in the MD runs snapshots were extracted at critical points and used for ab-initio calculations of two types. In one type seven water molecules were placed around the complex according to the snapshot and the bulk effect of water was simulated by the self consistent reaction field of Tomasi and coworkers [7]. In the other type of calculation the solute was surrounded by solvent molecules represented by point charges corresponding to SPC water [8].

The pmf function for the ionic complex exhibits two minima the deeper of which corresponds to CIP whereas the shallower one to SSIP. However, in terms of energy the SSIP turns out to be energetically prefered, what is in agreement with the ab-initio calculated energy. Concerning the PT both the MD and ab-initio calculations prefer the ionic form which become stabilized by the energy of the dipole-water interaction. One of the aims of these calculations was the testing of the GROMOS 37C4 force field [9] together with SPC water model. We found a good agreement with ab-initio results.

2. METHODS OF COMPUTATION

Molecular Dynamics simulations were performed by GROMOS-87 [9] package of programs implemented on a CONVEX C-220 computer. Ab-initio calculations were performed using Gaussian-86 suite of programs implemented on a MICROVAX-III computer.

2.1. MOLECULAR DYNAMICS

MD generates the ensemble of configurations on the basis of proposed molecular potentials by solving the equations of motions. Calculations of free energy and entropy are difficult, since these two quantities depend on the extent of configurational space available. On the other hand, the quantities like chemical equilibrium and rate constants depend directly on the relative free energies. Therefore, special methods have been developed to calculate the free energy differences. A review of methods for calculating free energy is given in References [10][11]. In this work we used themodynamic integration to evaluate the free enery differences. The structure and atom numbering of AMA is shown in Figure 1. Pmf is the free energy evaluated along the postulated reaction coordinate. In the study of association of AMA in aqueous solution the reaction coordinate was defined as the distance between the atoms C1 and N5. In the present work the pmf was evaluated by means of thermodynamic integration such that the atoms C1 and N5 were artificially restrained with a harmonic force of $2000 kcal/mol^2$. The equilibrium bond length of this virtual bond was gradually changed from its initial value corresponding to the state A toward its final value corresponding to the state B. In the GROMOS program PROMD [9] the bond length potential energy is written as following function of the coupling parameter λ if the force constants are equal for both states:

$$V(r_C, \lambda) = \frac{1}{2} K \left[r_C - (1 - \lambda) b_A + \lambda b_B \right]^2,$$

where r_C denotes the reaction coordinate and b_A and b_B are the equilibrium bond lengths for the states A and B respectively.

The ion pair AMA with a reaction coordinate of $7.5 Å$ was placed at the center of the rectangular simulation box with the requirement that the minimal solute to box wall distance is less than $11 Å$. The simulation box of AMA and 516 water molecules resulted. The spherical cutoff radius of $10 Å$ was used. No smooth continuation of the nonbonding interactions to zero was applied. MD was carried out with a coupling to a an external bath of 298 K and pressure of 1 bar [12]. The temperature coupling constant was of 0.05 ps in the equlibration and TI phase and 0.5 ps in the sampling phase. We applied the SHAKE procedure [13] for all the bonds involving hydrogen atoms and the leap-frog algorithm to integrate the equations of motion. Equilibration of 30 ps was preceded by 500 steps of energy minimization. Each subsequent TI was carried out for 10 ps in the course of MD in which the reaction coordinate was changed by $0.125 Å$. Each subsequent TI started from the coordinates of the previous one and additionally equilibrated for 1 ps. MD sampling runs of 60 ps at pmf minimum corresponding to the contact ion pair (CIP) and pmf minimum corresponding to the solvent separated ion pair (SSIP) were preceded by 30 ps equilibration phase.

The following interaction potential was used in MD simulations :

$$V_{tot} = V_{ww} + V_{sw} + V_s$$

The first term represents the water-water interaction energy in terms of the SPC (Simple Point Charge) model proposed by Berendsen et al. [8][14]. The second term corresponding to the solute-water interaction was modelled in terms of the 12-6-1 potential. The last term represents the ion-ion interaction. Nonbonding paramaters were taken from the standard GROMOS 37C4 force field [9]. The nonbonding parameters are collected in Table 1. One of the aims of the present study was to test the GROMOS force field that is widely used in the simulations of large systems. Although the authors of the GROMOS force field suggest for

Table 1: Parameters for the nonbonding interactions, used in the study of association of the ions methylammonium and acetate.
(a) in atomic units
(b) in $(\text{Å}^{12}\text{kcal/mol})^{1/2}$
(c) in $(\text{Å}^{6}\text{kcal/mol})^{1/2}$
(d) values corresponding to SPC water model

ATOM	CHARGE (a)	$(C12)^{1/2}$ (b)	$(C6)^{1/2}$ (c)
C1	0.270	898.0	23.65
O2	-0.635	900.0	23.25
O3	-0.635	900.0	23.25
C4	0.000	2500.0	46.06
N5	0.129	1500.0	24.13
C6	0.127	2500.0	46.06
H7	0.248	0.0	0.0
H8	0.248	0.0	0.0
H9	0.248	0.0	0.0
OW(d)	-0.82	793.3	25.01
HW(d)	0.41	0.0	0.0

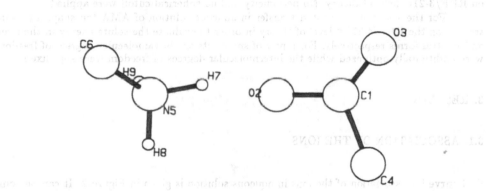

Figure 1: Structure and numbering of methylammonum acetate. Methyl groups C4 and C6 were treated as united atoms.

Lennard-Jones repulsion term between protonated amino nitrogen atoms and deprotonated carboxylate oxygen atoms a value of $1500(kcal/mol)^{1/2}Å^6$ instead of $900(kcal/mol)^{1/2}Å^6$ for oxygen atom Lennard-Jones parameters only the smaller value was used in the simulations. We estimated the influence of variation of this nonbonding parameter on the pmf.

In the study of proton transfer we deal with a chemical reaction in a strict sense. Since the molecular mechanics force fields are in general incapable of treatment of the chemical reactions, a considerable amount of informations from ab-initio calculations is necessary. We adopted the strategy that the difference in free energy of hydration of AMA is calculated by TI, while the intramolecular energy difference is calculated ab-initio.

In the study of proton transfer in aqueous solution of AMA the simulation box was a truncated octahedron and contained 253 SPC water molecules beside AMA. A spherical cutoff of $8Å$ was applied. The atom C1 was restrained to its initial position with a harmonic force of $10kcal/mol/Å^2$. MD was run at the pressure of 1 bar and temperature of 298 K. The equations of motion were integrated by a time step of 1 fs and the SHAKE procedure was applied for bonds involving hydrogen atoms. The system in ionic form was equilibrated for 120 ps.

2.2. AB-INITIO CALCULATIONS

From each long MD run corresponding to the CIP and SSIP respectively a snapshot was taken. By molecular building program INSIGHT[15] implemented on Silicon Graphics IRIS 4D50GT both ions and water molecules bridging the AMA and some forming the hydrogen bonds with the solute were selected. The cluster in the case of SSIP consisted of AMA and seven water molecules. Although in the cluster correspondig to the CIP all the water molecules did not fullfill the criterion, seven molecules were selected for the sake of energy comparison. Hydrogen atoms were added to both methyl groups treated in MD as united atoms. Clusters were treated on RHF level using the following basis sets: MINI-1, 4-31G and 3-21G including the solvent reaction field calculations [7]. In addition, the calculations were made where the solvent was included in calculations in the form of point charges corresponding to the SPC water values. The system consisted beside both solutes of 1548 point charges and was treated on RHF/3-21G level of theory. No periodicity and no spherical cutoff were applied.

For the study of the proton transfer in aqueous solution of AMA ten snapshots were treated on the RHF/4-31G* level of theory in order to evaluate the solute energy in the ionic and neutral forms respectively. For a pair of snapshots the intramolecular degrees of freedom were additionally optimized while the intermolecular degrees of freedom were kept fixed.

3. RESULTS

3.1. ASSOCIATION OF THE IONS

Pmf curve for association of the ions in aqueous solution is given in Figure 2. It can be compared to the energy profile in the gas phase(not shown) which is characterized by a deep contact minimum of -93 kcal/mol appearing at smaller value of the reaction coordinate. Although at the value of the reaction coordinate of $7.5Å$ the pmf curve is still slightly raising, its value was arbitrarily set to zero. The pmf is characterized by a deep minimum corresponding to CIP and a shallow second minimum corresponding to SSIP. In the free energy calculations convergence was better for the range of the reaction coordinate corresponding to smaller interionic distances. This was observed by monitoring the free energy changes in subsequent free energy calculations, when the coupling parameter λ was changed from 0 to 1 and from 1 to 0. One can rationalize the result in that the ion-water interaction energy as a function of interionic distance obeys two regimes. In the region of small interionic distances one can consider the ions as a large dipole and the leading term in solute-water interaction energy is the interaction between dipoles which falls off as r^{-3}. This can be contrasted with larger interionic distances where, with some exaggeration, one can claim that the leading terms are two ion-dipole interactions which fall off as r^{-2}. Since the number of water molecules is proportional to r^2, it is

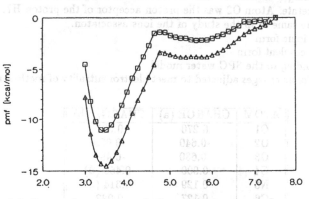

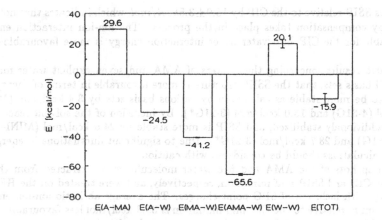

Figure 2: Potential of mean force for association of the ions methylammonium and acetate in aqueous solution.

Figure 3: Energy analysis from MD carried out at contanct and solvent separated ion pair. All energies are in kcal/mol relative to contact ion pair.

obvious that in the latter case one is much closer to the singularity of the interaction energy. In the studies of single ions only the water-water interaction energy and thermal disorder prevent the singularity in the ion-water interaction energy [16].

Authors of the GROMOS force field 37C4 suggest a value of $1500(kcal/mol)^{1/2}\text{Å}^6$ as an alternative to the Lennard-Jones repulsion term between protonated amino nitrogen atoms and deprotonated carboxylate oxygen atoms instead of $900(kcal/mol)^{1/2}\text{Å}^6$. Only the smaller value was used in this work. Obviously the former value introduces some additional repulsion between the ions that is of short range and one can expect that it will not influence the pmf curve beyond the barrier separating both minima. We assumed that in both cases the relative orientations of the ions is the same. In that case and with similar arguments as introduced in the umbrella sampling technique [17] we assumed that the CIP minimum occurs at about 0.2Å higher value of reaction coordinate and is about 3 kcal/mol less favourable. The result was obtained by adding the difference between both repulsive parts of the Lennard-Jones curves to the calculated pmf.

From the time averages of MD carried out at CIP and SSIP, respectively, averages of the energy components were calculated and are depicted in Figure 3. The latter analysis has shown that the average temperature has risen up to 10 K above the reference temperature. Hence the results should be accepted with some caution. In contrast to the free energy the

Table 2: Atomic charges used in the MD simulations of the proton transfer between methylammonium and acetate. Atom O2 was the proton acceptor of the proton H7. Lennard-Jones parameters were the same as in the study of the ions associaton.
(a) charges for the ionic form
(b) charges for the covalent form
(c) values corresponding to the SPC water model
(d) values of the atomic charges adjusted to meet electroneutrality of methylamine and acetic acid

ATOM	CHARGE (a)	CHARGE (b)
C1	0.270	0.321
O2	-0.640	-0.620
O3	-0.630	-0.498
C4	0.000	-0.458 (d)
N5	0.129	-0.314 (d)
C6	0.127	0.022
H7	0.294	0.339
H8	0.225	0.146
H9	0.225	0.146
OW(c)	-0.82	-0.82
HW(c)	0.41	0.41

energy favours SSIP relative to the CIP by $15.9 \pm 3.5 kcal/mol$, which indicates that noticeable energy-entropy compensation takes place in the process. The ion-ion interaction energy is more favourable for the CIP but water-water interaction energy is more favourable for the SSIP.

Ab-initio results concerning the clusters of AMA and seven explicit water molecules show, with all basis sets, that the SSIP structure is more favourable in terms of energy. SSIP is predicted to be more stable as calculated by various basis sets by 3.4 kcal/mol (MINI-1), 12.5 kcal/mol (4-31G) and 13.0 kcal/mol (3-21G*). By inclusion of the solvent reaction field the SSIP is additionaly stabilized, and SSIP is more stable by 24.6 kcal/mol (MINI-1), 28.2 kcal/mol (4-31G) and 28.7 kcal/mol (3-21G*). Due to significant fluctuations in energies the single point calculations should be considered with caution.

Ten snapshots of the AMA and 516 water molecules were extracted from the MDs carried out at CIP and SSIP pmf minimum, respectively, and were treated on the HF/3-21G level of theory with inclusion of SPC point charges. The average ab-initio ion-ion energy at the SSIP minimum relative to the CIP pmf minimum is 33.9 kcal/mol less favourable, in good agreement to MD average of 29.6 kcal/mol. The average ab-initio interaction energies AMA-water point charges of -131 kcal/mol and -189 kcal/mol for CIP and SSIP are also in good agreement with MD averages of -131.5 kcal/mol and -197.1 kcal/mol. The good agreement between ab-initio and MD results advocates the combination of GROMOS 37C4 force field with the SPC water model.

3.2 PROTON TRANSFER IN METHYLAMMONIUM ACETATE

Since the results of free energies of hydration are very much dependent on the nonbonding parameters, specially on charges, we treated them with special care. The choice of the nonbonding parameters is obviously less problematic for the ionic form the parameters of which were taken from the standard GROMOS 37C4 force field. In contrast to the study of association of ions we arbitrarily assigned the atoms O2-H7-N5 participating in hydrogen bonding. 37C4 atomic charges were scaled according to Mulliken charges from the HF/4-31G* calculations on the methylammonium formate with included solvent reaction field [2][3]. Additional constraints were the electroneutrality of both species in the covalent form and net charges of +1 and -1 for the ionic form of methylammonium and acetate, respectively. Lennard-Jones parameters were taken from GROMOS 37C4 force field[9] and were used unchanged in the simulations of the association and proton transfer in AMA. The nonbonding parameters are collected in Table 2. Bonding parameters as well as bonding parameters that we introduced instead of nonbonding and were applied in the TI procedure are collected in Table 3. The critical quantitiy is the interionic distance. For the ionic form we adopted the CIP pmf minimum,

Table 3: Values of the bonding parameters used in the thermodynamic integration when ionic
form was changed into covalent.
All distances are in $\mathring{A}$ and all angles in degrees.
Values of the force constants for the bonds are in kcal/mol/$\mathring{A}^2$ and for the bond angles and
improper dihedrals in kcal/mol/rad^2.

Bond	$b_0(ion)$	$k_f(ion)$	$b_0(cov)$	$k_f(cov)$
1-2	1.247	1000.	1.322	1000.
1-3	1.229	1000.	1.210	1000.
1-4	1.500	800.	1.500	800.
5-6	1.469	900.	1.471	900.
2-5	2.840	900.	3.129	900.
1-5	3.640	900.	3.928	900.
2-7	0.992	0.	0.992	1000.
5-7	1.007	1000.	1.007	0.
5-8	1.000	1000.	1.000	1000.
5-9	1.000	1000.	1.000	1000.
Angle	$\Theta(ion)$	$k_\Theta(ion)$	$\Theta(cov)$	$k_\Theta(cov)$
2-1-3	126.5	100.	126.5	100.
2-1-4	117.2	95.	117.2	95.
6-5-7	109.47	90.	109.47	0.
1-2-7	105.	0.	105.	90.
6-5-8	109.47	90.	109.47	90.
6-5-9	109.47	90.	109.47	90.
8-5-9	109.47	90.	109.47	90.
Impr. dihedral	$\xi(ion)$	$k_\xi(ion)$	$\xi(cov)$	$k_\xi(cov)$
4-1-2-3	180.	40.	180.	40.
1-2-7-5	-140.	20.	-140.	10.
3-1-2-7	-112.	30.	-112.	10.

Table 4: Relative free energies of hydration of methylamine and acetate when ionic form ($\lambda = 0$) is transformed to covalent ($\lambda = 1$).
The errors are estimated as absolute values of the hysteresis.
(a) Thermodynamic integration carried out in 50 ps.
(b) Thermodynamic integration carried out in 100 ps in such a way that λ is changed from 0 to 0.1 in 50 ps and then from 0.1 to 1 in next 50 ps.
(c) Thermodynamic integration carried out in 200 ps.

$\lambda_{initial}$	λ_{final}	t/ps	ΔA	ΔA
0.	1.	50.	54.13	
1.	0.	50.	38.20	46.2 ± 15.9 (a)
0.	0.1	50.	36.51	
0.1	0.	50.	36.86	
0.1	1.	50.	8.14	
1.	0.1	50.	5.83	43.7 ± 2.7(b)
0.	1.0	200.	39.75	
1.	0.	200.	37.82	38.8 ± 1.9(c)

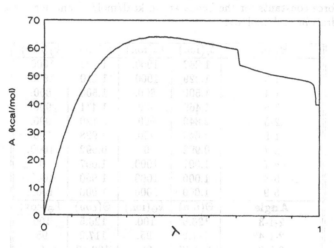

Figure 4: Free energy of hydration as a function of coupling parameter λ when ionic form of the methylammonium acetate ($\lambda = 0$) was changed into covalent form ($\lambda = 1$) in 200 ps of MD.

and for the covalent form we added the elongation of the C1-N5 distance from the 4-31G* calculation with methylammonum formate transformed from the ionic form to the covalent one. We would like to emphasize that this is a weak point of the calculation, since in order to know the exact value we should calculate the contact pmf minimum for the covalent form.

In the long MD simulations for ionic and covalent forms the artificial bonding interactions were replaced by nonbondig, except for the distance C1-N5 upon which energy components are dependent. Results of the free energies of hydration are collected in Table 4. Results are corrected by the Onsager correction for dipole-continuum interaction beyond the cutoff distance. Free energy as a function of a coupling parameter is shown in Figure 4. The best uncorrected TI result indicates that the free energy of hydration favours the ionic form by 38.8 ± 1.9 kcal/mol if the energetics of the proton transfer is excluded. Using correction formula for the contribution of the solvent beyond the cutoff distance of 8Å we obtained the corrections of 3.6 kcal/mol and 0.2 kcal/mol for ionic and neutral form of AMA respectively. The best corrected value for the free energy difference is 42.2 ± 1.9 kcal/mol.

For ten snapshots extracted from the trajectory for ionic and covalent forms ab-initio HF/4-31G* calculations were carried out without geometry optimizations. The average energy for ionic and covalent forms were -322.591 hartree and -322.674 hartree, respectively, with standard deviations of 5.8 kcal/mol and 3.9 kcal/mol, respectively. This corresponds to the by 52.5 kcal/mol more stable covalent form. By adding the free energies of hydration one can conclude that the covalent form is more stable by about 10 kcal/mol. By inclusion of the solvent reaction field with $\epsilon = 80$ in order to take into account the free energy of solvation the same basis set predicted the ionic form to be on the average by 3.7 kcal/mol more stable.

For one ionic and one covalent form the geometry optimization was performed by retaining fixed all the intermolecular degrees of freedom. The energy difference dropped and the covalent form was found to be by 35.5 kcal/mol more stable.

By combination of the free energy of hydration and the ab-initio energy difference for partially geometry optimized AMA one can conclude that the ionic form is more stable by 6.7 ± 1.9 kcal/mol. The error bar in free energy difference is probably larger, since the error bar originating from the ab-initio energy difference was not taken into account. The result is in agreement with pK_a results indicating the ionic form to be of 7.9 kcal/mol more stable [18]. The numbers can not be compared directly, since the experimental result includes free energy of association of ionic species and dissociation of covalent species from contact to infinity.

From MD carried out for the ionic and covalent forms the average energy components and their distributions were calculated. For ionic and covalent forms the average water-water interaction energies were found to be -3167.8 and -3203.6 kcal/mol, respectively. On the other hand the more favourable ionic AMA-water interaction energy of -148.5 kcal/mol can be contrasted with -33.2 kcal/mol for the covalent form. By adding the ab-initio energy difference of 35.5 kcal/mol one can conclude that in terms of energy the ionic form is favoured by 44.0 kcal/mol.

4. DISCUSSION

The result which indicates that the ionic associated form of AMA is in terms of free energy the most favourable looks at first glance quite reasonable. It supports the traditional scheme of drug-receptor recognition viewed as lock and key model where the molecular electrostatic potentials play an important role. In addition, the more stable folded form of the proteins is thus supported, although other interactions also are important in the process of protein folding. Interaction potentials used in the calculations seem to be quite reasonable, since they are in close agreement with ab-initio results. Boudon et al. [19] studied the pmf for association of the ions guanidinium acetate in aqueous solution. They found CIP and SSIP minimum to be of about equal depth and separated by a barrier of about 10 kcal/mol. Our pmf can be compared with the results of Ciccotti et al. [20] simulating the association of the opposite charged ions in model solvent with dipole solvent of 2.4 D. They found the CIP minimum to be of about $30kT$ more stable relative to SSIP which was predicted as a shallow minimum with a depth of about $2kT$.

We found the CIP minimum to be roughly 9.8 ± 5.0 kcal/mol more stable relative to the SSIP. Relative large error bar in the free energy difference between CIP and SSIP is indicative of the convergence was not being achieved and longer simulations would be benefitial. Nevertheless the predicted stability of the CIP relative to the SSIP as well as relative to the separated ions is probably exaggerated. Warshel [21] concluded from the ΔpK_a values of amino acids that the free energy changes are less than -3 kcal/mol if the charged amino and carboxylic groups are drawn in aqueous solution from 6Å to 3Å.

A possible reason for the exaggerated stability of CIP is hidden in the application of the rigid water model in the simulation by the SHAKE procedure. It was shown by ab-initio studies [22] that the flexible water model additionaly favours SSIP due to the nuclear polarizability.

Huston et al.[23] studied the pmf for association of the sodium and phosphate ions in aqueous soluton. They reported that the results depend very much on the treatment of longe range interactions. In contrast to spherical cutoff treatment the minimum image treatment gave the pmf that was repulsive practically in the whole range of interionic distances. The same phenomenon has been advanced by Friedman and Mezei [24] in the study of pmf for the association of the ion pair NaCl in aqueous solution.

A noticeable energy-entropy compensation is present in the association of the AMA ions as well as in the proton transfer process. A considerable reorganization of water molecules takes place in both cases, as demonstrated in the changes of water-water interaction energy. Yadav et al.[25] stressed the important differrence between the environment relaxation in enzymatic reactions relative to reactions in aqueous solution.

We think that in the near future it will be necessary to study systematicaly the effect of various treatments of long range interactions including the Ewald summation on the pmf of ions association to find an explanation for the differences.

The results of free energy calculations for the proton transfer are in an agreement with the expectation that the ionic form is more favourable. The result is quite close to the predicted difference from the pK_a values [18]. Nevertheless the reoptimization of the intermolecular degrees of freedom was necessary to obtain the agreement which may be taken as evidence that the MD force field was not perfectly tunned to ab-initio calculations. It is evident, that significant energy-entropy compensation was predicted by the simualations. Experimentaly the free anergy changes associated with processes such as carboxylic acid dissociations in aqueous solutions originate mostly from the entropic contribution [18].

Significant solvent reorganization was present in both processes and manifested itself through the changed the water-water interaction and radial distribution functions (not shown here).

Previously ab-initio calculated energy difference for ionic and covalent forms of the methylammonium formate gives evidence[2][1] that the results are very much dependent on the basis set and the presently used basis sets are still far away from the Hartree-Fock limit [2]. Mezei arrived to the same conclusion by the study of tautomerism in aqueous solution of N-methyl acetamide [26]. Therefore the modeling of the chemical reactions involving molecules of moderate size in the gas phase is not easy. Theoretical treatment of the same reaction in solutions becomes considerably more difficult.

With the increasing power of supercomputers and steady improvement in the quality of the force fields[27] one expects the modelling of the biological systems to become more accurate and reliable. On the other hand the methods to study kinetics of the proton transfer in condensed matter are developed [5][6] and ready to be applied.

ACKNOWLEDGEMENTS

We thank Drs. Mezei and Hynes for the preprints. This work was financed by the Ministry of Science and Technology of the Republic of Slovenia.

REFERENCES

[1] A. M. Sapse, C. S. Russel, J. Am. Chem. Soc., **107**, 174 (1985)

[2] M. Hodošček and D. Hadži, J. Mol. Struct., **198**, 461 (1989)

[3] M. Hodošček, PhD. Thesis, University of Ljubljana, Ljubljana 1989.

[4] G. S. Denisov, N. S. Golubev, J. Mol. Struct. **75**, 311 (1981)

[5] D. Borgis, J. T. Hynes, J. Chem. Phys., **94**, 3619 (1991)

[6] J. Timoneda, J. T. Hynes, Submitted to J. Phys. Chem.

[7] S. Miertuš, E. Scrocco, J. Tomasi, Chem. Phys., **55**, 117 (1981)

[8] H. J. C. Berendsen, J. P. M. Postma, W. F. van Gunsteren, J. Hermans, in Intermolecular Forces, B. Pullman, Ed., Reidel Publ., 1981, p. 331

[9] W. F. van Gunsteren, GROMOS, Groningen Molecular Simulation Program Package, Biomos B.V., Nijenborgh 16, 9747 AG Groningen, The Netherlands, 1987

[10] W. F. van Gunsteren, v Computation of Free Energy for Biomolecular Systems, W. F. van Gunsteren and R. K. Weiner, Eds., Escom Science Publishers, Leiden, The Netherlands, 1989, pp. 27

[11] M. Mezei, D. L. Beveridge, Free Energy Simulations, Ann. Acad. Sci. New York, **482**, 1 (1986)

[12] H. J. C. Berendsen, J. P. M. Postma, W. F. van Gunsteren, A. DiNola, J. R. Haak, **81**, 364 (1984)

[13] J. P. Ryckaert, G. Ciccotti, H. J. C. Berendsen, J. Comp. Phys. **23**, 327 (1977)

[14] W.F. van Gunsteren, H.J.C. Berendsen, J. Mol. Biol., **176**, 559 (1984)

[15] BIOSYM Technologies Inc., 9605 Scranton Rd., Suite 101, San Diego, CA, U.S.A.

[16] S. Bratos, Private Communication

[17] G. N. Patey, J. P. Valleau, J. Chem. Phys., **63**, 2334 (1975)

[18] N. S. Isaacs, Physical Organic Chemistry, Longman Scientific and Technical, Belfast, (1987)

[19] Boudon, G. Wipf, in Modelling of Molecular Structures and Properties, J. L. Rivail, Ed., Studies in Physical and Theoretical Chemistry Vol. 71, Elsevier Science Publishers, The Netherlands, 1990, pp. 203

[20] G. Ciccotti, M. Ferrario, J. T. Hynes, R. Kapral, J. Chem. Phys., **93**, 7137 (1990)

[21] A. Warshel, J. Phys. Chem., **83**, 1640 (1979)

[22] J. Mavri, To be published

[23] S. E. Huston, P. J. Rossky, D. A. Zichi, J. Am. Chem. Soc., **111**, 5680, 1989

[24] R. Friedman, M. Mezei, Private Communication

[25] A. Yadav, R. M. Jackson, J. J. Holbrok, A. Warshel, J. Am. Chem. Soc. **113**, 4800 (1991)

[26] M. Mezei, This volume

[27] M. J. Field, P. A. Basch, M. Karplus, J. Comp. Chem, **11**, 700 (1990)

CRYSTAL OF ADENINE AND DISSOCIATED AND NONDISSOCIATED CHLORO-ACETIC ACID - A PROMISING MODEL SYSTEM FOR THE STUDY OF PROTON TRANSFER PHENOMENA

Jan Florián[1], Vladimír Baumruk[1], Jana Zachová[1] and Jaroslav Maixner[2]

1/ Institute of Physics, Charles University
   Ke Karlovu 5, 121 16 Prague 2, Czechoslovakia
2/ Inst.Chem.Technology, Prague 6, Czechoslovakia

ABSTRACT

The structure of the single crystal of adeninium chloracetate chloroacetic acid solvate (adenine chloracetate) was determined by the X-ray diffraction technique [1]. The single crystal is stabilized by the net of N-H...O and N...H-O hydrogen bonds between adenine N1,N7,N9 atoms and NH2 group, and oxygen atoms of the molecules of chloracetate and chloroacetic acid. Adenine has been found to be protonated at N1 position. Tautomerism involving N1-H...O and N1...H-O structures is proposed to explain observed disorder of chlorine and oxygen atoms that for chlorine corresponds to the 13 degree rotation around the C-C bond of chloracetate.
The Raman spectrum of the single crystal shows at least 80% of adenine molecules to be protonated at N1 position.

INTRODUCTION

Theoretical studies [2] has shown a number of mechanisms for coupling conformational changes to proton pumping activity which is closely connected with such important biological events as spontaneous mutation or proton transfer across biomembranes. We present here a model system for an experimental study of these phenomena and also for the study of proton transfer between hydrogen-bonded nucleic acids bases.

EXPERIMENTAL

Crystals were grown by cooling of adenine solution in 1 mol/l chloroacetic acid saturated at 308K with aid of a crystal seed (cooling rate 1K/day). Twice recrystalized or sublimed adenine (Lachema) and twice redestiled chloroacetic acid (Lachema) were used. Crystals are colourless prismas, often twinned,

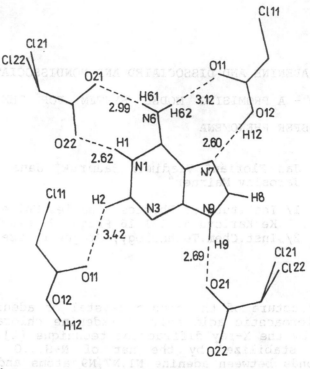

Figure 1: The net of hydrogen bonds observed in the adenine
chloracetate single crystal. The O...N distances
between atoms forming hydrogen bonds are indicated
in the figure, remaining geometry parameters can be
found in Table 2.

Table 1: Geometry parameters (given in degrees and angstroms)
of hydrogen bonds present in the adenine chloracetate
crystal (Fig.1). Standard deviations are given in
paranthesess.

X-H...Y	H...Y	X...Y	X-H...Y
N1-H1...O22[i]	1.85(6)	2.618(6)	175(2)
N9-H9...O21[iv]	1.92(6)	2.690(7)	169(2)
O12-H12...N7[v]	1.87(6)	2.598(7)	162(2)
N6-H61...O21[i]	2.28(6)	2.989(7)	164(2)
N6-H62...O11[ii]	2.42(6)	3.118(7)	165(2)

with maximal dimensions 7x1x1.5 mm.
Crystal structure: Orthorombic, space group Pccn, eight formula units of N1-protonated adenine, chloroacetic acid, and chloracetate in the unit cell [1].

THE STRUCTURE OF ADENINE CHLORACETATE SINGLE CRYSTAL (FIG.1) PROVIDES:

- A direct experimental evidence for adenine to be protonated more preferably at N1 position than at N7 position, the both nitrogens having the same chance, given by the geometry of the hydrogen bonded complex, to receive a proton from the molecule of chloroacetic acid.

- An experimental evidence for the energetic preference of the N9-H tautomeric form of N1-protonated adenine over its N7-H tautomeric form.

- A model system for a study of proton dynamic in hydrogen bonds, with high probability for proton tunneling phenomena. A proton tunneling in the N1-H...O22 and N9-H...O21 hydrogen bonds can easily induce proton tunneling in the N7...H-O12 hydrogen bond (Fig.1).

- A net of hydrogen bonds among molecules of monochloracetate, monochloracetic acid and protonated adenine that mimic the pattern of hydrogen bonds in a triple helix DNA where both the Watson-Crick and Hoogsteen types of hydrogen bonded base pairs are formed [3].

- A possibility for proton transfer in hydrogen bonds to be coupled to a conformational change in the molecule of chloracetate. Chlorine atom has been found to be disordered in the chloracetate molecules. Its two positions are occupied with 83% and 17% probability and they correspond to the 13 degrees rotation around the C21-C22 bond.

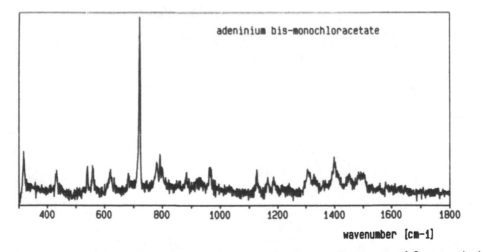

Figure 2: The Raman spectrum of the adenine chloracetate single crystal

RAMAN SPECTRA

To quantitatively evaluate the expected coexistence of more differently protonated forms of adenine in the single crystal we have measured its Raman spectrum (Fig.2).It was recorded on a modular UV-VIS Raman spectrometer [4]. The lower optical quality of the single crystals unfortunately did not allow us to obtain good signal/noise ratio Raman spectra. We can nevertheless conclude that the main observed spectral features correspond to the others previously studied crystals of N1-protonated adenine [5] so that at least 80% of adenine molecules, present in the crystal, are in the N1-protonated form. A more detailed Raman study of this system is in progress.

REFERENCES

[1] J.Maixner, J.Zachová, to be published in Collection Czech.Chem.Commun.
[2] S.Scheiner, P.Redfern, E.A.Hildebrand, Int.J.Quantum Chem. 29, 817 (1986)
[3] e.g. D.S.Pilch, R.Brousseau, R.H.Shafer, Nucleic Acids Research 18, 5743 (1990)
[4] J.Stepanek, V.Baumruk, P.Praus, J.Bok, Computer Phys. Commun. 50, 225 (1988)
[5] J.Zachová, V.Baumruk, J.Maixner, submitted for publ. in J.Cryst.Growth - Proceedings of the IV-th International Conference on Crystal Growth of Biological Macromolecules; V.Baumruk, PhD-thesis, 1991

AB INITIO HF SCF CALCULATION OF THE EFFECT OF PROTONATION ON VIBRATIONAL SPECTRA OF ADENINE

Jan Florián

Institute of Physics, Charles University

Ke Karlovu 5, CS-12116 Prague 2, Czechoslovakia

INTRODUCTION

Vibrational spectroscopy represents a powerful tool for studying the proton transfer phenomena [1,2]. It is highly desirable to interpret the observed spectral changes, which might be caused by protonation, on the basis of theoretical calculations. Such calculations must include vibrational spectra of series differently protonated, and often relatively large biological molecules, and take the effect of environment, mostly aqueous solution, into account. As a best suited for this purpose seems to be a combination of a low-cost ab initio HF SCF method, able to handle molecules up to 25 atoms, and normal coordinate calculation where the SCF force field is fitted by means of a few scaling factors to the observed frequencies. The resulting scaling factors implicitly involve the effects of electron correlation, weak intermolecular interactions, incompleteness of basis set, and anharmonicity. The transferability of scale factors among related molecules has been shown to be a good approximation [3,4].

We have used the above mentioned method, developed by Blom and Altona [5], and Pulay et al [6], to determine the scaled quantum mechanical (SQM) force field of neutral, N1-protonated, N3-protonated, and N1,N7-diprotonated forms of adenine as a counterpart of thorough polarized IR and Raman study on single crystals and aqueous solutions of nucleic acids (NA) bases carried out at Institute of Physics,Charles University, Prague [7]. Calculations of vibrational frequency shifts caused by protonation in the NA bases cytosine and guanine are in progress.

Protonation of NA bases and particularly adenine takes part in a number of biologically important processes, among them: spontaneous point mutations caused by the proton tunneling in hydrogen bonds between complementary NA bases in DNA (Lowdin's hypothesis [8]), base mismatches [9], self-association of nucleotides [10], formation of triple-stranded NA helices [11], and binding interaction between coenzymes NADH or NAD^+ and protein part of dehydrogenase enzymes [2].

Proton Transfer in Hydrogen-Bonded Systems
Edited by T. Bountis, Plenum Press, New York, 1992

COMPUTATIONAL METHOD

Ab initio HF SCF force field has been calculated in the minimal STO-3G basis set. The scaling procedure involved the simultaneous fitting of 12 scaling constants to the in-plane experimental vibrational frequencies of polycrystalline samples of adenine, N-deuterated adenine, and C8,N-deuterated adenine, in the internal coordinates given in Table 1. The resulting scale constants (Tab.1) were used in calculation of SQM STO-3G in plane frequencies of differently protonated molecules of adenine. The more detailed description of the computational method is given in the previous paper [3].

Table 1: In-plane internal coordinates of adenine and scale factors of the corresponding force constants

Description	Symbol	Definition	Scale const.
C-N stretch.	N1C2,N3C4,C6N10 C5N7,C8N9,N9C4	N1-C2, N3-C4, C6-N10 C5-N7, C8-N9, N9-C4	0.82
C-C stretch.	C5C6	C5-C6	0.72
C=N stretch.	C2N3,C6N1,N7C8	C2-N3, C6-N1, N7-C8	0.77
C=C stretch.	C4C5	C4-C5	0.69
6-memb.ring in-plane deformations	D6RTr	(C6-N1-C2)+(C2-N3-C4)+(C4-C5-C6)- (N1-C2-N3)-(N3-C4-C5)-(C5-C6-N1)	0.96
	D6RE1	2[(C2-N3-C4)+(C5-C6-N1)]-[(C6-N1-C2) +(N1-C2-N3)+(N3-C4-C5)+(C4-C5-C6)]	0.96
	D6RE2	(C6-N1-C2)+(N3-C4-C5)- (N1-C2-N3)-(C4-C5-C6)	0.96
C6N10 bend	DC6N10	(C5-C6-N10)-(N1-C6-N10)	0.96
5-memb.ring in-plane deformations	D5R=	0.63245(N7-C8-N9)+ (-0.51167)[(C5-N7-C8)+(C8-N9-C4)]+ 0.19544 [(C4-C5-N7)+(N9-C4-C5)]	0.88
	D5R|	0.37175 [(C8-N9-C4)-(C5-N7-C8)]+ 0.60150 [(C4-C5-N7)-(N9-C4-C5)]	0.88
NH2 sciss	DNH2S	- [(C6-N10-H11)+(C6-N10-H12)]+ 2(H11-N10-H12)	0.85
NH2 rocking	DNH2R	(C6-N10-H12)-(C6-N10-H11)	0.92
C2-H13 bend	DC2H	(N1-C2-H13)-(N3-C2-H13)	0.80
C8-H14 bend	DC8H	(N7-C8-H14)-(N9-C8-H14)	0.80
N9-H15 bend	DN9H	(C8-N9-H15)-(C4-N9-H15)	0.92

Interaction force constants:	stretch-stretch	0.58
	stretch-bend	1.26
	bend-bend	0.70

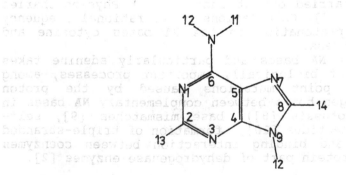

Figure 1: Atom numbering in the molecule of adenine

RESULTS AND DISCUSSION

The calculated interpretations of in-plane fundamental modes of differently protonated adenine are given in Tables 2-5. We do not present results for N-H and C-H stretching vibrations because these are strongly influenced by the intermolecular interactions so that the harmonic approximation and transferability of scale factors can not be assumed to be valid for them. Also interpretation of out-of-plane vibrational modes has been skipped in this study due to the lack of reliable set of experimental out-of-plane fundamentals for fitting scale factors, the unscaled STO-3G force field

Table 2: Calculated frequencies [cm^{-1}] and the potential energy distribution (PED) for in-plane vibrational modes of adenine

Exp.Freq.[13]		Our SQM-STO3G results		4-21G results [12]	
IR	Raman	Calc	Main % contrib. to the PED	Calc	Main % contrib. to the PED
-	327	278	DC6N1(53),D6RE2(16),D5R|(8)	265	DC6N10(51),D6RE2(15),D5R|(10)
-	530	530	DC6N1(35),D6RE2(20),D5R|(8)	506	DC6N10(27),D6RE2(23),C4N9(12)
543	535	563	D6RE1(76)	529	D6RE1(71)
622	623	634	C5C6(26),D6RE2(25),D5R|(17)	610	C5C6(26),D6RE2(23),D5R|(26)
723	723	716	N3C4(22),D5R|(22),D6RE2(17)	692	N3C4(25),C5N7(14),D5R|(12)
913	899	898	D6RTr(42),D6RE2(13)	877	D6RTr(41),D6RE2(19),C5N7(10)
951	952	954	D5R=(74),C5N7(10)	932	D5R=(67)
1025	1024	1024	C6N1(37),DNH2R(22),C6N10(9)	986	C6N1(35),DNH2R(32)
1126	1126	1120	DC8H(22),C8N9(19),D5R|(18)	1029	C8N9(72),DN9H(18)
1157	1163	1152	C8N9(42),DN9H(18),N9C4(8)	1101	C5N7(15),C4N9(11),N3C4(11)
1234	1235	1231	N1C2 (31),C6N1(16),C2N3(8)	1190	N1C2(27),C2N3(13)
1253	1249	1255	DNH2R(42),C5N7(19)	1211	DNH2R(34),C5N7(22)
1309	1307	1304	DC8H(40),C2N3(22),C5C6(8)	1267	C2N3(38),DC8H(25)
1335	1332	1334	C5N7(28),C2N3(15),C4C5(15)	1299	N1C2(21),DC8H(21),C5N7(11)
1368	1371	1390	DC2H(53),N1C2(11),DN9H(8)	1330	DC2H(17),DN9H(13),C6N10(13)
1421	1419	1432	DC2H(29),DN9H(16),C8N9(9)	1381	DN9H(23),C4N9(18),N7C8(14)
1451	1462	1467	N9C4(27),N7C8(15),C4C5(12)	1411	DC2H(27),C4C5(17),DN9H(13)
-	1482	1486	N7C8(27),D5R|(10),C5C6(8)	1477	DC2H(27),N1C6(20),C6N10(18)
1508	1514	1525	C6N10(16),DN9H(13),C8N9(12)	1490	N7C8(44),DC8H(19)
1604	1597	1589	C5C6(27),C6N10(27),D6RE1(8)	1580	C4C5(26),N3C4(11)
1638	-	1668	N9C4(19),N3C4(17),DN9H(14)	1594	C5C6(30),N3C4(12)
1673	1677	1694	DNH2S(86),C6N10(9)	1640	DNH2S(91)

being unable to provide acceptable results. This however does not represent a serious drawback in case of interpretation of vibrational spectra of aqueous solutions of NA bases, where the IR spectroscopy technique can can not be used and Raman intensities of the out-of-plane vibrational modes of NA bases are very low.
Protonation turned out to substantially change the force field of adenine, the most pronounced and unexpected being the increase of the C6N10 diagonal stretching force constant caused by protonation. The new fundamentals originating from the in-plane bending vibrations of "protons" were calculated to appear in the 1600-1720 cm^{-1} frequency region.

Table 3: SQM STO-3G frequencies [cm^{-1}] and the PED of in plane vibrational modes of the N1-protonated adenine

Exp.data		Calc	Main % contributions to the PED
HSA	HBr		
353	362	275	DC6N1(53),D6RE2(16),D5R| (8)
524	517	521	DC6N1(29),D6RE2(24),N9C4 (9),D5R| (8),N3C4 (6)
545	534	546	D6RE1(73),C4C5 (6),C5N7 (5)
638	612	637	C5C6 (31),D6RE2(24),D5R| (21)
722	716	690	N3C4 (30),D5R| (15),D6RE2(10),C5N7 (8),C4C5 (7),C6N1 (6)
902	895	918	D6RTr(45),D6RE2(14),C5N7 (9),D6RE1(8),C6N10(6),C2N3 (5)
950	947	955	D5R= (76),C5N7 (7)
1011	1016	1022	DNH2R(29),C6N1 (26),N1C2 (13)
1136	1136	1108	C8N9 (33),D5R| (22),DC8H (14),DN9H (11),N3C4 (6)
1175	1171	1141	C8N9 (30),N1C2 (19),DN9H (8),C5N7 (7),N9C4 (5),D5R= (5)
1197	1195	1203	N1C2 (30),DNH2R(15),C6N1 (10),DC8H (8),N7C8 (5)
1220	1241	1266	DC8H (25),DNH2R(22),DC2H (17),D5R| (7),C6N1 (5),D6RTr(5)
1304	1304	1292	DC8H (24),C4C5 (15),DC2H (12),C5C6 (7),D5R| (7),C5N7 (6)
1341	1334	1337	C5N7 (36),DC2H (16),C5C6 (8),N7C8 (8),D6RTr(7)
1387	1397	1389	DC2H (31),N7C8 (12),C4C5 (11),DN9H (11),N9C4 (8),C8N9 (8)
1416	1414	1412	N7C8 (15),C4C5 (10),C2N3 (8),C5N7 (8),D5R| (7),DN9H (7)
1430	1441	1469	N9C4 (22),N7C8 (20),C2N3 (14),N3C4 (11)
1479	1496	1499	C4C5 (18),DN1H (13),N7C8 (10),C8N9 (9),DN9H (7),C6N10(7)
1545	1546	1528	C6N10(21),DN9H (15),C6N1 (12),DN1H (11),D6RTr(8),C4C5 (8)
1582	1575	1610	C2N3 (48),DN1H (9),C5C6 (8),D6RE2(7),DC2H (6)
1624	1606	1649	N9C4 (24),N7C8 (15),DN9H (14),N3C4 (12),DN1H (6),DC8H (5)
1677	1671	1668	DNH2S(38),C6N10(25),DN1H (12),C6N1 (5)
1703	1700	1689	DNH2S(54),DN1H (21),C6N10(6),C5C6 (6),C6N1 (5)

HSA: adenine hemisulfate hydrate (HSA) [7]
HBA: adenine hydrobromide hemihydrate (HBA) [7]

Table 4: SQM STO-3G frequencies [cm^{-1}] and the PED of in plane vibrational modes of the N3-protonated adenine

Calc.	Main % contributions to the PED
278	DC6N1(51), D6RE2(17), D5R| (9), DNH2R(5)
517	DC6N1(32), D6RE2(22), N9C4 (7), D6RE1(7), N3C4 (7), D5R| (6)
536	D6RE1(63), D6RTr(7), C5N7 (6), C4C5 (5)
626	D6RE2(32), C5C6 (22), D5R| (20), N1C2 (6), C6N1 (6), C6N10(6)
713	N3C4 (19), D5R| (19), D6RE2(15), C4C5 (10), C5N7 (7), N9C4 (7)
926	D6RTr(40), D6RE2(12), C6N1 (11), D5R| (7), N1C2 (6), N9C4 (5)
952	D5R= (76), C5N7 (10)
995	C6N1 (41), DNH2R(18), C6N10(6), D6RTr(5)
1100	C8N9 (53), DN9H (15), D5R| (13), DC8H (5)
1136	C8N9 (18), C2N3 (17), D5R| (7), N9C4 (7), DC8H (7), C5N7 (6)
1207	C2N3 (33), DC8H (24), DNH2R(9), D6RTr(8), N9C4 (6), DN3H (5)
1255	DNH2R(38), C5N7 (16), DC8H (14), D5R| (9), C6N1 (5)
1286	C4C5 (25), DC8H (23), C5N7 (15), D5R| (6)
1322	DC2H (34), C5N7 (13), C2N3 (8), C5C6 (8), N7C8 (8), C4C5 (7)
1378	DC2H (35), N7C8 (14), N1C2 (9), C6N1 (7), DN9H (6), N9C4 (6)
1430	DN3H (24), DN9H (21), N9C4 (13), C2N3 (12), C8N9 (7)
1450	C6N1 (16), DN9H (14), C5C6 (10), DNH2R(8), N9C4 (7), C5N7 (6)
1510	N7C8 (40), N1C2 (16), N3C4 (9), D5R| (6), C2N3 (5)
1564	N1C2 (32), N7C8 (13), C5C6 (10), DC2H (7), C2N3 (6), DN3H (5)
1606	DN3H (31), DN9H (16), C4C5 (15), C6N10(13), C2N3 (8)
1626	C6N10(29), DNH2S(17), C5C6 (14), DN3H (13), C4C5 (5), D6RTr(5)
1673	DNH2S(71), C6N10(22)
1685	N3C4 (30), N9C4 (25), DN3H (7), DN9H (7), N7C8 (5)

Table 5: SQM STO-3G frequencies [cm^{-1}] and the PED of in plane vibrational modes of the N1,N7-diprotonated adenine

Exp[7] Calc. Main % contributions to the PED

Exp[7]	Calc.	Main % contributions to the PED
346	299	DC6N1(50), D6RE2(17), D5R| (10), C5C6 (7)
521	515	DC6N1(36), D6RE2(21), N9C4 (9), N3C4 (6), D5R| (6)
538	542	D6RE1(80)
612	616	C5C6 (34), D6RE2(19), D5R| (14), DC6N1(8), C6N10(5), N1C2 (5)
714	691	N3C4 (30), D5R| (20), D6RE2(17), C5N7 (6)
915	907	D6RTr(46), D6RE2(14), D6RE1(8), C6N10(7), N3C4 (6), C2N3 (6)
967	962	D5R= (79), C4C5 (7)
1055	1010	C6N1 (28), DNH2R(28), N1C2 (12)
1088	1077	DC8H (32), D5R| (18), N9C4 (9), N3C4 (7), D6RTr(6), C5N7 (6)
1138	1185	N1C2 (39), DC8H (15), C6N1 (14), N7C8 (6)
1207	1195	C8N9 (34), N7C8 (24), DN9H (24), DN7H (8)
1224	1217	DNH2R(28), DC8H (18), N9C4 (10), N7C8 (9), D5R| (7), N1C2 (7)
1307	1281	DC2H (37), DNH2R(11), C6N1 (8), N1C2 (8), C5C6 (5)
1343	1323	C5N7 (31), DC2H (15), D6RTr(11), DN7H (9), C5C6 (8)
1349	1373	DC2H (14), C5C6 (14), C2N3 (13), C6N1 (11), C4C5 (7)
1392	1390	N3C4 (24), N9C4 (17), DN9H (15), D5R| (7), C5N7 (7), DN7H (6)
1419	1425	C4C5 (24), DN9H (16), C8N9 (16), N9C4 (10), C2N3 (7), D5R= (7)
1489	1481	N7C8 (34), DN7H (14), C8N9 (14), D5R| (9), C4C5 (8), C5N7 (6)
1517	1516	DN1H (22), C6N10(17), C6N1 (10), C2N3 (9), DNH2R(6), N9C4 (6)
1578	1572	C2N3 (28), DN1H (15), DN7H (13), C6N10(13), D6RTr(8)
1602	1609	DN7H (14), C2N3 (14), C8N9 (12), C5N7 (9), D6RE2(8), C5C6 (7)
1625	1634	DN9H (22), C4C5 (13), N3C4 (13), N9C4 (11), N7C8 (8), DN1H (7)
1690	1667	DNH2S(44), C6N10(24), DN1H (11)
1718	1710	DNH2S(49), DN1H (18), C6N10(10), C5C6 (7), C6N1 (6)

ACKNOWLEDGMENT

The ab initio computations were performed during my stay at ICTP Trieste, and also on an IBM3090 computer in Prague as a part of the IBM Academic Initiative. The kind help of Dr.V.Baumruk and Dr.J.Štěpánek (Inst.Phys., Charles Univ., Prague) is gratefully acknowledged.

REFERENCES

[1] K.Gerwert, G.Souvignier, B.Hess, Proc.Natl.Acad.Sci.USA 87, 9774 (1989)
[2] J.C.Austin, C.W.Wharton, R.E.Hester, Biochemistry 28, 1533 (1989); H.Deng, J.Zheng, D.Sloan, J.Burgner, R.Callender, Biochemistry 28, 1525 (1989)
[3] J.Florián, J.Mol.Struct.(Theochem), in press
[4] G.Pongor, P.Pulay, G.Fogarasi, J.E.Boggs, J.Am.Chem.Soc. 106, 2765 (1984)
[5] C.E.Blom, C.Altona, Mol.Phys. 31, 1377 (1976)
[6] P.Pulay, G.Fogarasi, G.Pongor, J.E.Boggs, A.Vargha, J.Am.Chem.Soc. 105, 7037 (1983)
[7] J.Štěpánek, V.Baumruk, J.Mol.Struct. 219, 299 (1990) V.Baumruk, PhD-thesis; J.Štěpánek, unpublished results
[8] P.-O. Löwdin, in: Advances in Quantum Chemistry, ed. P-O.Lowdin, Vol.2, p.213, Acad.Press, New York 1965
[9] O.Kennard, in: Nucleic Acids and Molecular Biology, eds. F.Eckstein and D.M.Lilley, Springer-Verlag, Berlin 1987
[10] R.Tribolet, H.Sigel, Eur.J.Biochem 170, 617 (1988)
[11] P.Rajagopal, J.Feigon, Biochemistry 28, 7859 (1989)
[12] J.Wiorkiewicz-Kuczera, M.Karplus, J.Am.Chem.Soc. 112, 5324 (1990)
[13] A.Y.Hirakawa, H.Okada, S.Sasagawa, M.Tsuboi, Spectrochim Acta, 41A, 209 (1985)

STUDIES OF THE NATURE OF THE ORDERING TRANSFORMATION IN ICE Ih

J. W. Glen and R. W. Whitworth

School of Physics and Space Research
University of Birmingham
Birmingham, B15 2TT, UK

The stable phase of ice at low pressures with which we are familiar is the phase known as ice Ih (Fig. 1). In this phase the water molecules are bonded to each other tetrahedrally by hydrogen bonds, i.e. there is one proton on each O-O bond nearer to one of the two oxygen nuclei. These protons are however disordered; there is no regular crystallographic relation between these positions which are determined only by the so-called Bernal-Fowler rules - each oxygen has two protons near it and each bond has one proton on it. This disorder results in an entropy, which was first calculated by Pauling[1] to be approximately $k \ln 3/2$ per molecule.

This structure can have point defects resulting from breaches of the Bernal-Fowler rules. As in water, a small number of water molecules can be ionized, resulting in H_3O^+

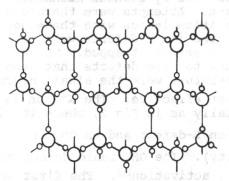

Fig. 1.   Arrangement of atoms in ice Ih. The larger circles represent oxygens, the smaller circles hydrogens (or protons). Note that the protons are not arranged crystallographically.

Proton Transfer in Hydrogen-Bonded Systems
Edited by T. Bountis, Plenum Press, New York, 1992

and OH⁻ ions, thus breaking the first rule, and a small
fraction of bonds can have no proton (which Bjerrum called L-
defects) or two protons (Bjerrum's D-defect). All four of
these defects have an associated electrical charge and are
potentially mobile. When they move they reorient water
molecules. Ionic defects move by a proton moving from one end
of a bond to the other, Bjerrum defects move by a proton
moving around an oxygen nucleus from one bond to another. Any
mobile defect can thus reorient water molecules and so can
enable such a reorientation to occur in the presence of an
electric field; it can thus be a mechanism for dielectric
relaxation. To obtain a direct electric current through ice
by proton migration, it is necessary for protons to move along
bonds and from bond to bond, so both an ionic and a Bjerrum
defect must be mobile.

At a sufficiently low temperature the disordered ice Ih
phase with its intrinsic entropy must cease to be the
thermodynamic equilibrium phase. An obvious possibility is
that it should change to a phase similar in structure but with
ordered protons. However as temperature falls the dielectric
relaxation time increases, indicating that the number of
defects or their mobility or both decreases with temperature,
as is to be expected if they are thermally activated, and
before the temperature has fallen to a value at which ice Ih
ceases to be the equilibrium phase, the rate of reorientation
becomes far too long for the reorientation to be observed on a
laboratory time scale.

Attempts have been made to lower the temperature at which
measurable relaxation can occur by using doped rather than
pure ice. It has been known for many years that acid or
alkaline impurities can give extrinsic defects. For example
HF dissolved substitutionally in ice produces L-defects and

H_3O^+ ions, while NH_3 produces D-defects and OH⁻ ions. At
higher temperatures the reorientation is produced most rapidly
by the Bjerrum defects, but they require an activation energy
to move, and so at low temperatures the ionic defects, which
appear to be able to move by quantum-mechanical tunnelling,
become more important. Attempts were therefore made to find a
disorder-order transition in ice with these dopants, but these
were unsuccessful; the ionic defects were either too tightly
bound to the impurity or were trapped. The impurities which
were eventually found to give defects that remained active at
the transition temperature were the alkali hydroxides, of

which KOH is the most effective. The K^+ ion is believed to
dissolve interstitially as in Fig 2, where it will be seen to

be associated with an L-defect and an OH⁻ ion (this preserves

electrical neutrality). The OH⁻ ionic defect can be shown to
move without thermal activation[2,3]. The first observations of
the dielectric properties of KOH-doped ice were made by

Kawada[4], but the demonstration that this was producing a more
ordered phase was made by Suga and coworkers[5]. Suga[6,7] gave
this phase the name ice XI.

The evidence for this transition comes from a combination of
calorimetric and dielectric measurements and neutron

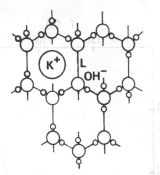

Fig. 2. Diagram to indicate how KOH
dissolved in ice is thought
to introduce defects. The
K^+ ion is interstitial, the

OH^- ion substitutes for a
water molecule leaving one
bond without any protons -
an L defect.

diffraction studies. From calorimetry it can be determined
that up to 68% of the entropy of disorder is released[8]. There
could be several explanations for the entropy of disorder not
being fully released: (i) only part of the ice may be
electrically active and hence able to transform; (ii) the
transition may be incomplete and so only partial ordering may
be achieved, even though the equilibrium phase would be
ordered; (iii) the new equilibrium phase may itself not be
fully ordered, which would of course imply that further
ordering should in principle occur at some still lower
temperature.

From dielectric measurements we can determine some features
of the process which reorients water molecules. In the
simplest case with one mobile defect and no traps we would
have a simple Debye relaxation whose strength would tell us
about the amount of reorientation in equilibrium, while the
single, well-defined relaxation time would tell us about the
dynamics. A fully ordered phase should have lost this
relaxation completely. In fact dielectric experiments[9] on
polycrystalline ice doped with KOH do show a relaxation whose
strength varies little with temperature, and whose relaxation
time also varies little until the temperature falls below
120K, consistent with a tunnelling process. Below 120K the
relaxation time rises (Fig. 3). This is attributed to the
trapping of defects on traps about 0.17 eV deep. Below the
transition temperature T_c (72K), however, the strength
decreases with time and the relaxation time rises, as the
material transforms and so changes to an ordered phase,
incapable of dielectric relaxation. On heating the material
reverts to its previous dielectric behaviour (Fig. 4). The
full study of the dielectric properties above and below the
transition temperature are complicated by the fact that there
is not a simple single dielectric relaxation[10,9]. This implies
there is more than one relaxation process involved, and the
relative strengths of the different processes change with

241

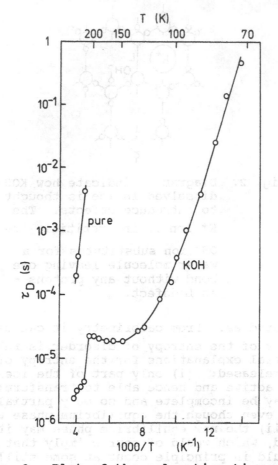

Fig. 3. Plot of the relaxation time τ_D
for dielectric relaxation in
ice doped with KOH as a
function of $1/T$. Note that
there is a temperature range
for which τ_D does not vary
much with temperature. This
is thought to be due to

extrinsic OH⁻ ions moving by
quantum-mechanical tunnelling.

temperature and, below the transition temperature, with time.
The detailed interpretation of this is still not certain, but
the overall behaviour of the material on annealing below T_C is
consistent with the disappearance of a large part of the
relaxation in the transformed material.

Dielectric measurements on single crystals of KOH-doped
ice[11] show that the transition can be obtained in single
crystal specimens and that above the transition temperature T_C
the dielectric relaxation strength follows a Curie-Weiss law
$\varepsilon = C/(T-\Delta)$ with different values of the parameters parallel

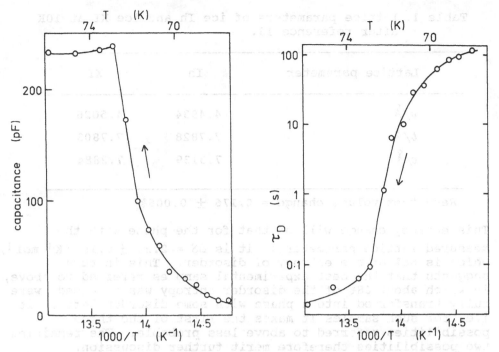

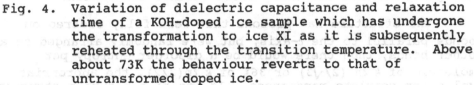

Fig. 4. Variation of dielectric capacitance and relaxation time of a KOH-doped ice sample which has undergone the transformation to ice XI as it is subsequently reheated through the transition temperature. Above about 73K the behaviour reverts to that of untransformed doped ice.

and perpendicular to the c-axis. The anisotropy is that $\varepsilon_{||} > \varepsilon_{\perp}$ and $\Delta_{||}$ is between 10 and 30K while $\Delta_{\perp}$ is between −50 and −80K.

In an attempt to obtain structural information about this transition, neutron diffraction measurements have been made on untransformed and transformed powder specimens[12,13]. Howe and Whitworth[13] have shown that in a transformed specimen the lattice parameters of part of the specimen have changed: ice XI has a shorter c-axis (Table 1). The volume change deduced from these lattice parameters together with the pressure dependence of the transition temperature reported by Yamamuro and others[14]

$$\frac{dT_c}{dp} = 0.015 \pm 0.001 \text{K MPa}^{-1}$$

allows us to deduce an entropy change using the Clausius-Clapeyron equation

$$\frac{dT_c}{dp} = \frac{\Delta V}{\Delta S}.$$

Table 1. Lattice parameters of ice Ih and ice XI at 10K after reference 13.

Lattice parameter	Ih	XI
$a/\text{Å}$	4.4934	4.5026
$b/\text{Å}$	7.7828	7.7803
$c/\text{Å}$	7.3139	7.2884

Resultant volume change = 0.176 ± 0.006%

This entropy change will be that for the phase with the measured lattice parameters; it is ΔS = 2.25 ± 0.15 J K^{-1} mol^{-1}, which is 66% of the entropy of disorder. This in turn suggests that the best experimental samples referred to above, in which about 68% of the disorder entropy was released, were fully transformed into a phase with some disorder left in it, i.e. for such samples it makes the first of the three possibilities referred to above less probable. The remaining two possibilities therefore merit further discussion.

If the structure had protons that were fully ordered on bonds parallel to the c-axis, but were randomly arranged on all other bonds, then there would be a disorder entropy per molecule of $k \ln (2/\sqrt{3})$ or 36% of $k \ln(3/2)$, the appropriate value for complete randomness. If ice XI had such a structure it would thus have lost 64% of the original disorder entropy, which is quite close to the loss of entropy deduced above, and also to that released in the best cases. Of course a small amount of short-range order among the remaining bonds could result in further entropy loss. Dr Rachael Howe (private communication) has calculated the relationship between entropy S and order parameter f', and this is shown in Fig. 5. Here f' is a parameter relating to the oblique bonds only, being 1/3 for the random case and 1 for complete ordering.

However the other possibility is that the thermodynamic equilibrium phase is completely ordered but that final ordering is very difficult to achieve. This may indeed be the case because of the mechanism by which ordering is accomplished. The model we have used depends on the movement of electrical point defects which must move along a linear path in the crystal reorienting water molecules at each step. However, when only a relatively small amount of disorder remains, this disorder must consist of "strings" of incorrectly oriented water molecules[15] as shown in Fig.6. The defect has then to move exactly along one of these strings to produce order. Thus in Fig. 6, order can be produced by an

OH$^-$ ion moving along the path indicated by the arrow, so moving each proton backwards along each bond. However, at each

step there are two possible paths for the OH$^-$ ion to take, and although the one which produces order is energetically favoured, the other option will have a finite, though lower, probability. Since there are a large number of such choices,

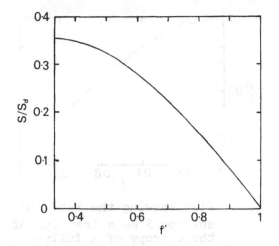

Fig. 5. Plot showing the ratio of
entropy S as a fraction of
the entropy of a fully
disordered sample S_d for ice
in which all bonds parallel

to the c-axis are ordered but
in which the remaining bonds
are partially ordered with an

order parameter f' which
varies from 1/3 when they are
completely random to 1 when
they are completely ordered.

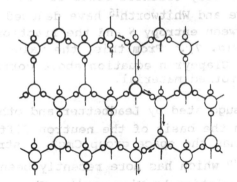

Fig. 6. Diagram to show ordered
ice with a string of
disordered water
molecules. Order can only
be achieved if an

OH⁻ ion traverses the path
indicated by the arrows.
As it does so protons move
backwards along each bond.

At each step the OH⁻ ion
has two paths it could
possibly take.

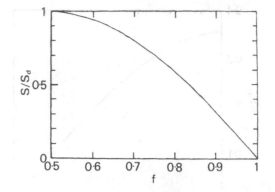

Fig. 7. Plot showing the ratio of entropy S as a fraction of the entropy of a fully-disordered sample S_d as a function of an order parameter f which measures the fraction of bonds which are correctly oriented, thus $f = \frac{1}{2}$ represents randomness and $f = 1$ complete order; all bonds are considered equivalent.

the probability that the ion will move along exactly the right path will be very small, so we can expect that strings of incorrectly oriented bonds will remain in the crystal. The fraction of incorrectly oriented bonds can be estimated from the entropy. Howe and Whitworth[15] have deduced the relationship between entropy S and the fraction of correctly ordered bonds f (Fig. 7). From this, the value of ΔS deduced from the Clausius-Clapeyron equation above corresponds to $f \approx 0.87$ in the transformed material.

The structure suggested by Leadbetter and others[12] and Howe and Whitworth[13] on the basis of the neutron diffraction from powdered samples has the space group $Cmc2_1$ a structure first predicted by Kamb[16] which has more recently been predicted in a thermodynamic calculation by Minagawa[17]. This structure is fully ordered and ferroelectric parallel to [001], the original [0001] direction of the untransformed ice Ih. However, the experimental evidence for this interpretation is not strong, and recently single-crystal neutron diffraction studies have been made[18]. A KOH-doped single crystal has been successfully transformed, and neutron diffraction data obtained from it. These fail to confirm the postulated structure, although they do show changes from the untransformed ice including some new diffraction peaks. These results are also inconsistent with the $Pna2_1$ anti-ferroelectric structure predicted by Owston[19] on the basis of minimizing the electrostatic energy of water molecule pairs, and with any of the other fully ordered structures so far suggested. Further work is proceeding to try to interpret the neutron diffraction data.

The investigation of the ordering transition in ice Ih has thus revealed that the ordering depends on proton movement along hydrogen bonds, and has shown how such motion can occur at appreciable rates at temperatures below 72K in KOH-doped ice. In turn, the hydrogen-bond reorientation process may itself make difficult a complete transformation of material into the ordered state.

ACKNOWLEDGEMENTS

The authors would like to thank Dr Rachel Howe for providing the data for Figs. 3 and 4 and also making available her unpublished calculations which form the basis of Fig. 5.

REFERENCES

1. L. Pauling, The structure and entropy of ice and of other crystals with some randomness of atomic arrangement, J.Am.Chem.Soc., 57:2680 (1935).
2. A. V. Zaretskii, V. F. Petrenko, and V. A. Chesnakov, The protonic conductivity of heavily KOH-doped ice, Phys.Stat.Sol.(a), 109:373 (1988).
3. R. Howe and R. W. Whitworth, The electrical conductivity of KOH-doped ice from 70 to 250K, J.Phys.Chem.Solids, 50:963 (1989).
4. S. Kawada, Dielectric dispersion and phase transition of KOH doped ice, J.Phys.Soc.Jap., 32:1442 (1972).
5. Y. Tajima, T. Matsuo, and H. Suga, Phase transition in KOH-doped hexagonal ice, Nature, 299:810 (1982).
6. H. Suga, [Phase diagram of ice and the discovery of phase XI], Kotai Butsuri, 20:125 (1985).
7. T. Matsuo, Y. Tajima, and H. Suga, Calorimetric study of a phase transition in D_2O ice Ih doped with KOD: ice XI, J.Phys.Chem.Solids, 47:165 (1986).
8. Y. Tajima, T. Matsuo, and H. Suga, Calorimetric study of phase transition in hexagonal ice doped with alkali hydroxides, J.Phys.Chem.Solids, 45:1135 (1984).
9. A. V. Zaretskii, R. Howe and R. W. Whitworth, Dielectric studies of the transition of ice Ih to ice XI, Phil.Mag.B, 63:757 (1991).
10. S. Kawada, Acceleration of dielectric relaxation by KOH-doping and phase transition in ice Ih, J.Phys.Chem.Solids, 50:1177 (1989).
11. M. Oguro and R. W. Whitworth, Dielectric observations of the transformation of single crystals of KOH-doped ice Ih to ice XI, J.Phys.Chem.Solids, 52:401 (1991).
12. A. J. Leadbetter, R. C. Ward, J. W. Clark, P. A. Tucker, T. Matsuo and H. Suga, The equilibrium low-temperature structure of ice, J.Chem.Phys., 82:424 (1985).
13. R. Howe and R. W. Whitworth, A determination of the crystal structure of ice XI, J.Chem.Phys., 90:4450 (1989).
14. O. Yamamuro, M. Oguri, T. Matsuo, and H. Suga, High pressure calorimetric study on the ice XI-I_h transition, J.Chem.Phys., 86:5137 (1987).

15. R. Howe and R. W. Whitworth, the configurational entropy of partially ordered ice, J.Chem.Phys., 86:6443 (1987).

16. B. Kamb, Crystallography of ice, in: "Physics and chemistry of ice", E. Whalley, S. J. Jones, and L. W. Gold ed., Royal Society of Canada, Ottawa, 28 (1973).

17. I. Minagawa, Phase transition of ice Ih-XI, J.Phys.Soc.Jap., 59:1676 (1990).

18. M. Oguro, R. W. Whitworth, and C. C. Wilson, Dielectric and neutron diffraction studies of the transformation of ice Ih to ice XI in KOH-doped single crystals, in: "International Symposium on the Physics and Chemistry of Ice, Sapporo, Japan, 1-6 September 1991", to be published.

19. P. G. Owston, La position des atomes d'hydrogène dans la glace, J.Chim.Phys., 50:C13 (1953).

DEFECT ACTIVITY IN ICY SOLIDS FROM ISOTOPIC EXCANGE RATES:

IMPLICATIONS FOR CONDUCTANCE AND PHASE TRANSITIONS

J. Paul Devlin
Department of Chemistry
Oklahoma State University
Stillwater, Oklahoma 74074

Introduction

This paper considers the results of several recent studies of icy thin films that have been focused on a) revealing the behavior of ionic and orientational point defects under a variety of conditions and b) the identification of phenomena that are dependent on defect activities. Since the absolute mobilities and concentrations of the defects are generally unknown, it is necessary to invoke the concept of defect activity which, throughout these studies, is identified with a rate of proton-deuteron isotopic exchange within isotopically nonequilibrium samples.

Historically, the functioning of point defects has been highlighted in models that describe protonic conductivity in ice and the orientational relaxation rates of icy substances.[1] However, detects have also been implicated in the mechanism of sheer (dislocation) propagation in crystalline ice[2], proton transport through biomembranes[3] and fast-ion conductors[4], the elementary steps of charge separation that lead to electrical storms[5], and the molecular process by which proton ordering can be induced in various ices[6,7].

Much of our particular fascination with defect activities derives from the premise that many low-temperature phase transformations in icy substances, such as the crystallization of amorphous ice and the direct vapor deposition of crystalline ice and clathrate hydrates at temperatures below ~150 K, require some minimum level of orientational point defect activity in the growing phase.[8] An interesting view is one of Bjerrum L defects scurrying about behind the advancing phase boundary to impose the Bernal-Fowler rules, ultimately through recombination with D defects. Confirmation of this premise would yield insights to the control of ice formation of potential value in cryobiology and other areas of technology, including the transport of natural gas at subfreezing temperatures.[9]

Our method of characterizing the level of either natural or induced defect activity of an ice depends on an ability to isolate D_2O molecules intact within an H_2O sample, an ability that has been developed for clusters and thin films of ice I, amorphous ice and a variety of clathrate hydrates.[10] Because the conversion of an intact isolated D_2O molecule to isolated HOD proceeds by two steps, with the rate of the first step determined by proton activity and of the second by orientational-defect activity, it is possible to independently monitor the activities of the ionic and orientational defects spectroscopically.

For example, a procedure that is used to deduce the L-defect activity of an icy substance may be appreciated by considering the two steps of isotopic scrambling outlined in Fig. 1. Proton hopping can only convert D_2O to neighbor-coupled HOD (i.e., $(HOD)_2$), since the deuteron is locked between two neighbor oxygens, while L-defect migration through the site can free a deuteron from a particular H-bond and thus, ultimately, convert $(HOD)_2$ to isolated HOD. It is clear that the orientational activity can best be characterized by observations on samples rich in $(HOD)_2$, but thermal proton hopping commonly onsets at a higher temperature than L-defect migration so that, with step (2) occurring more rapidly than step (1), samples do not become enriched in $(HOD)_2$.

This unfavorable situation can be surmounted using methods that artificially enhance the proton hopping activity, necessary for Step 1, at temperatures below the onset of the intrinsic L-defect activity. Methods that have led to high nonequilibrium concentrations of $(HOD)_2$ and

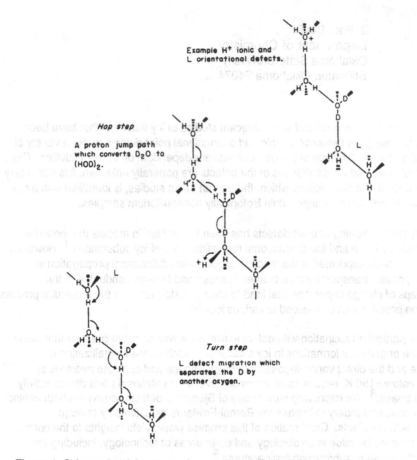

Figure 1. Schematic of the proton hop step and the L-defect turn step of proton transport.

thus to useful exchange-based observations on L-defect activity of thin films include MeV electron-beam bombardment,[11] self-ionization of electronically excited napthol dopant,[12] the photoconversion of the dopant nitrobenzaldehyde to self-ionizing benzoic acid and the single-photon vibrational-overtone excitation of pure ice with dense visible laser radiation[13].

In addition to the ease of manipulation of the relative rates of the hop and turn steps of proton transport, the molecular level observations of defect activities have also capitalized on the unique infrared spectra of the three key molecular units of Fig. 1: namely D_2O, HOD and $(HOD)_2$ at low concentrations in H_2O ice. It is clear from Fig. 1 that rate studies based on the isotopic

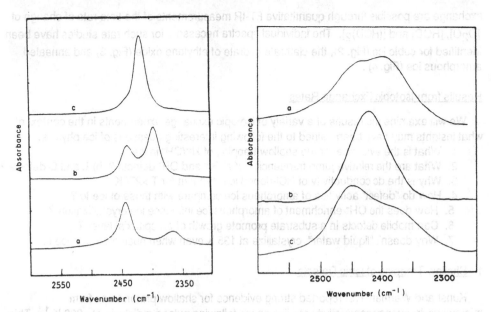

Figure 2. Infrared spectra of H_2O isotopomers isolated in H_2O ice Ih at 90 K: (a) D_2O; (b) $(HOD)_2$; (c) HOD.

Figure 3. Infrared spectra of isotopomers of H_2O isolated in the ethylene oxide clathrate hydrate: (a)$(HOD)_2$; (b) HOD; (c) $(HOD)_2$.

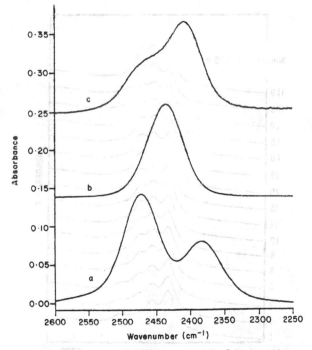

Figure 4. Infrared spectra of isotopomers of H_2O isolated in annealed amorphous ice at 80 K: (a) D_2O; (b) HOD and (c) $(HOD)_2$.

exchange are possible through quantitative FT-IR measurements of the time rate of change of [D_2O], [HOD] and [(HOD)$_2$]. The individual spectra necessary for such rate studies have been identified for cubic ice (Fig. 2), the clathrate hydrate of ethylene oxide (Fig. 3) and annealed amorphous ice (Fig. 4) .

Results from Isotopic Exchange Rates

We will examine the results of a variety of isotopic exchange experiments in the context of what insights may have been gained to the following interesting questions of ice physics:

1. What is the evidence for the shallow trapping of H^+/OH^- in ice.
2. What are the relative jump frequencies of a) H^+ and OH^- defects ? b) L and D defects ?
3. Why is the dc conductivity of HCl-doped ice invariant for T >270K.
4. How do "defect" activities of amorphous ice compare with those of ice Ic ?
5. How does the OH^- enrichment of amorphous ice influence ice crystallization ?
6. Can mobile defects in a substrate promote growth of an epitaxial layer ?
7. Why doesn't "liquid water" crystallize at 135 K even when nucleated with ice Ic?

1. Shallow Trapping of Ionic Defects

Kunst and Warman first reported strong evidence for shallow proton traps from microwave-frequency conductivity studies on ice following pulsed radiolysis at ~260 K.[14] This association of protons with shallow traps in crystalline ice has been confirmed by isotopic-exchange studies following the uv photolysis of ice samples lightly doped with 2-napthol.[12] Protons that were released to the ice from the 2-napthol excited state at 90 K were quickly immobilized in shallow traps as established by the subsequent onset of Step I (Fig. 1) during warming of the ice to near 110 K.

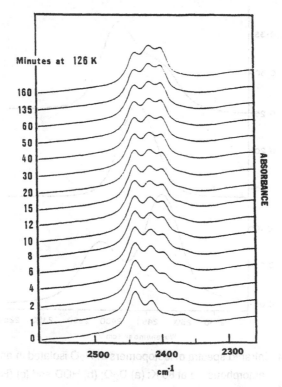

Figure 5. Spectra showing formation of (HOD)$_2$ vs. time at 126K from isolated D_2O in cubic ice after photoionization of 2-napthol dopant.

The spectra in Fig. 5 show this proton hopping activity at 126 K, a temperature at which no proton activity is observable in the absence of the photolysis stage. Measurement of the temperature dependence of the activity of protons released from shallow traps, together with an assumption of an "equilibrium" between trapped and free protons, revealed an association energy between the L defect and the trap of ~10 kcal. Following Kunst and Warman, the proton shallow trap has been tentatively identified with the negatively charged (0.37 e) L defect.

The presence of shallow traps in undoped crystalline ice has also been established using the self ionization of water molecules in vibrational overtone states at 90 K.[13] Dense blue (4880 A) or green (5145 A) argon laser radiation was shown to generate protons which, like those described above, were initially shallowy trapped but became active upon warming.

The shallow trapping of ionic defects in <u>amorphous</u> ice has also been observed by monitoring the onset of isotopic exchange for low-temperature deposits containing KOH or HNO_3. Data for KOH-doped amorphous-ice deposits prepared at 60 K are presented in Fig. 6. Samples containing significant quantities of KOH were obtained by codepositing water with molecular beams of potassium atoms.[15] For dense K-beams, extensive isotopic exchange occured during deposition at 60 K. For the particular case depicted in Fig. 6, the exchange was significant, but incomplete.

The near completion of exchange after the sample was warmed to 110 K (curve c) indicates that OH^- ions are released from shallow traps at temperatures well below the onset temperature (>140 K) of intrinsic ion-defect activity of amorphous ice.

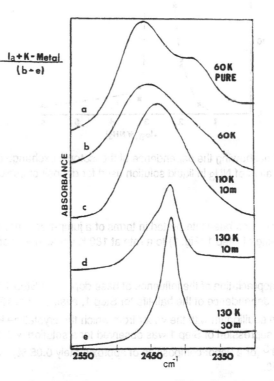

Figure 6. Spectra showing influence of dopant KOH on isotopic exchange (b and c) and crystallization (d and e) of amorphous ice. Curve (a) is for undoped amorphous ice at 60 K.

Also, the direct conversion from D_2O to HOD, without evidence of $(HOD)_2$ accumulating, indicated that Step 2 is more rapid than Step 1 (Fig. 1) for KOH-doped unannealed amorphous ice. This is not surprising since the concentration of L defects is enhanced by the presence of the hydroxide ions, and the intrinsic orientational defect activity of unannealed amorphous ice is presumed to be greater than that of crystalline ice (vide infra). The latter point may be an important one, since HNO_3-doped unannealed amorphous ice shows very similar behavior.

2. Relative Jump Frequencies of Ionic Defects:

It is generally accepted that the proton is the main agent of charge transport for pure crystalline ice though, from classical conductivity measurements, a considerable range of values have been attributed to the magnitude of the ratio of the mobility of the proton vs. the hydroxide ion.[1] Isotopic exchange spectroscopic data, for ice contaminated with a trace of the

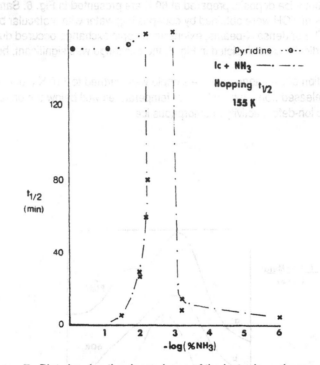

Figure 7. Plot showing the dependence of the isotopic exchange on the concentration of NH_3 in liquid solution used for deposit of cubic ice.

organic base 7-azaindole, has been interpreted in terms of a jump-frequency ratio greater than 10^2, based on the slowing of Step 1 (Fig. 1) to a rate at 180 K equal to that of pure ice at 140 K.[16]

A semiquantitative appreciation of the influence of base doping on Step 1 can be obtained from Fig. 7 which shows the dependence of the half-life for Step 1, evaluated at 155 K, on NH_3 content of an aqueous phase in equilibrium with the vapor from which the crystalline ice samples were grown. The maximum supression of Step 1 was observed for a solution with NH_3 mole % slightly greater than 10^{-3} (or a vapor composition of approximately 0.05 %), with a supression factor of ~50.

For higher $[NH_3]$, the mechanism that results in isotopic exchange must involve hydroxide ion "hopping" (i.e., via proton hopping from H_2O to OH^- sites). For vapor compositions much greater than 0.1 % in NH_3, the isotopic exchange rate accelerates rapidly and eventually exceeds that of pure ice (Fig. 7). Even for samples that displayed the maximum supression, it is likely that the $[OH^-]$ was orders of magnitude greater than the $[H^+]$ of pure ice. This suggests that, at 155 K, the proton jump frequency is many orders of magnitude greater than that of the hydroxide ion. However, these samples are enriched with both D defects and cations (NH_4^+), so the low activity may reflect hydroxide-ion trapping at positive centers.

It cannot be deduced from Fig. 5, but observations during the same set of experiments also led to a similar conclusion for the L defect mobility with respect to that of the D defect. It was observed that the orientational step of Fig. 1 occurs at the same rate for an ice sample formed from a vapor containing ~0.5 % NH_3 as for pure ice. The incorporation of each NH_3 molecule in the crystalline ice presumably introduces a D defect. Although most of the NH_3 is not incorporated in the crystalline ice,[17a] the D defect concentration for such samples is expected to be several orders of magnitude greater than for pure ice. This implies that, at 155 K, the jump frequency of an L defect is orders of magnitude greater than that of a D defect although, again, the trapping of the D defect by the abundant hydroxide ions may be a consideration[17b].

3. The invariant dc conductivity of HCl-doped ice:

The insensitivity of the conductivity of acid-doped ice to temperature variations slightly below the freezing point has been attributed to a combination of a dominant proton conductivity, the complete ionization of the acid dopant above some minimum temperature, and a temperature-independent mobilty of the proton.[18] If proton shallow-trapping is as significant as suggested by the results of Kunst and Warman and our low-temperature isotopic-exchange results (above), it is more reasonable to view the observed insensitivity of the dc-conductivity to temperature as the result of an accidental near cancellation of factors: that is the activation energy for L-defect formation (~ 8 kcal) nearly balances the dissociation energy of the proton from the L defect (~10 kcal) and, combined with a small negative activation energy for the proton mobility,[14] results in a mobile proton concentration that is relatively insensitive to temperature.

4. Defect activities of pure annealed amorphous ice:

Amorphous ice that is formed by deposition of water vapor at $T < 90$ K is microporous in nature.[19] Besides being a source of a large surface area as revealed through adsorbtion studies, this structure is apparent from the observation of distinct infrared bands for OH groups that "dangle" from the surface of the micropores (Fig. 8).[20,21] The charge transport/relaxation properties of this substance are undoubtedly dominated by the unusual mobility of the incompletely bonded surface groups. However, annealing of the deposits near 130 K results in a collapse of the pore structure so that annealed amorphous ice can be visualized as having a sufficiently complete network of near tetrahedral H-bonded molecules that the concept of defects, and a defect activity that dominates the transport and dielectric relaxation properties, may be useful.[22]

Our study of the defect activities of annealed amorphous ice paralleled that of crystalline ice[16] with one major exception. The proton-hopping activity of pure amorphous ice is not observable below the crystallization temperature of ~145 K, so isotopic exchange studies must depend on artificially induced proton activity. In an ongoing study, proton activity has been generated by 80 K photolysis of the dopant 2-napthol.[23] Semiquantitative results, based on the standard spectra of Fig. 3, indicated a) that the photogenerated protons were shallowly trapped at 80 K and b) that both the intrinsic L defects and the shallowly-trapped protons were activated upon warming to near 120 K. The exchange halflife (Step 2), of nearly one hour at 120 K, matched that for crystalline ice at ~135 K, confirming that the L-defect activity of amorphous ice does exceed that of crystalline ice.

5. Crystallization of hydroxide-rich amorphous ice:

It is well known that the doping of crystalline ice with KOH results in sufficient orientational mobility at 72 K for proton disordered ice to undergo a phase transition resulting in domains of ordered hexagonal ice.[6] Prompted in part by the unique ease with which the L-defect-rich crystalline clathrate hydrate of ethylene oxide forms from the vapor phase near 100 K and the fact that amorphous ice crystallizes at ~145 K, the temperature at which the L-defects of ice Ic become thermally activated on a laboratory time scale,[12,16] we have conjectured that the formation of crystalline ices at low temperature is also dependent on a mobility derived from the L defects within the growing phase.

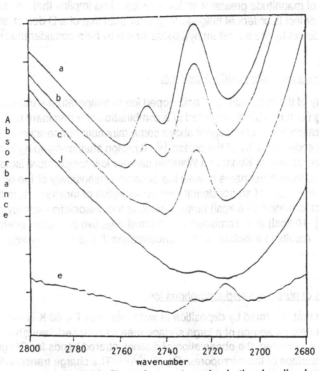

Figure 8. Infrared spectra of thin films of amorphous ice in the dangling-bond OD stretching-mode region: (a) pure D_2O at 15 K, (b) unexchanged 50% D_2O/50% H_2O at 15 K, (c) sample of (b) annealed 10 m at 60 K, (d) sample of (b) annealed 10 m at 120 K and recooled to 15 K, and (e) 15 K deposit of isotopically scrambled 20% D_2O mixture after 60 K anneal.

We have conducted several series of investigations to test the validity of this conjecture. In one series, 60 K codeposits of H_2O with varying amounts of potassium metal were monitored for a) isotopic exchange rates vs temperature and b) the temperature of onset of crystallization[15]. The exchange rates were useful as an indication of the relative KOH content of the samples, since the base content was otherwise difficult to establish. An inverse correlation was found between the minimum crystallization temperature and the base content, with samples for which rapid exchange was noted below 110 K invariably crystallizing in a matter of minutes at 130 K. The bottom two curves of Fig. 6 are representative of the spectroscopic monitoring of this crystallization at 130 K.

These data suggest that the temperature of crystallization of amorphous ice can be significantly reduced (15 - 20 K) by the presence of KOH, the presence of which is presumed to be accompanied by an increased L-defect activity in the growing cubic ice phase. Unfortunately the occasional crystallization of blank samples which contained no potassium metal, at comparably reduced temperatures, points to a need for skepticism of this KOH effect.

6. Can mobile defects in a substrate promote growth of an epitaxial layer?

It is known that glycopeptides can suppress the freezing point of body fluids of polar fish and there is some evidence that this occurs through the attachment of the peptides to the surface of ice crystal nuclei. This behavior implies a long-range control of crystal growth which might operate through peptide influence on defect activity. We have attempted to test defect activity-at-a-distance by examining the growth, in vacuum, of the L-defect-poor structure-I clathrate hydrate of CO_2 epitaxially to the L-defect-rich hydrate of ethylene oxide.[24] (Here the estimates of the relative defect concentrations are based on orientational relaxation rates)[25].

It is impossible to prepare the hydrate of CO_2 through a simple low-temperature vapor deposition in a near-vacuum environment. However, the CO_2 hydrate readily grows epitaxially to the EO hydrate for temperatures ranging from 110 to 150 K (Fig. 9, top). This fact alone might be explained in terms of hydrate nucleation, but it is not so easy to understand the

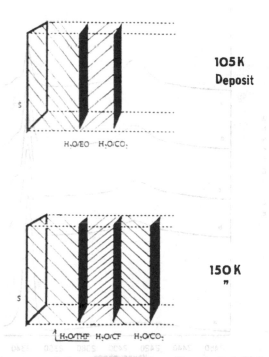

Figure 9. Schematic representation of the formation of layers of clathrate hydrate epitaxially to ether (EO and THF) hydrates: top structure I hydrate; bottom- structure II hydrates.

observation that the formation of the CO_2 hydrate is limited in thickness to ~ 1.0 micron with the limiting thickness a function of the thickness of the EO-hydrate substrate layer.

One explanation of this limited induced hydrate formation, consistent with our conjecture about crystal growth, is that the CO_2 hydrate does not form independently because of a shortage of orientational mobility[25] stemming from a modest L-defect concentration, particularly at 110 K, while the defect-rich EO-hydrate substrate can effectivly enhance the concentration of defects in the epitaxial layer to provide the necessary orientational mobility for growth. This enhancement is expected to extend over limited macroscopic distances. Although the average distance between terminating collisions of L and D defects is macroscopic, the space charge that accompanies preferential L-defect migration into the second layer may act to limit the spatial range of the enhancement.

In a second related series of epitaxial growth studies, it has been shown that, whereas one can grow micron-thick films of the structure-II hydrate of CO_2 epitaxially to the hydrate of tetrahydrofuran, that thickness is reduced by the insertion of a layer of the hydrate of chloroform between the THF and CO_2 hydrate layers (Fig. 9, bottom). This is demonstrated in Fig. 10, which shows that, as the deposit time of the chloroform-hydrate layer was increased in stages from 0 to 9 min, the spectrum of the outer layer, which alone contained isolated HOD, progessively converted from the broad O-D band of the hydrate to a much sharper feature denoting crystalline ice. Since the acceptor H-bonding tendency of the encaged $CHCl_3$ molecules can be envisioned to promote D-defect enrichment, the chloroform hydrate is expected to be

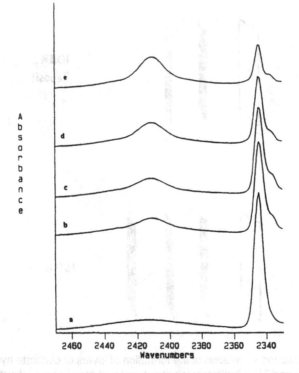

Figure 10. Spectra reflecting the influence of an insulating layer of chloroform clathrate hydrate on the growth of the type II CO_2 hydrate, epitaxially to the THF hydrate at 150 K. Insulating layer is from (a) 0.0 m, (b) 0.5 m, (c) 3 m, (d) 6 m and (e) 9 m deposits.

particularly lacking in orientational L defects and may inhibit migration of L-defects into the outer CO_2 hydrate layer.

7. Why can water remain liquid at 140 K (even when nucleated by Ic) ?

Amorphous ice undergoes a glass transition near 124 K to a "liquid" phase which does not crystallize until near 145 K. Similar behavior has been reported for vitreous ice containing some measure of ice crystallites.[26] If this is, in fact, a liquid phase it would be expected to homogeneously nucleate, so its metastability near 140 K requires an explanation.

One explanation, based on our crystal-growth conjecture, is that nucleation alone is not sufficient. Rather, the growing phase must also possess orientational mobilty to gaurantee rapid fulfullment of the Bernal-Fowler rules of the cubic ice. Significantly, the isotopic-exchange defect-activity studies have shown that the L-defect activity of cubic ice is such that the average time between passage of a defect through a water- molecule site is approximately thirty minutes at 140 K.[12,16] Even with the heat of crystallization available for defect activation, the associated mobility appears to be inadequate to maintain a crystallization front.

Summary

The monitoring of isotopic exchange rates, using the intensities of the O-D stretching-mode infrared bands as a measure of the concentrations of isolated D_2O, HOD and $(HOD)_2$, has enabled the characterization of the ionic-defect and orientational-defect activities of a variety of icy substances. These data have been used to show a) the importance of ion-defect trapping, b) the relatively much greater activity of the proton and the L defect vs. the hydroxide ion and the D defect in cubic ice and c) L-defect activity that correlates with the known T_g values of cubic and amorphous ice and d) justification of an alternate interpretation of the conductivity of acid-doped ice near the melting point. Related data have also been presented to further probe the role of orientational defects in low-temperature phase transformations. In particular, the crystallization temperature of amorphous ice has been reduced by 15-20 K through doping with KOH, while the growth of the clathrate hydrates (I and II) of CO_2, epitaxially to ether hydrates, is shown to be consistent with a dependence on L-defect activity associated with the unusual abundance of L-defects present within the substrate hydrates.

Acknowledgment

Support of this research by the National Science Foundation under Grants CHE-8719998 and CHE-9023277 is gratefully acknowledged.

References

1. P. V. Hobbs, Ice Physics (Clarendon, Oxford) (1974).
2. J. W. Glen, Phys. Condens. Matter, 7, 43 (1968).
3. J. F. Nagle and S. Tristam-Nagle, J. Membrane Bio., 74, 1 (1983).
4. V. H. Schmidt, J. E. Drumheller, and F. L. Howell, Phys. Rev. B, 4, 4582 (1971).
5. G. W. Gross, J. Geophys. Res., 87, C9, 7170 (1982).
6. Y. Tajima, T. Matsuo, and H. Suga, Nature, 299, 810 (1982).
7. T. Matsuo, Y. Tajima, and H. Suga, J. Phys. Chem. Solids, 47, 165 (1986).
8. P. J. Woolridge, H. H. Richardson, and J. P. Devlin, J. Chem Phys., 87, 4126 (1987).
9. P. B. Dharmawardhana, W. R. Parrish, and E. D. Sloan, I&EC Fundamentals, 19, 410 (1980).
10. J. P. Devlin, International Reviews in Physical Chemistry, 9, 29 (1990).
11. J. P. Devlin and H. H. Richardson, J. Chem. Phys., 81 3250 (1984).

12. P. J. Woolridge and J. P. Devlin, J. Chem. Phys., **88**, 3086 (1988).

13. J. P. Devlin, J. Phys. Chem., **92**, 6867 (1988).

14. M. Kunst and J. M. Warman, J. Phys Chem., **87**, 4093 (1983).

15. G. Ritzhaupt and J. P. Devlin, unpublished work (1988).

16. W. B. Collier, G. Ritzhaupt, and J. P. Devlin, J. Phys. Chem., **88**, 363 (1984).

17. a) G. W. Gross, C. Wu, L. Bryant, and C. McKee, J. Chem. Phys., **62**, 3085 (1975); b) As noted by J. Nagle in a comment following this workshop paper.

18. I. Takei and N. Maeno, J. Chem. Phys., **81**, 6186 (1984).

19. E. Mayer and R. Pletzer, Nature, **319**, 298 (1986).

20. B. Rowland and J. P. Devlin, J. Chem. Phys., **94**, 812 (1991).

21. V. Buch and J. P. Devlin, J. Chem. Phys., **94**, 4091 (1991).

22. J. L. Green, A. R. Lacey, and Sceats, M. G., J. Phys. Chem., **90**, 3961 (1986).

23. M. Fisher and J. P. Devlin, unpublished work (1991).

24. F. Fleyfel and J. P. Devlin, unpublished work (1990).

25. D. W. Davidson, S. K. Garg, S. R. Gough, R.E. Hawkins and J. A. Ripmeester, in Inclusion Compounds II, edited by J. L. Atwood, J. E. D. Davies and D. D. MacNicol (Academic, New York, 1984).

26. G. P. Johari, A. Hallbrucker, and E. Mayer, Nature, **330**, 552 (1987).

SINGLE-MOLECULE DYNAMICS AND THE "OPTICAL LIKE" COLLECTIVE MODES IN LIQUID WATER.

D. Bertolini[1], A. Tani[2] and D. Vitali[3].

[1] IFAM-CNR, via del Giardino 7, 56100 Pisa (Italy)
[2] Dip. di Chimica, Univ. di Pisa, 56126 Pisa (Italy)
[3] GNSM-CNR, Piazza Torricelli 2,56100 Pisa (Italy)

1.INTRODUCTION

This paper presents molecular dynamics results on the TIP4P [1] model of water in the supercooled region and is mainly focused on the connection between its single-molecule and collective dynamics. This subject has recently received much attention in studies of models for the dynamic and dielectric properties of pure molecular liquids and mixtures [2].

In the self part of the spectrum of the hydrogen current at low k (Fig. 1) one can observe three bands. The two bands at the lowest frequencies (8 and 45 THz) are the same as that visible in the spectrum of the autocorrelation function of the center of mass velocity, while the broad band between ~90 and 170 THz is determined by librational motions. The latter band can be decomposed into contributions from hindered rotations around the three principal axes of inertia, (90, 130, 165 THz). Collective effects, whose contribution increases in size in the low-k region, affect the longitudinal and transverse component of the current in a quite different way. At low frequency, the transverse component shows a viscoelastic peak, and the longitudinal component the acoustic mode. The band at 45 THz is no longer visible in either spectrum. As for the librational band, one can observe that collective effects enhance the intermediate frequency (130 THz) component in the transverse current and the highest frequency one in the longitudinal spectrum, to a much larger extent. In a series of papers [3], we have carried out a detailed analysis of collective effects on the librational dynamics stressing their relation with orientational (dipolar) correlations and the dielectric behavior of liquid water. Two main conclusions emerged: a), the bulk of the collective effects is due to the first and second shell of neighbors, and, b), the "zero-order" assumption that the cross terms of the collective time correlation function are equal to the time-propagated self parts is able to qualitatively account for the shape of the above mentioned bands of the spectra of transverse and longitudinal currents. The time delay is different for the three modes and the ratio of the transverse and longitudinal time shifts for the librational region has been linked to dielectric features in the same frequency region [3c]. However, the time shifts have been empirically determined and, more important, this simple picture cannot explain the large increase of intensity visible, e.g., in the longitudinal spectrum.

We shall show in what follows that a linear, unidimensional, harmonic model [4] can reproduce satisfactorily the collective effects outlined above for the low-frequency part of the spectrum, though this is still to be proved for the high-frequency librational region. Moreover, our results point out that current theories [5,6,7] that relate single-molecule and collective dynamics need to be improved in order to account for the lattice-like behavior especially important in highly-structured hydrogen bonded liquids.

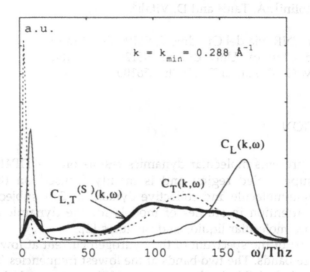

Fig.1   Spectra of the transverse (T) and longitudinal (L) hydrogen current at 245 K. The self contribution (S) is equal for (T) and (L) components at this k.

2. RESULTS AND DISCUSSIONS

a) MD results and time-propagated functions

The comparison between the autocorrelation function (acf) of the velocity of the center of mass and its cross counterpart, defined as follows

$$\Phi(t)^{(C)} = <\vec{v}_1^{(CM)}(0) \sum_j' \vec{v}_j^{(CM)}(t)> - \Phi(t)^{(S)} \qquad (2.1)$$

is shown in Fig.2.

The sum in Eq.2.1 is extended to the molecules which at t=0 are found at a distance $R \leq 3.27Å$ (first minimum of g(r)) from the central molecule 1, which is included in the sum.

Fig.2 clearly shows that the two functions have very similar shapes. (Note that the acf has been shifted in time exploiting its time-reversal parity). Thus, from the time-domain relation

$$[\Phi(t)]^{(C)} = [\Phi(t-\tau)]^{(S)} \qquad\qquad (2.2)$$

one obtains for the spectra

$$[\Phi(\omega)]^{(C)} = e^{-i\omega\tau} [\Phi(\omega)]^{(S)} \qquad\qquad (2.2')$$

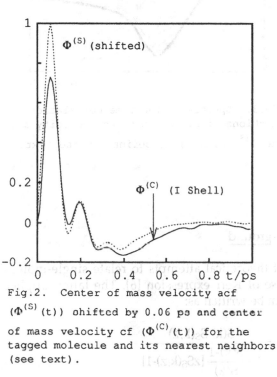

Fig.2. Center of mass velocity acf
($\Phi^{(S)}(t)$) shifted by 0.06 ps and center
of mass velocity cf ($\Phi^{(C)}(t)$) for the
tagged molecule and its nearest neighbors
(see text).

Actually, one can see (Fig.3) that $[\Phi(\omega)]^{(C)}$ is qualitatively reproduced by Eq. 2.2' and, even better, by the spectrum of $[\Phi(t-\tau)]^{(S)}$, with $\tau=\tau_{tr}=0.06$ ps. In this case, the self and cross time correlation functions describe the exchange of linear momentum between the tagged molecule and its nearest neighbors. The same procedure has ben applied to the acf of the current of the center of mass, shown in Fig.4 at the lowest k accessible in our simulation, ~0.3 Å^{-1}. The band at ~45 THz, visible in the spectrum of the self part of the acf, disappears in that of both the longitudinal and the transverse component, and this effect is reproduced by assuming, also in this case, that $[\Phi(t)]^{(C)}= [\Phi(t-\tau)]^{(S)}$, with $\tau=\tau_{tr}=0.06$ ps. In the case of the librational dynamics one needs two different values of time shift for the longitudinal and transverse component, namely $\tau_L=0.038$ ps and $\tau_T=0.048$ ps. Their ratio has been shown to be given by the amplitude of the band in the same frequency region of the dielectric spectrum. Fig.5 shows that the shape of the MD spectra is qualitatively reproduced.

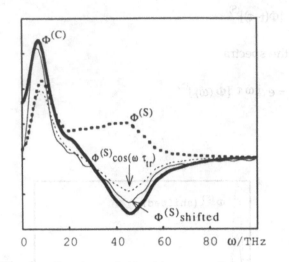

Fig.3. Spectra of the time correlation functions of Fig.2 and of the velocity acf.

The $\Phi^{(S)}\cos(\omega\tau_{tr})$ is defined in the text.

b) Theoretical background

Basically, most theoretical attempts to relate single-molecule and collective dynamics make use of Kerr expression [6]. The latter, applied to the dynamic structure factor, can be written as:

$$S(k,z) = \frac{S(k)\,S_S(k,z)}{1+\dfrac{S(k)-1}{S(k)}[zS_S(k,z)-1]}$$

(2.3)

where $S(k,z)$ is the dynamic structure factor and $S_S(k,z)$ its single-molecule, "self", part and $S_D(k,z)$ the distinct or "cross" part. Kim and Nelkin's [8] memory function approach provide a fairly straightforward route to this result. One can use a single-component or a multi-component description, and both will be very briefly reviewed here.

i) One-component description

With the usual definition of density correlation function, a Langevin equation for the self part and one for the distinct part are obtained. Taking their correlation function one can write, in the frequency domain:

$$z\,S_S(k,z)-1+m_S(k,z)S_S(k,z)=0$$

$$zS_D(k,z)-[S(k)-1]+m_S(k,z)S_D(k,z)=<f_S(k,z)\,|\,\rho_D(k)>$$

(2.4)

$f_S(k,t)$ is the random force on the tagged molecule and $\rho_D(k,t)$ is the distinct part of the Fourier component of the density fluctuation. In general, the r.h.s of

Eq.2.4 does not vanish, because the random force can be correlated with position and momentum of the molecules nearby. Neglecting this term and eliminating the memory function $m_S(k,z)$ from Eq. 2.4 one recovers Vineyard's relation [5]. More generally one obtains:

$$S_D(k,z)=[<f_S(k,z) \mid \rho_D(k)>+S(k)-1]S_S(k,z) \tag{2.5}$$

ii) Two-component description

The two-component description is obtained writing a system of four equations similar to Eq.2.4 with four memory functions:

$$z\,S_S(k,z)-1 + m_{11}(k,z)S_S(k,z)+m_{12}(k,z)S_D(k,z)=0$$

$$zS_D(k,z)-[S(k)-1]+m_{21}(k,z)S_S(k,z)+ m_{22}(k,z)S_D(k,z)=0$$

$$zS_D(k,z)-[S(k)-1]+m_{11}(k,z)S_D(k,z)+m_{12}(k,z)N\,S(k,z)=0$$

$$N[zS(k,z)-S(k)]+ m_{21}(k,z)S_D(k,z)+m_{22}(k,z)NS(k,z)=0 \tag{2.6}$$

If N>>1, one obtains from the last of Eqs.2.6

$$zS(k,z)-S(k)+m_{22}(k,z)S(k,z)=0 \tag{2.7}$$

and from second and third

$$m_{12}(k,z)= \frac{(m_{22}(k,z)-m_{11}(k,z))}{N} = \frac{m_{21}(k,z)}{N} \tag{2.8}$$

so the first becomes

$$zS_S(k,z)-1+m_{11}(k,z)S_S(k,z)=0 \tag{2.9}$$

Eqs.(2.7) and (2.9) contain two memory terms and Vineyard and Kerr approximations are obtained from the following relations between the memory functions:

$$m_{11}(k,z) = m_{22}(k,z) \qquad \text{(Vineyard)} \tag{2.10}$$

$$m_{11}(k,z) = S(k)\, m_{22}(k,z) \qquad \text{(Kerr)} \tag{2.11}$$

Kerr result can easily be extended to the currents through the well-known relation:

$$C_L(k,z) = zS(k) - z^2\, S(k,z) \tag{2.12}$$

Substituting Eq. 2.12 and its self counterpart into Eq. 2.3 one has:

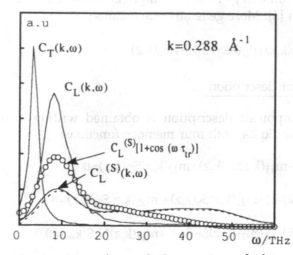

Fig.4. Comparison of the spectra of the
longitudinal (L) and transverse (T) current
of the center of mass with the spectra of
the self (S) term , of the time-propagated
self and the Kerr results (Eq.2.13).

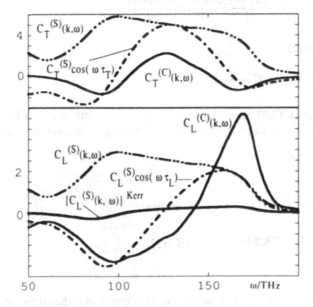

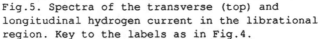

Fig.5. Spectra of the transverse (top) and
longitudinal hydrogen current in the librational
region. Key to the labels as in Fig.4.

$$C_L(k,z)^{Kerr} = \frac{z \, C_L(k,z)^{(S)}}{z + \dfrac{[1-S(k)]}{S(k)} C_L(k,z)^{(S)}}$$

(2.13)

Kim and Nelkin [8] extended their approach to include four and six components. This improves the description of the short-time regime, but we are not going to discuss it here.

The results for the current of the center of mass and of the hydrogens, computed according to Eq. 2.13, are shown in Figs. 4 and 5. The agreement with the MD spectra is clearly worse than that obtained assuming that the cross functions are given by time-propagated self-terms. The spectra given by Eq. 2.13 are similar to that relevant to the self part, which confirms that Kerr approximation is fairly accurate only at high k and ω.

Eqs. 2.7 and 2.9 can be written, using Eq. 2.12 :

$$zC_L(k,z)-C_L(k)+\frac{C_L(k)\, C_L(k,z)}{S(k)} z^{-1}+ M_C(k,z)C_L(k,z)=0$$

(2.14)

$$zC_L(k,z)^{(S)}- C_L(k)+C_L(k)C_L(k,z)^{(S)}z^{-1}+M_S(k,z)C_L(k,z)^{(S)}=0$$

(2.15)

where

$$m_{11}(k,z) = \frac{C_L(k)}{z + M_S(k,z)}$$

(2.16)

$$m_{22}(k,z) = \frac{C_L(k)/S(k)}{z + M_C(k,z)}$$

(2.17)

and

$$C_L(k) \equiv C_L(k,t=0)$$

(2.18)

Kerr approximation, equivalent to Eq. 2.11, is :

$$M_S(k,z)=M_C(k,z)$$

(2.19)

As a consequence, it is easy to show that the assumption of time-propagated functions leads to a non-linear relation between the memory kernels $m_{11}(k,z)$, $m_{22}(k,z)$ and $M_C(k,z)$ and $M_S(k,z)$.

We can assume that the relation between total current and its self part becomes, with the model of time-propagated functions (see Eq. 2.2'):

$$C_L(k,z)=C_L(k,z)^{Kerr} [1+B(k,z)e^{-z\tau}]$$

(2.20)

where $B(k,z)$ is a function that approaches one at low k's and ω's and zero at high k's and ω's.

Then, substituting Eq. 2.20 into 2.14 and eliminating $C_L(k,z)^{(S)}$ from Eqs. 2.14 and 2.15 we obtain:

$$M_S(k,z)=M_C(k,z)\,[1+B(k,z)e^{-z\tau}]\quad+\quad B(k,z)e^{-z\tau}\,[z+\frac{C_L(k)}{zS(k)}])$$

(2.21)

or

$$m_{11}(k,z) = \frac{m_{22}(k,z)\,S(k)}{1+B(k,z)\,e^{-z\tau}\,[1+\frac{m_{22}(k,z)}{z}]}$$

(2.22)

Hence, Vineyard and Kerr results can be obtained if a linear relation between the memory kernels is assumed (see Eqs. 2.10 , 2.11 and 2.19). Conversely, the hypothesis of time-propagated functions appears to require a non-linear dependence between the memory kernels (see Eqs. 2.21 and 2.22). It should be assessed whether this means that also the effective potential is non linear.

3. DYNAMICS OF ONE-DIMENSIONAL HARMONIC LIQUIDS

A clue to the answer to this question is provided by a one-dimensional harmonic model of a liquid [4]. In this case, the analytic solution for both self and cross part of the dynamic structure factor and currents is known.
The relevant formulae are:

$$S_s(k,t)=\exp[-\frac{(k\sigma)^2}{4}\,f_0(2ct/d\,)]$$

(3.1)

and

$$S_D(k,t)=\sum_{n=1}^{\infty} 2\cos(nkd)\exp\{[-\frac{(k\sigma)^2}{4}\,f_{2n}(2ct/d)]-\frac{n(k\sigma)^2}{2}\}$$

(3.2)

where

$$f_{2n}(2ct/d)=\int_0^{2ct/d} (2ct/d-s)\,J_{2n}(s)\,ds$$

(3.3)

and

$$(\sigma/d)^2 = \frac{k_BT}{mc^2}$$

(3.4) .

In Eq. 3.4 c is the velocity of sound, m is the mass of the particles, d the nearest neighbor distance in the lattice and s a coupling parameter. From Eqs. 3.1 and 3.2 the corresponding currents can be obtained through the relations:

$$C_S(k,t)=-\frac{d^2S_s(k,t)}{dt^2}$$

(3.5)

$$C_D(k,t) = -\frac{d^2 S_D(k,t)}{dt^2}$$
(3.6).

$C_S(k,t)$ and $C_D(k,t)$ and the relevant spectra have been computed at various k's, with parameters appropriate to water at 245 K, namely c=15 Å/ps, $(\sigma/d)^2$ =0.0502 and d=3 Å.

The spectra of the current are shown in Fig.6 at four values of k. One can notice that the functions corresponding to k=0.2 and 0.45 Å^{-1} exhibit analogous effects to those found in the MD spectra of the simulated water, i.e. some spectral region of the cross term are equal in magnitude and opposite in sign to the corresponding region of the self term, so that they cancel each other in the total spectrum. When k=0.8 Å^{-1} the situation resembles what observed in TIP4P water at k=k_{min}=0.2877Å^{-1} (Fig.4), where the collective acoustic mode occur at (roughly) the same frequency as the low-frequency band of the self spectrum. No other high frequency band is visible in the spectra of the unidimensional harmonic liquid.

Despite the same value of velocity of sound (c=15 Å/ps), the frequency, ω_{max}, of maximum spectral intensity is 2-3 times higher in water than in the harmonic model, in the low-k region. This is a consequence of the strong dispersion of the acoustic mode in water [9].

In the harmonic model ω_{max} is in accord with the hydrodynamic limiting low ω_{max}= c k up to k~0.4Å^{-1}. At higher k's, ω_{max} is always smaller than c k (see Fig.6 at k=1.2 Å^{-1}) due to the increase of S(k), i.e. of static correlations, which is peaked at k~2 Å^{-1}, where S(k)=2.16. Also water has a peak in S(k) at k~2 Å^{-1} [10].

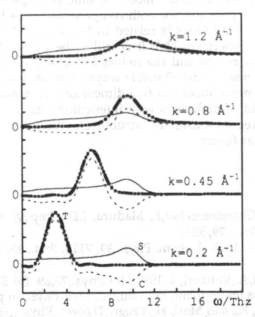

Fig.6. Spectra of the currents relevant to the unidimensional harmonic model of liquid (see text)

4. CONCLUSIONS

Existing theories that aim at relating single-molecule and collective dynamics do not provide satisfactory predictions mainly in the low-k, low-ω regime [11]. Kerr approximation, for instance, gives a total dynamic structure factor which is close to its self part, which is only correct when collective effects play a secondary role. The idea of cross terms equal to the corresponding time-propagated self terms, although can be seen as a heuristic approach, leads to encouraging results. However, this treatment cannot be extended to hydrodynamic modes, like the sound mode. As can be seen from Eqs. 2.21 and 2.22, this approach involves a non-linear, oscillatory, relation between the memory kernels relevant to the self and cross functions, while both Vineyard and Kerr approximations assume a linear relation. This might suggest that a non-linear effective potential is required to observe cancellation effects like those found in the MD spectra of water. However, the analysis of the unidimensional harmonic model, where we have linear interactions, has shown an overall behavior that qualitatively parallels that of simulated water, if values corresponding to that of water are given to the parameters. The main difference concerns the strong dispersion of the sound mode at low k's in water [9], which is missing in the harmonic model. On the other hand, qualitatively new features are observed in the librational region of the spectra of longitudinal and transverse hydrogen current. We have shown [3c] that they are related to the orientational correlations determined by the hydrogen bonds. These orientational correlations give a different weight to the cross contributions to the time correlation functions of the angular velocity. The longitudinal spectrum shows a strong enhancement of the highest frequency mode and an attenuation of the other two. In the transverse case, the intermediate frequency band is increased, but to a much lesser extent (see Fig.1 and 5).

With two different delay times, the model of time-propagated functions gives qualitatively satisfactory results, for both components of the hydrogen current. The ratio of the two delay times is related to the amplitude of the librational band of the dielectric spectrum [3c]. It is still to be assessed if the enhanced intensity of a spectral region and the reduced intensity of some other can be accounted for by a linear model. Possible ways to an answer require either an extension of the harmonic model to two dimensions or a molecular dynamics simulation of a liquid of molecules with interactions similar to that of water, but a different three-dimensional "structure". We plan to pursue both approaches in the near future.

REFERENCES

1) Jorgensen, W.L., Chandrasekhar,J., Madura, J.D., Impey, R.W., and Klein, M.L., 1983, J. Chem. Phys. **79**, 926.
2) Wei, D., Patey, G.N., 1990, J. chem. Phys., **91**, 7113; ibid., **93**, 1399; **94**, 6785; **94**, 6795.
3)(a)Bertolini,D.,Tani,A.,Vallauri,R.,1991,Mol.Phys.,**73**,69. (b) Bertolini, D., Tani, A., Mol. Phys.,in press.(c) Bertolini, D., Tani, A., Mol. Phys., in press.
4) Yoshida, T., Shobu, K., and Mori, H., Progr. Theoret. Phys. , 1981, **66**, 759.

5) Vineyard,G.H., Phys. Rev., 1958, **110**, 999.

6) Kerr,W. C., Phys. Rev., 1968, **174**, 316.

7) (a) Singwi K. S.,Skold, K.,Tosi,M.P.,Phys.Rev.Lett. ,1968, **21**, 881; Phys. Rev. ,1970, **A1**, 454 (b) Ortoleva, P., and Nelkin, M., Phys. Rev., 1970, **A2**, 187.

8) Kim, K., and Nelkin, M., Phys. Rev., 1970, **A4**, 2065.

9) (a) Wojcik, M., and Clementi, E., J. Chem. Phys.,1986,, **85**, 6075; (b) Ricci, M.A., Rocca, D., Ruocco, G., Vallauri, R., 1989, Phys. Rev. A40, 7226. Other authors invoke a high-frequency acoustic mode : see for example (c) Kowall, Th, Mausbbach, P., and Geiger, A., Ber. Bunsenges. Phys. Chem. , 1990, **94**, 279. On the other hand, the neutron scattering experiments show only one peak with the same dispersion as found in computer simulation results: see for example (d) Teixeira, J., Bellisent-Funel, M.C., Chen, S.H., and Dorner, B., Phys. Rev. Lett., 1985, **54**, 2681.

10) Geiger, A.,Mausbach, P., Schnitker,J., Blumberg, R.L., Stanley, H.E., 1984, J. de Phys., **C7**, 13; see also refs. 9(a) and 9(b).

11) Hansen, J.P., McDonald, I.R., 1986, 'Theory of simple liquids", 2nd ed., (Academic).

FREE-ENERGY COMPUTER SIMULATIONS FOR THE STUDY OF PROTON
TRANSFER IN SOLUTIONS

Mihaly Mezei*

Department of Chemistry and Center for Study
in Gene Structure and Function
Hunter College and the Graduate Center, CUNY
New York, NY 10021, USA

INTRODUCTION

The understanding of the effect of a solvent on proton
transfer requires the elucidation of the solvent contribution to
the free energy of solvation. Due to the high dimensionality of
the configuration space involved in the problem, the solvent
effect can only be modeled via computer simulation for a complex
polyatomic solute. While simulation of rather large systems have
recently become relatively routine, the methods for the
calculation of free energy from computer simulation has been
still recognized as a computationally exacting task[1-4]. The
purpose of this paper is to give a brief overview of the free
energy simulation methodology and demonstrate its capabilities
on the study of the thermodynamics of the formation of the
glycine zwitterion in water.

FREE-ENERGY SIMULATION METHODOLOGY

Unless the calculations are performed in the
grand-canonical ensemble, free energy computer simulation
techniques are characterized by the path used to connect the two
systems between which the free-energy difference is to be
calculated in the configuration space and by the quantity
chosen whose Boltzmann average is related to the free energy.
The path is described by the introduction of a coupling
parameter λ into the energy function. The various choices of the
coupling parameter generally fall into the following two
categories[3]:

$$E(\lambda,\underline{X}^N) = \lambda^k * E_1(\underline{X}^N) + (1-\lambda)^k * E_o(\underline{X}^N) \quad \text{or} \quad (1)$$

$$E(\lambda,\underline{C},\underline{X}^N) = E(\lambda*\underline{C}_1+(1-\lambda)*\underline{C}_o,\underline{X}^N) \quad (2)$$

Here E_o and E_1 are the energy functions for the two systems

---

*Present address: Department of Physiology and Biophysics, Mount
Sinai School of Medicine, CUNY, New York, NY 10029.

Proton Transfer in Hydrogen-Bonded Systems
Edited by T. Bountis, Plenum Press, New York, 1992

and λ is chosen in such a way that $\lambda=0$ and $\lambda=1$ in equations (1) and (2) describe systems with energy function E_0 and E_1, respectively. Mathematically speaking, equations (1-2) describe a homeomorphism between E_0 and E_1. The path described by equation (1) is linear for k=1 and 'nearly linear' for k>1. The introduction[3] of k>1 is a generalization of the integral transform introduced by Abrams et al.[5] The symbol $\underline{c}$ in equation (2) stands for the collection of potential parameters (including molecular geometries) whose values differentiate systems 0 and 1. Moving along the path described by equation (1) the system 0 is gradually "fading away" while system 1 is simultaneously being "turned on" - a kind of Chesire cat approach. Equation (2), on the other hand, describes a path where system 0 is continuously deformed into system 1.

As the free energy is a state function, several different paths can be used to calculate a given free-energy difference. A creative example for this is the use of a thermocycle by McCammon and coworkers to replace the "obvious" but computationally unfavorable path with a seemingly more complex but computationally tractable one[6]. The nonlinear path defined by Equation (2) was introduced in conjunction with the perturbation method by Jorgensen[7]. The possibility that a nonlinear path can be singularity-free for thermodynamic integration was noted by Mezei and Beveridge[3] and by Cross[8].

Once the path is specified, the free energy difference between the two states can be obtained in various ways. Thermodynamic integration uses the expression of Kirkwood[9]:

$$\Delta A = A_1 - A_0 = {_0\int^1} \partial A(\lambda)/\partial \lambda \ d\lambda \qquad (3)$$

$$= {_0\int^1} <\partial E(\lambda)/\partial \lambda>_\lambda \ d\lambda \qquad (4)$$

where k is the Boltzmann constant, T is the absolute temperature and the symbol $<>_\lambda$ stands for the Boltzmann average of the quantity enclosed using $E(\lambda)$ in the Boltzmann factor. The integration can be carried out with a quadrature[5,10,11] (preferably Gaussian) or using the slow-growth method[12] where λ is continuously varied during the simulation. The quadrature method with the exponent k in Equation (1) set to 4 was found to perform reliably and efficiently for the calculation of the free energy of liquid water[13] and for the solvation free energy of lithium and sodium ions[14]. The larger than 1 value for k is required to eliminate singularities in the integrals of equations (3,4) that would otherwise arise from the repulsive core of the potential[3]. The slow growth method was used recently to obtain the free-energy profile (i.e. to calculate the potential of mean force) of the proton transfer in aqueous solution between formic acid and methylamine by Mavri and Hadic[15]. While this latter technique ensures that all λ values will be sampled it was found that for this system there are serious ergodic difficulties ('large hysteresis').

The perturbation method[6,16,17] is based on the expression

$$\Delta A = -kT \ \ln \ <\exp[-(E_1-E_0)/kT]>_0 \ , \qquad (5)$$

where the symbol $<>_0$ stands for the Boltzmann average of the quantity enclosed using E_0 as the energy in the Boltzmann factor. While the perturbation method is widely used, the presence of the exponential in Equation (5) warrants caution for larger changes in λ since exponentiation drastically enlarges statistical fluctuations and thereby can easily introduce large numerical errors[3,18]. Finite difference thermodynamic integration[19] combines these two methods: the integral of equation (3) is evaluated by approximating the integrand with a finite difference ratio over a small λ interval. The small change in the free energy, needed in the finite difference ratio is calculated with the perturbation method. As only small changes are required in λ, the perturbation method results will be reliable.

The solvation free energy can also be related to various probabilities. These methods include the acceptance ratio method of Bennett[16], the overlap ratio method developed by Jacucci and Quirke[20] (based on the ideas put forward by Bennet[16] and shown to perform well in aqueous systems[21]) and the probability ratio method.

The probability ratio method was originally developed for the determination of the potential of mean force[21], and first applied to the determination of free energy differences by Mezei, Mehrotra and Beveridge[23]:

$$\Delta A = - kT \ln [(P(\lambda)_{\lambda=1}/V_1)/(P(\lambda)_{\lambda=0}/V_0)] , \qquad (6)$$

where $P(\lambda)$ is the Boltzmann probability of the system to be at the intermediate stage λ when λ is also a variable during the simulation and V_0, V_1 represent the configuration space volume corresponding to the $\lambda=0$ and 1 state, respectively. A familiar example for $P(\lambda)/V_\lambda$ is the radial distribution function $g(r)$ where V is the well known $4\pi r^2$ factor. Valleau, Patey and Torrie have recognized that equation (6) translates small free energy differences into large ratios in the probability of sampling and thus this method requires non-Boltzmann sampling with a modified Hamiltonian, $E'(\underline{X}^N,\underline{R}(\lambda))$ (usually referred to as "umbrella sampling")[17,24] to sample λ values whose probability is small:

$$E'(\underline{X}^N,\underline{R}(\lambda)) = E(\underline{X}^N,\underline{R}(\lambda)) + E_W(\lambda) . \qquad (7)$$

The Boltzmann average $\langle Q \rangle_B$ of any quantity Q can be recovered as

$$\langle Q \rangle_B = \langle Q \, w(\lambda) \rangle_W / \langle w(\lambda) \rangle_W \qquad \text{where} \qquad (8)$$

$$w(\lambda) = \exp [E_W(\lambda)/kT] \qquad (9)$$

and $\langle \rangle_W$ implies configurational average using the modified Hamiltonian given by equation (7). Most previous calculations determined $E_W(\lambda)$ empirically, either in tabular or in analytical form. The fact that the best choice for $E_W(\lambda)$ is $W(\lambda)$ suggested iterative approaches that not only provided computational efficiency but resulted in a method that is inherently self-checking. Paine and Scheraga[25] obtained the gas-phase conformational free energy map of the alanine dipeptide and Mezei recalculated the free energy difference between the C_7 and α_R conformations of the alanine dipetide in aqueous solution[26,18]. For the aqueous system several technical difficulties had to be overcome: matching of iterations with large statistical noise, recognition of equilibration phase, guiding the simulation to undersampled regions and others. It turned out to be important that the normalization factors of the estimated probability distributions be continually redetermined through the solution of a nonlinear minimization problem as the calculation proceeds (the "matching" problem of the previous studies). This iterative scheme is called adaptive umbrella sampling. In recent work on the dimethyl phosphate anion the adaptive umbrella sampling method proved to be significantly more reliable than the use of the harmonic weighting function[27]: on a thermodynamic cycle consisting of three distinct solute conformations the closure error was 8 kcal/mol for the harmonic method and 0.6 kcal/mol with the adaptive method. The potential of mean force between sodium and chloride ions in water has also been calculated with the adaptive method[28]. It is thus expected to perform well also for the study of a proton transfer free-energy profile.

GLYCINE ZWITTERION FORMATION

Glycine forms zwitterion in water at neutral pH. As described in the Appendix the free energy of the zwitterion formation can be estimated as -11 kcal/mol, favoring the zwitterionic form. The gas-phase energy difference has been calculated by ab initio methods most recently by Langlet, Caillet, Evleth and Kassab[29] at the 6-31G* level with geometry optimization (see references therein for earlier ab initio work). As they estimated that at the best level the neutral form is more stabile by about 20 kcal/mol, the solvation free energy must favor the zwitterionic form by about 30 kcal/mol. In the study reported here ab initio calculations were performed for the estimation of the gas phase contribution up to 2nd order Moller-Plesset level and the difference between the solvation free energies was calculated by thermodynamic integration. A major limitation of the calculations reported here is the use of a prefixed geometry for both the neutral and the zwitterionic form. Thus the results of the present calculation could be refined by exploring the effect of intramolecular conformations, requiring both additional ab-initio studies and potential of mean force calculation along the various torsion angles describing the two molecules.

Ab Inito Calculations

The molecular geometries were taken form the work of Clementi and coworkers[30,31]. Calculations were done at the STO-3G, 6-31G, 6-311G++ and 6-311G** levels[32]. The 6-311G** calculation also included correlation contributions at the MP2 level. The calculated energies are given in Table 1. The highest level calculations confirm the estimate of Langlet et al. The fractional charges on the atoms obtained by Mulliken population analysis of the ab-initio calculations are given in Table 2.

Free Energy Simulations

The water-water interactions were described by the TIP4P potential[33] and the solute-water interactions used the AMBER parameter set[34]. The atomic charges were taken from the Mulliken population analysis described above. As the fractional charges are strongly basis set dependent, calulations were performed with both the charges from the STO-3G and the 6-311G** results.

Table 1

Calculated ab-initio energies

Basis set	Glycine	Zwitterion	Difference a.u.	kcal/mol
STO-3G	-279.1029	-278.9666	0.1363	85.5
6-31G	-282.6697	-282.6424	0.0273	17.1
6-311G**	-282.8862	-282.8515	0.0347	21.8
6-311G++	-282.7537	282.7288	0.0249	15.6
6-311G**/MP2	-283.7546	-283.7182	0.0364	22.8

Table 2

Fractional charges obtained by Mulliken population analysis

	Glycine				Glycine zwitterion		
	STO-3G	6-31G	6-311**		STO-3G	6-31G	6-311**
O(H)	-0.2786	-.6935	-0.4249	O	-.4850	-.7233	-0.6246
O(C)	-0.2607	-.5718	-0.4741	O	-.4294	-.6757	-0.5804
C(O)	0.2951	.7663	0.5230	C(O)	.2406	.7269	0.4882
C(A)	-0.0407	-.2001	-0.0361	C(A)	-.0483	-.2775	-0.1428
N	-0.3855	-.8107	-0.5242	N	-.3583	-.8788	-0.3194
H(N)	0.1476	.3182	0.1940	H(N)	.3326	.4789	0.3160
II(N)	0.1584	.3326	0.2038	II(N)	.2901	.4325	0.2736
H(O)	0.2161	.4368	0.2761	H(N)	.2901	.4325	0.2736
H(C)	0.0617	.1916	0.1188	H(C)	.0837	.2422	0.1578
H(C)	0.0865	.2307	0.1434	H(C)	.0837	.2422	0.1578

The glycine was surrounded by 215 waters in a cell corresponding to the FCC close packing (truncated octahedron).

The simulations used the path described by equation (1) and evaluated the free energy differences by thermodynamic integration, i.e. using equation (4). The integrals were approximated by Gaussian quadratures. At first, the free-energy difference between the two glycine forms was calculated using the charges derived from the STO-3G calculations with a 5-point quadrature and λ exponent k=4. Next for both glycine forms the free energy differences between the models using the STO-3G and 6-311G** charges were calculated with 3-point quadratures and λ exponent k=1 (as the change from one model to an other did not involve the introduction of new repulsion centers). The diagram below gives the calculated free energy differences calculated from the simulations as well as the free energy differences derived form them and the ab initio calculations, in kcal/mol. The error estimates were obtained by the method of batch means[35,36] and represent 95% confidence intervals (2 S.D.).

```
    Glycine           -32.0±2.3  Glycine zwitterion
 (STO-3G charges)  ───────────>   (STO-3G charges)

      │ -3.8±0.2                     │ -9.2±0.2
      V                             V
    Glycine           -37.4±2.3  Glycine zwitterion
(6-311G** charges) ───────────>  (6-311G** charges)
                     (indirect)
```

Total free energy difference: -9.2±2.3 (STO-3G charges)
                                -14.6±2.3 (6-311G** charges)

Considering the approximations made in describing the
intermolecular interactions as well as the lack of complete
geometry optimization, the final result compares rather well
with the experimental estimate of -11 kcal/mol, showing that the
methods desribed here are capable of a reasonable treatment of
the thermodynamics of proton transfer.

Appendix

Estimate of the experimental free energy
of the zwitterion formation

The comparison of the pK_a of several amines show surprisingly
little variation[37]:

$NH^+_4 \rightleftharpoons NH_3 + H^+$ $\qquad\qquad pK_a=9.26$

$N^+H_3(CH_2)_2OH \rightleftharpoons NH_2(CH_2)_2OH + H^+$ $\quad pK_a=9.50$

$N^+H_3CH_3 \rightleftharpoons NH_2CH_3 + H^+$ $\qquad pK_a=10.72$

$N^+H_3CH_2CH_3 \rightleftharpoons NH_2CH_2CH_3 + H^+$ $\quad pK_a=10.67$

$H_3N^+CH_2COO^- \rightleftharpoons H_2NCH_2COO^- + H^+$ $\quad pK_a=9.6$

Thus, it is reasonable to assume that

$H_3N^+CH_2COOH \rightleftharpoons H_2NCH_2COOH + H^+$ $\quad pK_a=9.6$.

Combining this estimate with

$H_3N^+CH_2COOH \rightleftharpoons H_3N^+CH_2COO^- + H^+$ $\quad pK_a=2.35$

gives (by dividing the two equilibrium constant expressions)

$[H_3N^+CH_2COO^-]/[H_2NCH_2COOH] = 10^{-8.2}$.

Using $\Delta A = -kT \ln K$ results in $\underline{\Delta A = -11\ kcal/mol}$.

Acknowledgements

This work was supported under an RCMI grant #SRC5G12RR0307 from
NIH to Hunter College and a CUNY/PSC grant. Computing resources
were provided by the City University of New York, University
Computing Center. The author is grateful for the support of Nato
making the attendance of the workshop possible.

References

1. A. Pohorille, and L.R. Pratt, Methods in Enzymology, Biomembranes. 127:64 (1986).
2. D. Frenkel, In Molecular-Dynamics Simulation on Statistical Mechanical Systems, Soc. Italiana di Fisica, Bologna, (1986).
3. M. Mezei, and D.L. Beveridge, Ann. Acad. Sci. N.Y. 482:1 (1986).
4. D.L. Beveridge and F.M. DiCapua, Annu. Rev. Biophys. Chem. 18: 431 (1989).
5. M.R. Mruzik, F.F. Abraham, D.E. Schreiber, and G.M. Pound, J. Chem. Phys. 64:481 (1976).
6. J.A. McCammon, Science 238:486 (1987).
7. W.L. Jorgensen, and C. Ravimohan, J. Chem. Phys. 83:3050 (1985).
8. A. Cross, Chem. Phys. Letters 128:98 (1986).
9. J.G. Kirkwood, In Theory of Liquids. B.J. Alder, Ed. Gordon and Breach. New York, NY, (1968).
10. M. Mezei, Molec. Phys. 47:1307 (1982); Erratum: Ref. 13.
11. M. Mezei, S. Swaminathan, and D.L. Beveridge. J. Am. Chem. Soc. 100:3255 (1978).
12. H.J.C. Berendsen, J.P.M. Postma, and W.F. van Gunsteren, in Molecular Dynamics and Protein Structure, edited by J. Hermans, (Polycrystal Book Service, Western Springs, Illinois, 1985) p43. Chem. Soc. 17:55 (1982).
13. M. Mezei, Molec. Phys. 67:1205 (1989) (erratum to Ref. 10).
14. P.V. Maye and M. Mezei, submitted.
15. J. Mavri and D. Hadic, J. Comp. Chem. in print; this volume.
16. C.H. Bennet, J. Comp. Phys., 22:245 (1976).
17. G.M. Torrie, and J.P. Valleau, J. Comp. Phys. 23:187 (1977).
18. M. Mezei, Molecular Simulation 3301 (1989).
19. M. Mezei, J. Chem. Phys. 86:7084 (1987).
30. G. Jacucci, and N. Quirke, Molec. Phys. 40:1005 (1980).
21. M. Mezei, Molec. Phys. 65:219 (1988).
22. A. Ben-Naim, Water and Aqueous Solutions, Plenum Press, New York (1974).
23. M. Mezei, P.K. Mehrotra, and D.L. Beveridge, J. Am. Chem. Soc. 107:2239 (1985).
24. G. Patey and J.P. Valleau, J. Chem. Phys. 63:2334 (1977).
25. G.M. Paine, and H.A. Scheraga, Biopolymers 24:1391 (1985).
26. M. Mezei, J. Comp. Phys., 68:237 (1987).
27. B. Jayaram, M. Mezei, and D.L. Beveridge. 1988. J. Am. Chem. Soc. 110:1691 (1988).
28. R. Friedman and M. Mezei, to be published.
29. J. Langlet, J. Caillet, E. Evleth, and E. Kassab, in 'Modeling of Molecular Structures and Properties", J.L. Rivail, Ed., Studies in Physical and Theoretical Chemistry 71:345 (1990), Elsevier, Amsterdam.
30. E. Clementi, F. Cavallone, and R. Scordamaglia, J. Am. Chem. Soc. 99:5531 (1977).
31. L. Carozzo, G. Corongiu, C. Petrongolo, and E. Clementi, J. Chem. Phys. 68:787 (1978).
32. GAUSSIAN-82 from the laboratory of J.A. Pople, adapted to IBM by S. Topiol and R. Osmond.
33. W.L. Jorgensen, J. Chandrashekar, J.D. Madura, R. Impey, and M. Klein, J. Chem. Phys. 79:926 (1983).
34. S.J. Weiner, P.A. Kollman, D.A. Case, U.C. Singh, C. Ghio, G. Alagona, S. Profeta, Jr., and P.K. Weiner, J. Am. Chem. Soc. 106:765 (1984).

35. R.B. Blackman, and J.W. Tuckey, The Measurement of Power Spectra. Dover (1958).
36. W.W. Wood, Physics of Simple Liquids. H.N.V Temperly, F.S. Rowlinson, and G.S. Rushbrooke, Eds. North-Holland (1968).
37. D.G. Peters, J.M. Hayes and G.M. Hieftje, Chemical Separations and Measurements, Saunders, Philadelphia (1979).

A KINETIC MODEL FOR PROTON TRANSFERS IN SOLUTIONS

Luís G. Arnaut

Chemistry Department
University of Coimbra
3000, Coimbra, Portugal

INTRODUCTION

Modelling is a key feature of scientific methodology. It is an essential tool to extract the predominant physical parameters determining the behavior of complex systems. The study of reactivity in solution is one of such systems, where theoretical models are most needed. Recently it even became fashionable to interpret kinetic data in solution by means of two independent models: one for changes taking place in the reactive bonds and the other for the changes in solvation that accompany the reaction.[1, 2] A frequent outcome of such approaches is the description of reactivity by means of two "orthogonal" coordinates, separately translating the events taking place in the reactants and in the solvent molecules. The authors of such models are certainly well aware of the 104 years old statement by Menshutkin that "a solvent cannot be separated from the reaction",[3] but found it conceptually useful to develop kinetic models where such separation is carried out. It is important to realize that alternative intellectual constructs are also possible, and that the effect of solvation may be inseparable from the reaction coordinate describing the bond-breaking bond-forming process.

The Intersecting State Model, ISM, was designed to provide a framework for the calculation of rate constants both in the gas phase and in solution, using a single reaction coordinate, RC.[4, 5] The rationale behind ISM follows from the usefulness to relate, via a simple kinetic model, the information readily available for the stable reactants and products of a given reaction, with the properties of the transition state, TS, that connects them.

Ideally, a kinetic model should first make approximations leading to a physically meaningful and mathematically tractable RC. Then, it should use structural crystallographic and spectroscopic data from the potential energy regions around the reactant and product minima, and thermochemical data from the energetic difference between these minima, to provide estimates of the height of the energy barrier to the reaction. In practice, theoretical work in solution needs to invoke adjustable parameters, which admit constant values for closely related reactions, but cannot be quantitatively determined *a priori*. With the inclusion of such parameters, the models acquire some empirical character. Simple, empirical models can be very useful to rationalize experimental data, but they only become valuable theoretical tools if they can go beyond phenomenological interpretations and can predict the behavior of unknown systems. In order to do so, they need to meet the following criteria: i) the value assigned to each empirical parameter is consistent with its physical meaning, which has a quantum-mechanical analog; ii) the constancy of the empirical parameters leads to chemically meaningful reaction series, i. e., the reactants differ by substitution in positions leading to small perturbations in their transition states; iii) the values of the empirical parameters follow systematic trends with the structure of the reactants and nature of the medium, and can be extrapolated to new systems.

ISM is a serious attempt to provide a kinetic model to carry computations subscribing to the above conditions. It incorporates structural and vibrational information from the reactants and products, the free energy of the reaction and two intuitive partially adjustable parameters, to calculate activation free energies for a large number of systems. In this paper, after a brief description of the model and the physical meaning of its parameters, we consider some of its applications to proton transfer reactions, emphasizing the patterns followed by the adjustable parameters. We discuss reactions occurring both in the ground and in excited states, and give special attention to acid-base catalysis and solvent effects.

INTERSECTION STATE MODEL

Let us consider, for the sake of brevity, the prototype proton transfer reaction

$$AH + B^- \underset{k_p}{\overset{k_d}{\rightleftharpoons}} A^- + HB$$

Its RC can be divided into three consecutive regions. The first refers to configurations close to that of the reactants, where the energetic variations of the system are adequately described by the change in the AH bond length. In the intermediate region there is a strong interaction between the AH bond that is broken and the HB bond that is formed. Finally, the last region is the analogue for the first around the HB bond equilibrium configuration. It is reasonable to describe the energy variations in the first and last regions by the force constant of the AH and HB bonds respectively, f_r and f_p. However, a quantitative representation of the intermediate region is an exceedingly difficult problem.

The objective of ISM, as a kinetic model, is to provide estimates of the free energy of activation for the reaction, $\Delta G^{\ddagger}$, rather than to mimic the variation of the free energy along the whole RC. Therefore, the model deliberately avoids the difficulties associated with the intermediate region, and provides a technique to calculate $\Delta G^{\ddagger}$, using data from the first and last regions and a suitable RC. This RC is given by the sum of the reactants and products bond extensions, which are assumed to have a harmonic behavior, and it is appropriately scaled to compare the reactivity of bonds of different equilibrium bond lengths (l_{AH} and l_{HB}).

The sum of the reactant and product bond extensions at the TS (Δl), is lower than that of the isolated reactants and products ($\Delta x_r + \Delta x_p$).[6] This difference increases with the reaction energy, ΔG^0. ISM utilizes the concept of bond orders given by Pauling, to relate the sum of the bond extensions, $d = \Delta x_r + \Delta x_p$, to ΔG^0. To a good approximation it has been shown that d has a quadratic dependence on ΔG^0.[4, 5]

$$d = \eta \ (l_{AH} + l_{HB}) \tag{1}$$

$$\eta = (a' \ln 2 \ / \ n^{\ddagger}) + (a'/2) \ (\Delta G^0/\lambda)^2 \tag{2}$$

Given that a' is a constant (a'=0.156), the free energy dependence is weighted by an empirical parameter λ, called "mixing entropy". It accounts for the shape of the PES in the intermediate region. In a steep PES a given geometric change will give rise to a larger activation energy than in a more gradual PES (Figure 1), and a smaller λ is required to reproduce the energy barrier. The calculation of the reduced bond extension, η, requires another adjustable parameter: the transition state bond order $n^{\ddagger}$. In a symmetric proton exchange, $\Delta G^0 = 0$, one single bond is broken and another is formed, thus $n^{\ddagger} = 0.5$. Apparently no empiricism should be necessary to obtain the energy barriers for symmetric reactions. However, such an approach neglects the quantum mechanical resonance in the intermediate region, which leads to the splitting of reactant and product surfaces near the configuration where they intersect. The extent of the resonance will depend on the electronic nature of the reactants, and will be susceptible to dipolar field/inductive (F), to π electron delocalization/hyperconjugation (R) and to polarizability (P) effects.[6] Resonance splitting leads to a lower energy TS (Figure 2), which, in terms of ISM, is energetically equivalent to a TS with formal enhanced bonding, i. e., with a larger $n^{\ddagger}$. Thus, $n^{\ddagger}$ can be treated as an empirical parameter larger than 0.5 for bond-breaking bond-forming processes, and is related to the amount of resonance splitting in the intermediate region.

Both λ and $n^{\ddagger}$ are solvent dependent.[7] As the polarity of the solvent increases, its influence on the RC of proton transfers will also increase. Qualitatively, the mixing of lower frequency solvent modes into the RC, will transform it into a smoother RC, which must be reflected by an increased λ. On the other hand, in polar solvents, the electronic wavefunctions describing reactants and products are spread over larger, and different, dimensions and their overlap will be smaller. This

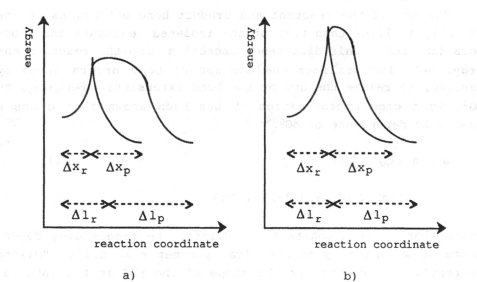

a)   b)

Figure 1. Effect of the PES shape in the intermediate region on the parameter λ. a) Gradual PES, translatable by a larger λ because the dependence on ΔG^0 is smaller. b) Steep PES, reflected by a smaller λ, required by the higher $\Delta G^{\ddagger}$.

leads to less resonance at the TS, translated by ISM as a $n^{\ddagger}$ closer to 0.5.

Given the values of λ and $n^{\ddagger}$, a model for the free energy profile of the reaction can be constructed (Figure 3) and simple relations are obtained to calculate $\Delta G^{\ddagger}$:

$$(1/2) \ f_r \ x^2 = (1/2) \ f_p \ (d - x)^2 + \Delta G^0 \tag{3}$$

$$\Delta G^{\ddagger} = (1/2) \ f_r \ x^2 \tag{4}$$

In summary, the rate constant for a proton transfer in solution, k, can be calculated from the knowledge of the AH and HB equilibrium force constants and bond lengths, the reaction free energy, the value of d given by eqs 1-2, and the use of an Eyring type of expression

$$k = (k_B T/h) \ C^{1-m} \ \exp(-\Delta G^{\ddagger}/RT) \tag{5}$$

where C=1 M is the standard concentration and m=2 is the molecularity of the reaction.

It is convenient to make statistical corrections to compare proton transfers between different acids and bases

$$\Delta G^0 = -R \ T \ [2.303 \ pK_{BH} + \ln(p_B/q_B) - 2.303 \ pK_{AH} - \ln(p_A/q_A)] \tag{6}$$

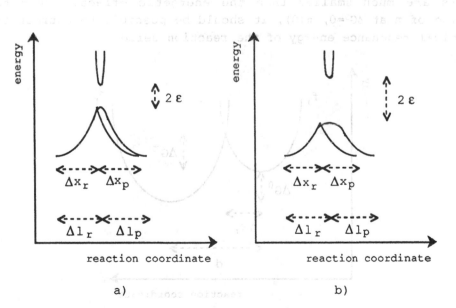

Figure 2. Effect of the resonance splitting, ε, in the intermediate region on the parameter $n^{\ddagger}$. a) Small resonance, leading to $n^{\ddagger} \approx 0.5$. b) Large resonance, implying $n^{\ddagger} > 0$.

$$k_d' = k_d / p_A / q_B \qquad (7)$$

$$k_p' = k_p / p_B / q_A \qquad (8)$$

where p_A (p_B) is the number of equivalent acidic protons in AH (BH) and q_A (q_B) is the number of equivalent basic sites in A^- (B^-).

The applications of ISM described in the remaining part of this work use preliminary knowledge of the rate constants to extract typical values of λ and $n^{\ddagger}$ for a large variety of systems. The results obtained can be used to rationalize and classify the experimental data, and provide the parameters necessary to apply ISM to new systems. In these applications we will use $[H_2O]=55.5$ M, which leads to $pK_a(H_3O^+)=-1.74$ and $pK_a(H_2O)=15.74$. Other parameters used in these calculations are $f_{OH}=420$, $f_{NH}=380$, $f_{CH}=290$, $f_{FH}=580$ J mol^{-1} pm^{-2}, and $l_{OH}=97$ (95.8 for water), $l_{NH}=101$, $l_{CH}=107$, $l_{FH}=97.1$ pm.[8]

APPLICATIONS

The value of λ and of $n^{\ddagger}$ for each reaction series can, in principle, be obtained from the η values that reproduce the experimental rate constants of proton transfers involving structurally related acids and bases, applying eq. 2. This assumes that, within the series, the variations in the electronic factors are much smaller than the energetic effects. From the value of η at $\Delta G^0=0$, $\eta(0)$, it should be possible to extract the typical resonance energy of the reaction series.

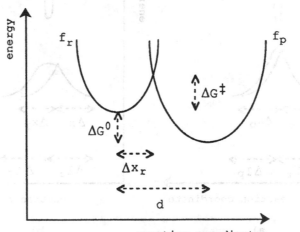

Figure 3. Representation of the reaction coordinate according to ISM; Δx_r is the reactants bond extension, and d the total bond extension to the TS.

In Figure 4 we present the correlations obtained for the protonation of different carbon bases by carboxylic acids in aqueous solutions at 298 K. The relevant parameters for these reactions are collected in Table 1. Such correlations neglect variations in $n^{\ddagger}$ with different substituents either in the carbon base or in the carboxylic acids. We have discussed extensively the nature and extent of the electronic effects induced by substituents in nitroalkanes.[6] We were able to show that they are significant and are quantitatively related to field/inductive and delocalization/hyperconjugation effects. This advises against gathering together different nitroalkanes in η vs $(\Delta G^0)^2$ plots. We also provided evidence illustrating that the deprotonation of series of substituted ketones admit smaller variations in $n^{\ddagger}$ than series of substituted nitroalkanes, and related this to the lower electron affinity of the former. Thus, it seems reasonable to plot together different CO- or CN-substituted carbon bases. It is important to realize that in the nitronate ion the negative charge is localized in the oxygen atoms and the nitro group contributes to the RC. This yields effective force constants and bond lengths with contributions from four bonds. This also happens with the ketones, but to a smaller extent, and the best description of their reactivity is intermediate between no participation and

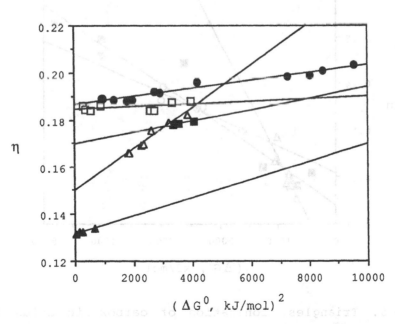

Figure 4. Protonation of carbon bases by carboxylic acids in aqueous solutions, at 298 K. Filled triangles, $(NO_2)_2CH_2$; open triangles $(CN)_2CHR$; filled squares, $(CH_3CH_2SO_2)_2CHCH_3$; open squares, $(RCO_2)_xCH_{4-x}$; circles, CH_3COR.

Table 1. Optimized λ and $n^{\ddagger}$ values for the protonation of carbon bases by carboxylic acids $A^- + RCOOH \rightarrow AH + RCOO^-$ in aqueous solution, at 298 K.[a]

AH	λ (kJ/mol)	$n^{\ddagger}$
$(RCO_2)_xCH_{4-x}$	377	0.585
$(CH_3CH_2SO_2)_2CHCH_3$	180	0.636
$(CN)_2CHR$	94	0.723
CH_3COR	218	0.578[b]
$(NO_2)_2CH_2$	143	0.824

[a]Kinetic and equilibrium acidities from Ref.[9].
[b]Using f_r=608, f_p=584 J/(mol pm^2) and l=239 pm,[6] one obtains λ=315 kJ/mol and $n^{\ddagger}$=0.901.

equal participation of the CO bond with the CH bond in the RC.[6]

The role played by the electronic and energetic factors in the series of carboxylic acids can be assessed by studying their deprotonation in water, Figure 5. The correlations with

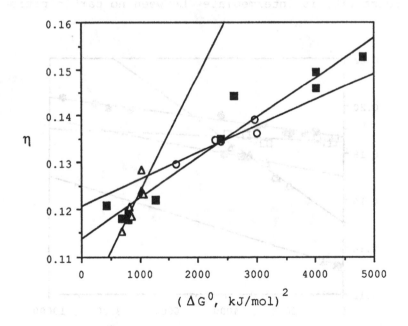

Figure 5. Triangles, ionization of carboxylic acids (λ=56 kJ/mol, $n^{\ddagger}$=1.097); squares, ionization and hydrolysis of amines (λ=95 kJ/mol, $n^{\ddagger}$=0.851); circles, ionization of phenols (λ=117 kJ/mol, $n^{\ddagger}$=0.897). Kinetic and equilibrium acidities from Ref.[10].

the energetic effects are encouraging. However, the relationship of RCH_2COOH (R=H, CH_3, CH_3CH_2, Ph) with electronic effects[11] gives,

$$n^{\ddagger}(RCH_2COOH) = 0.842 + 0.155 \ F - 0.082 \ R \qquad (9)$$

with a correlation coefficient of 0.995, This coefficient drops
to 0.901 for the series RCOOH, and the field effect decreases
although the proximity of the substituent to the reaction center increases, eq. 10. This does not seem reasonable, as the
importance of field effects in carboxylic acids acidity has already been shown.[12] Probably these correlations obtained with
electronic effects are largely coincidental, resulting from the
small range of substituents studied. The 2-hydroxy substituent
seems to lead to specific interactions and was not included in
these correlations.

$$n^{\ddagger}(RCOOH) = 0.936 + 0.094 \ F + 0.174 \ R \qquad (10)$$

It is important to note that $n^{\ddagger}$ the values reported in Table 1
are much smaller than those obtained for carboxylic acids. They
seem to be essentially determined by the nature of the carbon
acid, which will render negligible, in the catalysis of carbon
bases, the eventual variation of $n^{\ddagger}$ with the substituent in the
carboxylic acid.

In summary, we cannot exclude some electronic effects in
the series of oxygen or nitrogen acids and bases, but their use
as catalysts of carbon bases and acids seems to define series
with essentially constant λ and $n^{\ddagger}$ values.

In symmetric proton exchanges λ has no effect on η. Table 2
presents some results of proton exchanges between oxygen and
nitrogen acids/bases, in some cases mediated by water
molecules. The pre-exponential factor for the H_3O^+ and OH^- catalyzed exchanges may be larger than the others, because a
Grotthus type of mechanism may operate.[13] A correction for this
would result in $n^{\ddagger}$ values closer to 1.0. The exchanges between
amines and ammonium ions are subject to both electronic and
steric effects. The former tend to increase $n^{\ddagger}$ with alkyl substitution, while the latter will increase η, specially in direct proton exchanges. The exchange of the proton in acetic
acid seems to involve a ring formed by the two oxygens of
acetic acid and two water molecules. This system is not directly comparable with the bimolecular rate constants of Table
2. Fortunately, a very good $\eta(0)$ value for proton exchange in
carboxylic acids can be obtained from the proton transfer from
acetic acid to carboxylic bases in aqueous solutions. The re-

Table 2. Proton exchange in symmetric systems.

	k_d' (M^{-1} s^{-1})	η	$n^\ddagger$	ε (kJ/mol)
$H_3O^+ - H_2O$	3.3×10^9	0.0977	1.106	72.7[a]
$OH^- - H_2O$	1.7×10^9	0.1019	1.061	71.1[a]
$PhOH - H_2O - {}^-OPh$	7.1×10^8	0.1073	1.008	68.9[a]
$CH_3COOH - (H_2O)_n - {}^-OOCCH_3$			0.939	65.4[b]
$NH_4^+ - NH_3$	3.2×10^8	0.1131	0.9555	66.6[c]
$NH_4^+ - H_2O - NH_3$	1.2×10^7	0.1304	0.8289	57.9[c]
$CH_3NH_3^+ - NH_2CH_3$	1.4×10^8	0.1176	0.9191	64.0[c]
$CH_3NH_3^+ - H_2O - NH_2CH_3$	1.8×10^8	0.1160	0.9317	64.8[c]
$(CH_3)_2NH_2^+ - NH(CH_3)_2$	2.5×10^7	0.1270	0.8513	59.6[c]
$(CH_3)_2NH_2^+ - H_2O - NH(CH_3)_2$	5.0×10^8	0.1106	0.9773	67.1[c]
$(CH_3)_3NH^+ - H_2O - N(CH_3)_3$	3.6×10^8	0.1114	0.9705	66.8[c]

[a]Rate constants from Ref.[14].

[b]Obtained from best fit to data plotted in Figure 6, restricted to the condition $\lambda >> |\Delta G^0|$. The number of water molecules n is not known.

[c]Rate constants from Ref.[10].

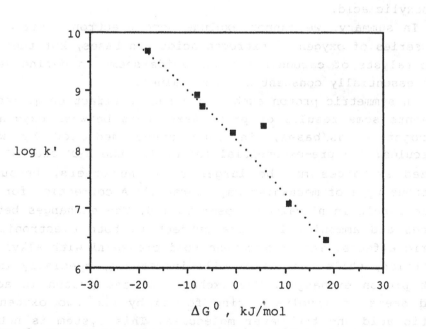

Figure 6. Proton transfer from acetic acid to bases in aqueous solutions at 293K, ionic strength 1.0 M. The bases are: $H_2PO_4^-$, $(CH_3)_2AsO_2^-$, $HO_2CCH_2CO_2^-$, $CH_3CH_2CO_2^-$, HCO_2^-, $ClCH_2CO_2^-$, $Cl_2CHCO_2^-$. Kinetic and equilibrium acidities from Ref.[15].

sults presented in Figure 6 were obtained with $f_r = f_p$, $\lambda \gg |\Delta G^0|$ and $n^{\ddagger} = 0.939$. Under these conditions the Brønsted plot is linear in the range amenable to experimental studies. The point that deviates is due to catalysis by $HCOO^-$. It is interesting to note that, of all the reactions studied here, only in this catalysis by $HCOO^-$ and in the deprotonation of 2-hydroxy substituted carboxylic acids by water, is the reverse rate constant (calculated with $\Delta G_r^{\ddagger} = \Delta G_f^{\ddagger} + \Delta G^0$ and eq. 4) not within a factor of two of the experimental value.

Results on the application of ISM to the deprotonation of several acids are presented in Table 3. Carbon acids have high λ values and approximate $n^{\ddagger}$ values can be calculated from η. It is encouraging to note that the extrapolated $n^{\ddagger}$ values for the deprotonations of Table 3 are, in general, in good agreement with the values obtained for the protonations presented in Table 1. Only the $n^{\ddagger}$ values for $(CN)_2CHR$ are significantly different, but this is certainly due to its lower λ value. The trend for higher $n^{\ddagger}$ values for amines, phenols and carboxylic acids is consistent in Figures 5 and 6, and in Tables 2 and 3.

Excited state proton transfers are characterized by an increase in the rates of protonation that may cover 15 orders of magnitude, and by pK_a changes up to 30 units.[17] Such variations with the electronic excitation make it very difficult for any model to interpret both ground and excited state proton transfers. Table 4 presents the optimized λ and $n^{\ddagger}$ values for series

Table 3. Optimized η values for the deprotonation of several acids $AH + H_2O \rightarrow A^- + H_3O^+$, at 298 K.[a]

AH	ΔG^0 (kJ/mol)	k_d' (M^{-1} s^{-1})	η	$n^{\ddagger}$	ε (kJ/mol)
$(CH_3CH_2CO_2)_2CH_2$	84.8	2.3×10^{-7}	0.1907	0.567	16.9
$(CH_3CH_2SO_2)_2CH_2$	78.5	1.4×10^{-4}	0.1689	0.640	29.3
$(CN)_2CH_2$	71.7	2.6×10^{-4}	0.1749	0.618	26.6
$(CH_3CO)_2CH_2$	59.7	1.4×10^{-4}	0.1894[b]	0.571	18.6
$(NO_2)_2CH_2$	29.3	7.5×10^{-3}	0.1314	0.823	119.2
NH_4^+	63.4	1.1×10^{-1}	0.1460		
$o-O_2NC_6H_5OH$	48.1	1.8×10^1	0.1392		
H_2S	50.2	3.9×10^1	0.1403		
CH_3COOH	32.5	1.4×10^4	0.1257		
HF	23.5	1.2×10^6	0.1056		

[a]See Ref.[6] for carbon acids data, and Ref.[16] for the other data. [b]Using $f_r = 584$, $f_p = 608$ J/(mol pm^2) and $l = 239$ pm,[6] one obtains $n^{\ddagger} = 0.8981$, $\varepsilon = 134.7$ kJ/mol.

Table 4. Optimized λ and $n^{\ddagger}$ values for the deprotonations of aromatic alcohols and ammmonium ions by H_2O, protonation of the carbonyl in naphthoic acids and methyl p-naphthyl ketone by H_3O^+, and deprotonation of 2-methylstyrene by H_2O, H_3BO_4, $H_2PO_4^-$, H_3PO_4 and H_3O^+.

	λ (kJ/mol)	$n^{\ddagger}$
p-methylstyrene	114	0.51[a]
$ArNH_3^+$	99	0.838[b]
$RC=O$	74	1.052[b]
$ArOH$	51	1.129[b]

[a]From Ref.[19].
[b]From the kinetic acidities reviewed in Ref.[17].

of excited state reactions. The trends found in the excited state reactions are close to those of the ground state ones. According to ISM, the most remarkable change brought about by electronic excitation in these molecules is the decrease in λ. This has been attributed to their increase in dipole moment,

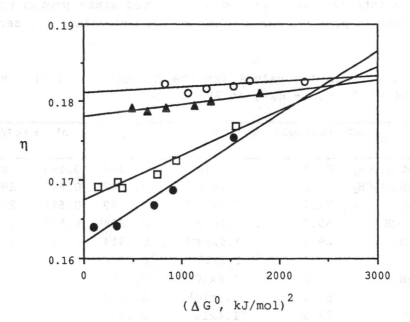

Figure 7. Solvent effects on the deprotonation of acetylacetone by carboxylate ions. Open circles, 100% H_2O (λ=326 kJ/mol, $n^{\ddagger}$=0.597); triangles, 50% H_2O-50% Me_2SO (λ=224 kJ/mol, $n^{\ddagger}$=0.607), squares, 10% H_2O-90% Me_2SO (λ=117 kJ/mol, $n^{\ddagger}$=0.646), filled circles, 5% H_2O-95% Me_2SO (λ=97 kJ/mol, $n^{\ddagger}$=0.668).

which leads to a tighter TS than their analogous ground state reactions in polar solvents.[17] We must recognize that there is an appreciable error in $\lambda(ArNH_3^+)$ due to the small range of ΔG^0 values experimentally available. Yates applied several kinetic models to photochemical proton transfer reactions and concluded that ISM is to be preferred.[18] The differences in the parameters obtained by Yates and those reported in Table 4 for the deprotonation of methylstyrene, result from the treatment of these reactions as adiabatic vs. diabatic, and do not invalidate the applicability of ISM.

Solvent effects can also be studied in the framework of this model. The deprotonation of acetylacetone by carboxylate ions in different mixtures of water/dimethyl sulfoxide (Figure 7) illustrates this effect.[7] As the reactants and products become more strongly solvated, λ increases and $n^{\ddagger}$ decreases. This is caused by the involvement of some of the electron density of the acid and base active sites in solvation.

Finally, the effect of ion pair formation can also be modelled by ISM. A very interesting system to study is the deprotonation of toluene by lithium and cesium cyclohexylamide (LiCHA, CsCHA) in cyclohexylamine, because the statistically corrected pK_a values of toluene and of the solvent are very close (41.2 and 41.6),[20] and energetic effects are thus negligible. The hydrogen exchange rate of toluene with LiCHA is 0.10 $M^{-1} s^{-1}$ at 298 K,[21] which leads to $n^{\ddagger}=0.525$. The exchange rate with CsCHA is 10^3-10^4 times faster,[22] thus $n^{\ddagger} \approx 0.60$. It is known that the small lithium cation requires extensive solvation, which is provided by a substantial interaction with the cyclohexylamide ion; on the other hand, the large cesium cation has much lower solvation requirements.[23] For proton transfers in apolar solvents, one would expect lithium salts to induce a behavior similar to that of a polar solvent, while with cesium salts the situation typical of apolar solvents is approached, as observed.

CONCLUSION

The Intersecting State Model is a kinetic model that relates the properties of reactants and products with some features of the transition state that connects them. ISM provides a framework to calculate reaction rates, making use of two adjustable parameters, the entropy of mixing (λ) and the transition state bond order ($n^{\ddagger}$). The former is related to the shape of the potential energy surface in the TS region, and is high for a gradual surface and low for a steep one. The latter is associated to the resonance splitting at the TS, and approaches the lower limit of 0.5 when the resonance is small (TS is not stabilized by an increased bonding character of the electrons present). From the application of the model to a wide range of systems, some general trends emerge. Some outstanding features exhibited by the model are: i) the higher $n^{\ddagger}$ values for oxygen and nitrogen acids than for carbon acids, and the appreciable dependence of the latter on field/inductive and π electron delocalization/hyperconjugation effects; ii) the relatively low λ values observed in oxygen and nitrogen acids, specially in excited state proton transfers; iii) the increase in $n^{\ddagger}$ and decrease in λ when the solvent is changed so as to decrease the extent of solvation; this effect is also observed in ion pair formation, when the coulombic interaction becomes looser with the increase of the cation size. These applications of ISM clearly show that it is a valuable tool to interpret and predict the rates of proton transfer reactions.

REFERENCES

1. W. J. Albery, C. F. Bernasconi and A. J. Kresge, J. Phys. Org. Chem., 1:29-31 (1988).
2. W. J. Albery, Ann. Rev. Phys. Chem., 31:227-267 (1980)
3. N. Menshutkin, Z. Physik. Chem., 1:611 (1887).
4. A. J. C. Varandas and S. J. Formosinho, J. Chem. Soc., Faraday Trans. 2., 82:953-962 (1986).
5. S. J. Formosinho, in: "Theoretical and Computational Models for Organic Chemistry," S. J. Formosinho, I. G. Csizmadia and L. G. Arnaut, ed., Kluwer, Dordrecht (1991), p. 159-206.
6. L. G. Arnaut, J. Phys. Org. Chem., in press:(1991)
7. L. G. Arnaut and S. J. Formosinho, J. Phys. Org. Chem., 3:95-109 (1990).
8. A. J. Gordon and R. A. Ford, "The Chemist's Companion," Wiley, New York (1972), p.107, 114.
9. A. Argile, A. R. E. Carey, G. Fukata, M. Harcourt, R. A. More O'Ferrall and M. G. Murphy, Isr. J. Chem., 26:303-312 (1985).
10. J. E. Crooks, in: "Comprehensive Chemical Kinetics," C. H. Bamford and C. F. H. Tipper, ed., Elsevier, Amsterdam (1977), p. 197-250.
11. C. G. Swain, S. H. Unger, N. R. Rosenquist and M. S. Swain, J. Am. Chem. Soc., 105:492-502 (1983).
12. M. R. F. Siggel, A. Streitwieser Jr. and T. D. Thomas, J. Am. Chem. Soc., 110:8022 (1988).

13. W. J. Moore, "Physical Chemistry," Prentice-Hall, New Jersey (1972), p. 435.

14. Z. Luz and S. Meiboom, J. Am. Chem. Soc., 86:4766-4768 (1964).

15. F. Hibbert, Adv. Phys. Org. Chem., 22:113-212 (1986).

16. S. J. Formosinho, J. Chem. Soc. Perkin Trans. 2., 61-66 (1987).

17. L. G. Arnaut and S. J. Formosinho, J. Phys. Chem., 92:685-691 (1988).

18. K. Yates, J. Phys. Org. Chem., 2:300-322 (1989).

19. L. G. Arnaut and S. J. Formosinho, to be published.

20. A. Streitwieser Jr. and F. Guibé, J. Am. Chem. Soc., 100:4532-4534 (1978).

21. A. Streitwieser Jr., P. H. Owens, G. Sonnischsen, W. K. Smith, G. R. Ziegler, H. M. Niemeyer and T. L. Kruger, J. Am. Chem. Soc., 95:4254-4257 (1973).

22. A. Streitwieser Jr., R. A. Caldwell, R. G. Lawler and G. R. Ziegler, J. Am. Chem. Soc., 87:5399-5402 (1965).

23. A. Streitwieser Jr. and R. A. Caldwell, J. Am. Chem. Soc., 87:5394-5399 (1965).

RAMAN INVESTIGATION OF PROTON HYDRATION AND STRUCTURE

IN CONCENTRATED AQUEOUS HYDROCHLORIC ACID SOLUTIONS

G. E. Walrafen, Y. C. Chu, and H. R. Carlon*

Chemistry Department
Howard University
Washington, D. C. 20059

ABSTRACT

Polarized X(ZZ)Y, depolarized X(ZX)Y, and isotropic X(ZZ)Y - (4/3)X(ZX)Y Raman spectra, all Bose-Einstein corrected, were obtained between ≈3 and 350 cm⁻¹ from concentrated aqueous HCl solutions having compositions of 32.1, 34.6, and 36.9 wt. %. These spectra were compared to the corresponding spectra from pure water and concentrated aqueous NaCl. Marked Raman intensity enhancements were observed for bands between ≈50 to ≈200 cm⁻¹ compared to the corresponding bands from liquid water, and these enhancements are thought to be related to proton tunnelling, which occurs because the hydrogen bonds of the concentrated HCl solutions are short, ≈2.5 Å, and nearly symmetric. Also, a very intense, broad, and asymmetric peak centered near 200 cm⁻¹ dominates the istoropic spectra from the HCl solutions, and a peak near 50 cm⁻¹ dominates the X(ZX)Y spectra. These features are readily modelled by the $H_9O_4^+$ ion, and intermolecular force constants for $H_9O_4^+$ were obtained from normal mode calculations. An isotropic Raman component observed near 120-135 cm⁻¹ is suggested to arise from $Cl^-(H_2O)_n$.

INTRODUCTION

Aqueous HCl solutions having compositions of 36.89, 34.60 and 32.12 wt. %; densities of 1.1832, 1.1721 and 1.1599 g-cm⁻³; and, stoichiometric [H₂O]/[HCl] mole ratios of 3.46, 3.83, and 4.28, respectively, were examined by means of low-frequency Raman spectroscopy in this work (1). A stoichiometric [H₂O]/[HCl] mole ratio of four corresponds to $H_9O_4^+$ which should be the major cationic species at this composition (2,3) However, another significant structure should be $Cl^-(H_2O)_n$, where n is a small number (4). Moreover, one might expect the $H_9O_4^+$ and $Cl^-(H_2O)_n$ ions to share water molecules in some sort of compact (albeit temporal) structure, formally, $H_3O^+(H_2O)_mCl^-$, where m is a small integer. This sharing of water molecules is suggested by the short O-Cl and Cl-Cl distances of 3.1 and 3.5 Å, respectively, as seen from the x-ray pair correlation function (5).

Proton Transfer in Hydrogen-Bonded Systems
Edited by T. Bountis, Plenum Press, New York, 1992

$H_9O_4^+$ has long been considered to be an energetically important structure of the hydrated proton in water (3). The case for and against other structures, e.g., $H_5O_2^+$, $H_7O_3^+$, etc., has been made in an excellent recent review (6). Whether or not $H_7O_3^+$ is a significant species in the solutions studied here in addition to $H_9O_4^+$ is a difficult question which we will not attempt to answer.

$H_9O_4^+$ is thought to have a central core of H_3O^+ which is trigonal, i.e., C_{3v} symmetry. Three water molecules are attached by linear (or nearly linear) hydrogen bonds to this core, resulting in a structure which is probably also trigonal in terms of the central oxygen atom and the three oxygen atoms in the first hydration shell of the H_3O^+. One of these three H_2O molecules is absent in $H_7O_3^+$, and two are absent in $H_5O_2^+$. However, a factor of paramount importance is that the hydrogen bonds to the H_3O^+ core, whether one, two, or three, are very short, and not too far from symmetric, i.e., the proton is not too far from being centered between its nearest-neighbor oxygen atoms.

The x-ray pair correlation function for an aqueous HCl solution of composition, $[H_2O]/[HCl] = 3.99$, displays a sharp, well-resolved peak near 2.52 Å (5). This value is just above the O-O distance of 2.4 Å, at, or below which, truly symmetric hydrogen bonds are thought to occur (7). A normal asymmetric hydrogen bond, in contrast, is much longer, $\approx 2.7 - \approx 2.9$ Å.

An ordinary asymmetric hydrogen bond, O-H$\cdots$O, such as that occurring in water, has an O-H distance of ≈ 1 Å, and an H$\cdots$O distance of ≈ 1.9 Å. A truly symmetric hydrogen bond, O-H-O, such as might occur when H_2O molecules are subjected to pressure above ≈ 500 kbar (8), should have an O-O distance of ≈ 2.4 Å, and an O-H distance of ≈ 1.2 Å.

The potential well for an ordinary asymmetric hydrogen bond is double, and the barrier to quantum mechanical tunnelling is large, and difficult to surmount (9). In contrast, the potential well for a symmetric hydrogen bond is single, and flat, which makes quantum mechanical tunnelling easy (9).

A rough sketch of the tunnelling situation, presented solely for perspective, results from approximating the order of magnitude of the velocity and the frequency of the proton tunnelling by treating the proton as a deBroglie wave constrained between its two nearby oxygen atoms.

Consider that proton tunnelling occurs freely within one of the short, nearly symmetric, O-H$\cdots$O units of the $H_9O_4^+$ ion. Further, assume that the deBroglie wave of the proton, wavelength λ_p, forms a standing wave within the 2.52 Å O-O distance, obtained from the x-ray data (5). The pertinent de Broglie relation for the proton velocity, V_p, is then:

$$V_p = h/m_p\lambda_p, \qquad (1)$$

where m_p is the proton rest mass, 1.673×10^{-24} g, $\lambda_p = 2.52 \times 10^{-8}$ cm, and h is Planck's constant, 6.6262×10^{-27} erg-sec.

A proton tunnelling velocity of $V_p = 1.57 \times 10^5$ cm-sec^{-1} results from eqn (1). Moreover, one obtains a tunnelling frequency, $\nu_p = 6.24 \times 10^{12}$ sec^{-1}, from $V = \lambda\nu$. These velocity and frequency values are only approximate, of course.

It is next useful to compare the proton tunnelling frequency with the intermolecular vibrational frequencies of pure liquid water.

One of the major intermolecular vibrations of liquid water involves restricted translations of two H_2O molecules along the line of the hydrogen bond which joins them (parallel restricted translation). An alternate description of this mode is as a longitudinal, dilatational p-wave of a group of hydrogen-bonded H_2O molecules (10). Another related intermolecular vibration of liquid water may be described as O-O-O bending of hydrogen-bonded molecules (perpendicular restricted translation). An alternate description in this case is as a transverse, shear, s-wave of a group of hydrogen-bonded H_2O molecules (10). The p-wave is centered in the general vicinity of ≈ 155 cm^{-1}, and the s-wave near $\approx 45-50$ cm^{-1} [this work, and (10)]. Both of these intermolecular modes give rise to very broad Raman bands whose breadth results from inhomogeneous broadening.

The point of the above description is that the vibrational frequencies of the restricted translational mode from pure liquid water, ≈ 4.7 x 10^{12} sec^{-1}, and its apparent analogue for concentrated HCl (isotropic peak, ≈ 200 cm^{-1}, this work), about 6.0 x 10^{12} sec^{-1}, are close to the deBroglie proton frequency. Therefore, it is reasonable to expect that a strong coupling might occur between the restricted translations, and the proton tunnelling, because of the near resonance of their frequencies, 4.7 to 6.0 x 10^{12} sec^{-1} compared to 6.2 x 10^{12} sec^{-1}.

The short, 2.52 Å hydrogen bond is the strong intermolecular interaction which makes the restricted translation within the $H_9O_4^{\cdot}$ ion, that is, O versus H_2O of an O-H-O unit, possible, and proton tunnelling should occur within this short 2.52 Å distance.

Such resonant interaction between the intermolecular phonon mode and proton tunnelling should have a major effect on several features of the intermolecular Raman spectrum from concentrated aqueous HCl. For example, proton tunnelling should enhance the low-frequency Raman intensity. Rapid proton tunnelling along the O-H$\cdots$O direction should increase the Raman polarizability derivatives via a close temporal approach between the proton and the polarizable O atoms. Indeed, we have found that marked intensity enhancements occur in the low-frequency Raman spectrum from the concentrated acid, compared to water. This effect, and other interesting effects in the Raman spectrum are now described, and the low-frequencies observed from the concentrated HCl solutions are also modelled according to the normal modes of $H_9O_4^+$ plus a contribution from $Cl^-(H_2O)_n$.

RAMAN SPECTRA

Low-frequency Raman spectra obtained between 3 and 350 cm^{-1} are shown for 34.6 wt. % HCl in Fig. 1. The X(ZZ)Y Raman spectrum is shown in (a), the X(ZX)Y Raman spectrum in (b), and the isotropic Raman spectrum, X(ZZ)Y - (4/3)X(ZX)Y, in (c). Corresponding Raman spectra from water are shown for comparison in (a) and (b), but the isotropic Raman spectrum from water is shown subsequently.

If the polarized, X(ZZ)Y, amplitudes from the HCl spectra in (a) near 50 and 170 cm^{-1} are compared with those from water, it is evident that the HCl spectra are roughly 3.5 times more intense than the water spectra, under the same conditions of excitation intensity and detection.

An intensity ratio of roughly 3.5 also results near 50 cm^{-1} for the depolarized Raman spectra, X(ZX)Y, shown in (b). However, the intensity ratio at 170 cm^{-1} is only about 2.5, because of the obvious polarization of the X(ZX)Y spectrum evident in the general

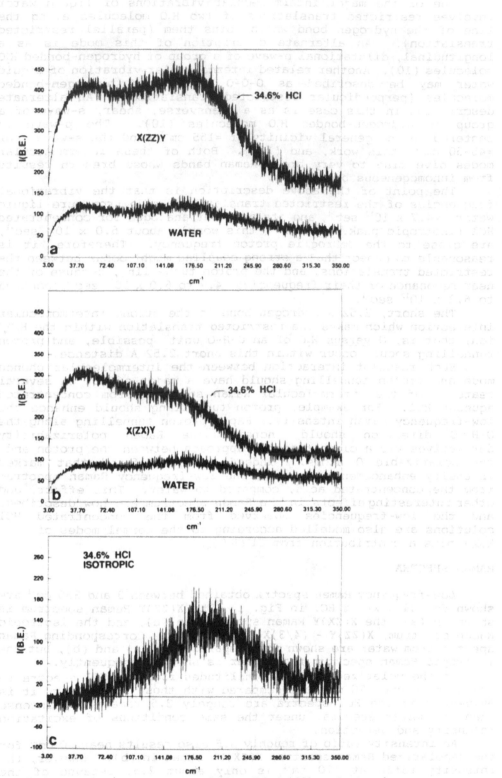

Figure 1. X(ZZ)Y, (a), top. X(ZX)Y, (b), middle. Isotropic, X(ZZ)Y - (4/3)X(ZX)Y, (c), bottom.

vicinity of 170 cm^{-1}. This polarization is emphasized by the isotropic spectrum shown in (c).

The (c) spectrum of Fig. 1 shows an intense peak near 200 cm^{-1}. However, the peak of the X(ZZ)Y spectrum occurs near 167 cm^{-1}, and the shoulder of the X(ZX)Y spectrum is centered in the vicinity of about 155 cm^{-1}, which means that the pure isotropic spectrum is shifted upward by about 30 to 45 cm^{-1}.

A difference of 45 cm^{-1} between the isotropic Raman spectrum and the shoulder position of the depolarized Raman spectrum means that two (or more) broad components are involved. One of these broad components occurs near 200 cm^{-1}, and it is highly polarized. The other broad component occurs near, and because of the overlapping 200 cm^{-1} neighbor component, well below, the depolarized shoulder position of ≈155 cm^{-1}. The position of this second component is about 120-135 cm^{-1}, but this is considered subsequently in relation to Gaussian deconvolution of the Raman spectra.

Polarized, X(ZZ)Y, plus depolarized, X(ZX)Y, (a), and isotropic Raman spectra, X(ZZ)Y - (4/3)X(ZX)Y, (b), are shown in Figs. 2 and 3 for 36.9 wt % and 32.1 wt. % HCl, respectively. Figs. 2 and 3 were obtained under essentially identical conditions of excitation intensity and detection. (These conditions differ from those of Fig. 1, for which the excitation intensity was smaller.)

Peaks occur near 50-51 cm^{-1} and near 164-169 cm^{-1} in the polarized spectra of Figs. 2 and 3, and in the depolarized spectra near 43 cm^{-1}, (a). Peaks occur near 202 cm^{-1} in the isotropic Raman spectra of Figs. 2 and 3. Nominal, 200 cm^{-1}.

Polarized and depolarized Raman spectra from water are shown in Fig. 4, (a). The corresponding isotropic Raman spectrum from water is shown in Fig. 4, (b). The isotropic Raman spectrum from water is very weak between 3 and 350 cm^{-1}, which indicates that the deporization ratios in this region are close to 3/4.

The contrast between the isotropic Raman spectrum from water, Fig. 4 (b), and the isotropic Raman spectra from the HCl solutions, Figs. 1, 2, and 3, emphasizes the very strong polarization near 200 cm^{-1} from the HCl solutions.

INTERPRETATION

Anions such as Cl$^-$, Br$^-$, and I$^-$ (11, 12) are known to produce very large effects in the intermolecular and intramolecular regions of the Raman spectrum of water. The effects of cations such as Li$^+$, Na$^+$, and K$^+$, on the other hand, are generally small.

Whether or not the hydrated proton falls into the normal category of small Raman cationic effects is a major question of the present interpretation. We believe, in short, that it does not, because of the presence of short, nearly symmetric hydrogen bonds.

Low-frequency Raman spectra were recently reported for NaI in water (11). An intense peak near 155 cm^{-1} was observed in the isotropic Raman spectrum. A shoulder near 45 cm^{-1} was also observed in the isotropic spectrum (11). (The isotropic spectrum from NaI, like those of Figs. 1 to 4 shown here, was corrected for the Bose-Einstein thermal population factor.)

The iodide ion may be surrounded by as many as six water molecules which form its first hydration layer (11). I$^-$···H-O

301

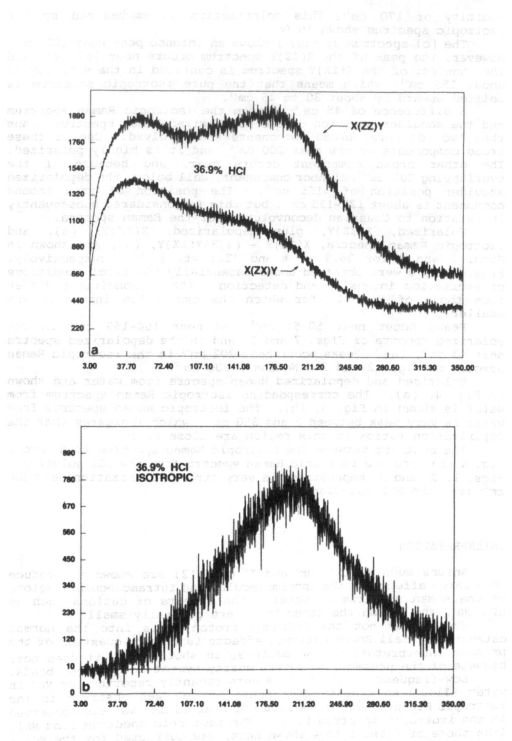

Figure 2. X(ZZ)Y, and X(ZX)Y, (a), top. Isotropic,
X(ZZ)Y - (4/3)X(ZX)Y, (b), bottom.

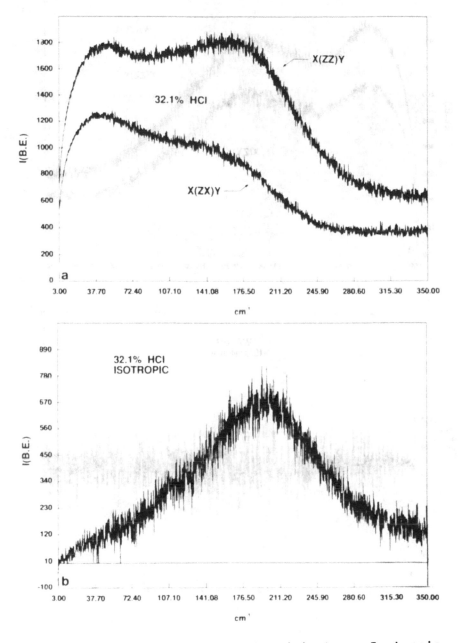

Figure 3. X(ZZ)Y, and X(ZX)Y, (a), top. Isotropic,
X(ZZ)Y - (4/3)X(ZX)Y, (b), bottom.

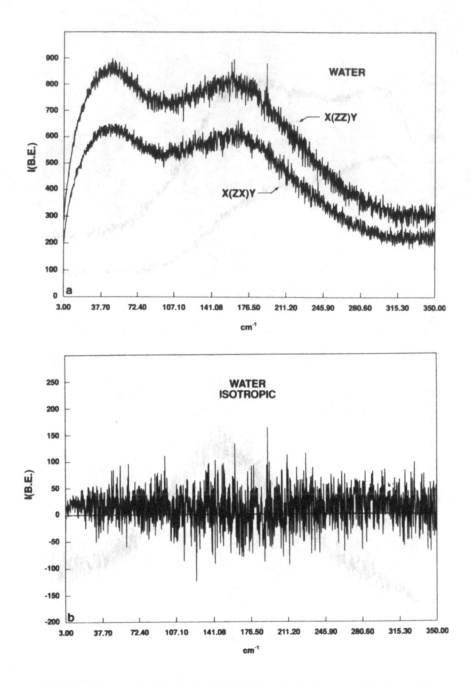

Figure 4. X(ZZ)Y, and (ZX)Y, (a), top. Isotropic,
X(ZZ)Y - (4/3)X(ZX)Y, (b), bottom..

hydrogen bonds are involved between the iodide ion and its waters of hydration. The I⁻ ion is large, and its polarizability is high. Vibration of the proton along the I⁻···H-O hydrogen bond results in a large isotropic polarizability derivative because of the large polarizability of the I⁻ ion (11).

The Cl⁻ ion also has first sphere waters of hydration, perhaps about four (4,12). However, the Cl⁻ ion is not so polarizable as the I⁻ ion, and thus a smaller isotropic Raman intensity might be expected for Cl⁻, compared to I⁻, as observed (12).

The ionic polarizability of I⁻ is a little bit more than twice that of the Cl⁻ ion (13).

Low-frequency Raman spectra were recently reported for aqueous chloride solutions (12). These spectra displayed an isotropic Raman peak at roughly 160 cm^{-1}. However, the authors actually quoted a value of 190 cm^{-1} (12), but they stated that this value resulted from computer deconvolution, as opposed to the actual frequency of the isotropic Raman intensity maximum.

We re-examined the polarized, depolarized and isotropic Raman spectra from an aqueous NaCl solution (saturated at 21°C, concentration about 5.3M) because we were uncertain about the accuracy of the published results (12). The shapes of the polarized and depolarized Raman spectra which we obtained are visually the same as those from the concentrated aqueous HCl solutions studied here. However, measurement of the X(ZZ)Y peak frequency indicates that it occurs at 145 cm^{-1}, compared to 167 cm^{-1} for the HCl solutions. Moreover, the peak of the isotropic Raman spectrum from the NaCl solution occurs at 168 cm^{-1}, compared to ≈200 cm^{-1} for concentrated HCl, and a very weak shoulder occurs near 46 cm^{-1}.

Shifts of some 22 to 32 cm^{-1} are about 10 times larger than the error in the peak frequency measurements. Hence, the low-frequency Raman spectra from the concentrated HCl solutions are clearly much different than those from the NaCl solution.

The HCl solutions studied here have Cl⁻ concentrations in the range of about 10 to 12 molar. Therefore, it is reasonable to expect that hydration of Cl⁻ should contribute significantly to the low-frequency Raman spectra. The problem is, however, that all of the currently available information about Cl⁻ hydration indicates that the isotropic peak observed here near 200 cm^{-1} for the HCl solutions occurs at much too high a frequency to be produced by hydrated Cl⁻. Hence, we a forced to the conclusion that the ≈200 cm^{-1} isotropic peak must come from another source, such as from hydration of the proton.

A weak isotropic feature near 120-135 cm^{-1} was mentioned previously. This feature might be associated with Cl⁻ hydration at concentrations between 10 to 12 molar, because it has been shown (12) that frequency decreases occur in the low-frequency Raman spectrum with increasing Cl⁻ concentration.

It is also shown subsequently that a frequency of 120-135 cm^{-1} does not arise from our normal coordinate calculations of the intermolecular vibrations of $H_9O_4^+$.

The most reasonable interpretation thus appears to be one in which the low-frequency Raman spectra from concentrated HCl are considered to contain major contributions from the hydrated proton, which at the concentrations studied here, should be $H_9O_4^+$, in addition to contributions from the hydrated Cl⁻ ion. Moreover, it appears that overlapping of the intermolecular vibrations from the hydrated proton and the hydrated chloride ion is considerable, although the hydrated chloride vibrations may have lower intermolecular Raman intensities. Despite these difficulties, we believe that we have been able to identify the main features of the $H_9O_4^+$ intermolecular Raman spectrum, and we also think that we have uncovered one intermolecular feature from the hydrated chloride ion.

We used the following model in this work. We considered the $H_9O_4^+$ ion to be pyramidal, C_{3v} symmetry, with the first hydration sphere H_2O molecules as single points of mass 18, and we further considered that these three mass 18 points vibrate against the central O atom of the H_3O^+ core, which is also treated as a single point, mass 16. This treatment corresponds to a 4-atomic C_{3v} model. Moreover, it ignores the masses of the protons of H_3O^+, because these protons are engaged in tunnelling, and hence contribute by similar amounts to the H_3O^+ core as well as to the outer H_2O molecules. We then carried out normal mode calculations using central force and valence force fields with this simple model.

Of course, we tried and finally rejected numerous variations on the above model. For example, we used a model in which the H_3O^+ core was treated as an atom of mass 19. In addition we tried numerous input vibrational frequencies. But the final solution involved the fact that the intense isotropic component has a frequency of about 200 cm^{-1}, and must be polarized, whereas the intense component of the X(ZX)Y spectrum has a peak near 50 cm^{-1}, which is entirely absent from the isotropic spectrum, and hence is depolarized. From this observation, one must conclude that the totally symmetric stretching, A_1, mode of our 4-atomic model must occur near 200 cm^{-1}, and that the asymmetric deformation, E, mode must occur near 50 cm^{-1}.

The central force and valence force calculations that we used involve two pairs of vibrational frequencies. For example, one pair involves the nondegenerate frequencies, that is, the totally symmetric stretching and totally symmetric deformation frequencies, whereas the other pair involves the degenerate frequencies, namely, the asymmetric stretching and deformation frequencies. It is not possible to start the calculations with one frequency from one pair and another frequency from the second pair, see below.

Before we could carry out normal mode calculations, however, it was essential to conduct Gaussian computer deconvolutions of the spectra from the HCl and NaCl solutions to obtain accurate component frequencies. The component frequencies from this procedure are listed in Table I.

The most reliable values from Table I for concentrated HCl are those at 50 cm^{-1} and 210 cm^{-1} from the X(ZX)Y and isotropic spectra, first and third columns. The corresponding values for NaCl are 45 cm^{-1} and 175 cm^{-1}, first and third columns. See (*). We first assumed that the symmetric and asymmetric deformation modes were either accidentally degenerate or unresolved and that both contributed to the 50 cm^{-1} feature of Table I. But when the 50 cm^{-1} and 210 cm^{-1} nondegenerate frequency pair was employed in the central force or valence force equations given by Herzberg (14), we obtained deformation force constants which were much too high, that is, they were about one-half of the stretching force constants.

We next assumed that the 50 cm^{-1} peak corresponds primarily to the degenerate E deformation mode, and that the 210 cm^{-1} isotropic component corresponds to the totally symmetric stretching A_1 mode. This assumption required that we would have to obtain values for the totally symmetric deformation mode by trial-and-error, because either a nondegenerate frequency pair, or a degenerate frequency pair is required to obtain stretching and deformation force constants, from which the remaining pair of frequencies may be calculated, as described above.

Table I. Gaussian Component Frequencies of Low-Frequency Raman Spectra from Concentrated HCl and NaCl Solutions (Bose-Einstein Corrected).

HCl	X(ZZ)Y	X(ZX)Y	Isotropic
	48	50*	---
	122	119	135
	188	187	210*

NaCl	X(ZZ)Y	X(ZX)Y	Isotropic
	45	45*	---
	88	88	96
	164	163	175*

Table II. Valence Force Field and Central Force Field Frequencies and Force Constants, F_1 Stretching, and F_2 Deformation, for $H_9O_4^+$ Considered as a 4-Atomic Pyramical XY_3 Molecule, with $M_X = 16$, and $M_Y = 18.02$. (1), (2), (3), and (4), refer to totally symmetric stretching, totally symmetric deformation, asymmetric stretching and asymmetric deformation, respectively. (1) and (2) are A_1 species, and (3) and (4) E species.

	Valence Forces					Central Forces				
(1)	190,	195,	200,	205,	210	190,	195,	200,	205,	210
(2)	27,	27,	27,	27,	27	27,	27,	27,	27,	27
(3)	269,	276,	283,	290,	297	261,	268,	275,	282,	290
(4)	51,	51,	51,	51,	51	50,	50,	50,	50,	50

F_1 2.96, 3.12, 3.28, 3.45, 3.62, Valence Forces, all units of 10^4 dyne/cm

F_1 2.81, 2.97, 3.14, 3.30, 3.48, Central Forces, all units of 10^4 dyne/cm

F_2 726, 725, 725, 725, 725, Valence Forces, all units of dyne/cm

F_2 2.18, 2.18, 2.17, 2.17, 2.16, Central Forces, all units of 10^3 dyne/cm

After several trials we obtained reasonable frequencies and force constants with a nondegenerate input pair of 27 cm^{-1} and 210 cm^{-1}. The results of this type of calculation for central force and valence force fields are listed in Table II.

We also used totally symmetric stretching frequency values ranging from 190 to 210 cm^{-1} in our calculations, but the (assumed) totally symmetric deformation value of 27 cm^{-1} was fixed, Table II. Such input pairs, yielded the stretching and deformation force constants from which the degenerate E mode frequencies were calculated, and from which a reasonable value of 50 to 51 cm^{-1} for the E deformation resulted, in agreement with observation. In addition, the calculated values of 27 cm^{-1} for the symmetric deformation mode, and 261-297 cm^{-1} for the asymmetric stretching E mode, place these vibrations in regions which would not be observable because of band overlap (deformation modes), or overlap and weakness (asymmetric stretching mode). See Figs. 1, 2, and 3.

Examination of Table II indicates that the main difference between the central force and valence force calculations involves the deformation force constant. The value of F_2 for the valence force field seems somewhat too small, relative to F_1, whereas the F_1 and F_2 values from the central force field seem reasonable.

F_1 values from the central force field calculations are about 35% higher than the intermolecular hydrogen bond stretching force constant from ice (15), and the same conclusion results from the valence force field F_1 values. Moreover, because the O-H···O hydrogen bonds of $H_9O_4^{\cdot}$ are short, 2.52Å, and nearly symmetric, it is entirely reasonable for their stretching force constants to be large, and thus the fact that the F_1 values exceed the value for ice is not unreasonable.

Another key conclusion which may be drawn from comparison of Tables I and II is that no calculated frequency comes close to the range of 119-135 cm^{-1}, second row, HCl, Table I. Nominal 120-135 cm^{-1}. The isotropic Raman component obtained in this range from Gaussian deconvolution occurs at 135 cm^{-1}; this spectrum must be polarized to occur in the isotropic spectrum. The absence of the 120-135 cm^{-1} component from both the central force and valence force results, suggests that this polarized component refers to another species, which we infer to be $Cl^-(H_2O)_n$. Of course, we realize that the basis for this assignment involves a very simple model, but it is nevertheless hard to understand how two polarized components, both in the range that most reasonably would correspond to intermolecular stretching, could arise from $H_9O_4^{\cdot}$. Moreover, if we were to assign the 135 cm^{-1} component to symmetric deformation, we would obtain a deformation force constant which would be much too large, and the degenerate bending and stretching frequencies (E modes) would also be too large. Hence, it would appear that we have little choice but to assign the 120-135 cm^{-1} region to $Cl^-(H_2O)_n$.

The assignment of the 120-135 cm^{-1} region to the hydrated chloride ion may not be without other problems, however. For example, examination of Table I clearly shows that the pattern of the polarized, depolarized, and isotropic frequencies for HCl is, apart from a general frequency shift of all modes, rather similar to the corresponding pattern for NaCl. Moreover, we do not have the option of assigning the 88-96 cm^{-1} NaCl region to another species, as we did for HCl.

The ideal situation, of course, would be to use a model of the formal type $H_3O^{\cdot}(H_2O)_mCl^-$. If the value of m were known, and if the structure and point group of this solvent shared entity were known, one could carry out a more detailed normal coordinate analysis. Under these circumstances, it might be worthwhile to include the intramolecular vibrations, and abandon the procedure used here, of considering H_2O as a single mass point, etc. Moreover, we might hope, at least, that a point group type of approach to the structure would be adequate. However, because the x-ray structural correlation length is roughly 8 to 9 Å, even the aforementioned "ideal" situation might be an approximation, that is, one cannot be certain about the extent of the collective excitations involved.

Despite the above caveats, we consider that the present model constitutes a rational approach to the data obtained in this work, even though our model treats the intermolecular vibrations of $H_9O_4^{\cdot}$ separately from those of $Cl^-(H_2O)_n$.

The frequencies of $H_9O_4^{\cdot}$ were first calculated from 4-atomic normal mode calculations, which yielded frequencies in reasonable agreement with experiment, except that one polarized vibration, observed in the intermolecular stretching region, was not predicted by the normal mode calculations. We therefore assigned this additional polarized vibration to a symmetric intermolecular stretching mode of $Cl^-(H_2O)_n$. But we also had good additional reasons for this assignment. The frequency of this additional mode is about what one would expect for the 10-12 M Cl$^-$ of the HCl solutions on the basis of present and reported (12) Raman studies of aqueous NaCl. Moreover, two polarized <u>stretching</u> vibrations <u>cannot</u> arise from a 4-atomic C_{3v} structure.

309

A significant success of our model would also seem to be that the calculated intermolecular stretching force constants are considerably larger ($\approx 35\%$) than the corresponding force constants for ice (15). These large force constants are in accord with the experimental fact that the x-ray pair correlation function for aqueous HCl shows a resolved peak at 2.52 Å (5). Such short O-O distances between the H_3O^+ core and the H_2O molecules of the first hydration shell demand that the strengths of the corresponding hydrogen bonds be unusually large. These short hydrogen bonds also make proton tunnelling very probable, as detailed in the INTRODUCTION, a circumstance very likely to enhance the intermolecular Raman intensity of the concentrated aqueous HCl solutions.

REFERENCES

1. Composition (wt. %) and density data for aqueous HCl solutions are presented in, "CRC Handbook of Physics and Chemistry", R. C. Weast ed., 66th edition, CRC Press, Inc., Boca Raton, Florida, 1985-86, see pg. D-232. From these data we obtained the following least squares equation, applicable above 10 wt. % HCl: $C = A_0 + A_1 D + A_2 D^2 + A_3 D^3$, where C = wt. %, and D = density in g-cm^{-3}. A_0 = -1474.18506, A_1 = 3653.28272, A_2 = -3113.59696, and A_3 = 934.18982. This cubic polynomial was used to determine composition from densities measured in this work.

2. $H_9O_4^+$ corresponds stoichiometrically to 4 H_2O per H^+. This in turn corresponds to $H_3O^+(H_2O)_3$.

3. M. Eigen and L. De Maeyer, Chapter 5 in, "The Structure of Electrolytic Solutions", W. J. Hamer ed., Wiley and Sons, Inc., New York, 1959, pg. 65.

4. K. Heinzinger and G. Palinkas, Chapter in, "Interactions of Water in Ionic and Nonionic Hydrates", H. Kleeberg ed., Proceedings of a Symposium in honor of the 65th birthday of W. A. P. Luck, Springer-Verlag, Berlin, 1987. Fig. 19, pg. 19, of this article shows that the highest probability for the first sphere hydration number of Cl^- is 5, followed closely by 4.

5. R. Triolo and A. H. Narten, J. Chem. Phys. 63, 3624(1975).

6. C. I. Ratcliffe and D. E. Irish, Chapter in, "Water Science Reviews 3", F. Franks ed., Cambridge Univ. Press., 1988.

7. F. H. Stillinger, private discussion.

8. A. Polian and M. Grimsditch, Phys. Rev. Lett. 52, 1312(1984).

9. J. C. Speakman, Chapter in, "MTP International Review of Science", J. M. Robertson ed., Butterworths, University Park, Baltimore, 1972.

10. G. E. Walrafen, J. Phys. Chem. 94, 2237(1990).

11. G. E. Walrafen, M. S. Hokmabadi, and Y. C. Chu, Chapter in, "Hydrogen-Bonded Liquids", J. C. Dore and J. Teixeira eds., NATO ASI series, Vol 329, Kluwer Academic, Dordrecht, 1991.

12. P. Terpstra, D. Combes, and A. Zwick, J. Chem. Phys. 92, 65(1990).

13. W. L. Jolly, "Modern Inorganic Chemistry", McGraw-Hill, New York, 1984. See Table 11.8, pg. 285.

---

*Adjunct Graduate Professor

14.   G. Herzberg, "Infrared and Raman Spectra of Polyatomic Molecules", Van Nostrand, New York, 1945.   See equations on pgs. 163, and 176.

15.   The intermolecular O-O stretching force constant for liquid water is about 1.9×10^4 dyne/cm, whereas the corresponding value for ordinary ice I_H is $\approx 2.4 \times 10^4$ dyne/cm.   See pg. 275 of Ref. (11).

PROTON CHARGE TRANSFER IN A POLAR SOLVENT

Yuri I. Dakhnovskii

Institute of Chemical Physics, USSR Academy of Sciences
Kosygin str.4, 117334 Moscow, USSR

ABSTRACT

The role of the temperature and solvent dynamics in tunneling phenomena is discussed. An expression for the tunneling action is exactly derived for an arbitrary classical anharmonic potential energy surface of a medium and for an arbitrary nonlinear coupling between a particle and the medium coordinates. It is concluded that the proton transfer rate is essentially increased by the interaction with polar modes. The dependence of the instanton action on temperature and the anharmonicity parameter is also studied.

Proton transfer reactions play a fundamental role in chemistry and biology [1-6]. The characteristic feature of these reactions is that they may appear at an activated and fall to an almost constant value at low temperatures, this constant value being greater than the one predicted by extrapolating the Arrhenius value to low temperatures. A feature of proton and other heavy atom or group transfer is the rapid variation of the proton's potential surface by the intermolecular motion, typically a vibration, which must be accounted for. The importance of the intermolecular contribution to proton transfer has been emphasized by many authors [7-16]. Hynes and co-workers [17] have pointed out the importance of the solvent dynamics. In their treatment, the solvent dynamics is treated classically while the intermolecular modes are allowed to be quantum. Our approach is very close to Hynes, except for the intermolecular mode which is considered to be classical. In our theory, we consider a bath, both as a set of harmonic oscillators and an arbitrary anharmonic function.

We consider a particle of mass M which has the relevant dynamics described by a variable q moving in a potential V(q). The particle is coupled with the solvent coordinates Q_α and the intermolecular mode $Q^{(i)}$ by an arbitrary interaction $F(q, Q_\alpha, Q^{(i)})$. The solvent modes are described by the potential energy surface $U(Q_\alpha, Q^{(i)})$. We assume the motion of the medium particles to be classical, i.e. the following condition is valid

$$\omega_\alpha \beta << 1 \quad , \quad \omega^{(i)} \beta << 1. \tag{1}$$

For an adiabatic problem of the escape from a metastable well, the rate constant can be evaluated from the imaginary part of the free energy, calculated via semiclassical theory or the metastable well [18-23]

$$k = (2/\hbar) \operatorname{Im} F \tag{2}$$

In the bounce analysis, the imaginary part of the free energy may be written in terms of the actions for the bounce path

$$k = A. \exp[-S_B]. \tag{3}$$

In this paper, we will only be interested in the bounce action S_B. For the system under consideration, which incorporates the solvent and intermolecular dynamics, the action functional has the following form

$$S = \int\limits_{\beta/2}^{\beta/2} dt \left[\frac{M\dot{q}^2}{2} + V(q) - F(q,Q_\alpha,Q^{(i)}) + \sum_\alpha \frac{m_\alpha \dot{Q}_\alpha^2}{2} + \frac{m_{(i)}\dot{Q}_{(i)}^2}{2} + U(Q,Q^{(i)}) \right],$$

(4)

where M and q is a proton mass and coordinate respectively; $V(q)$ is a tunneling potential; m_α and $m_{(i)}$ are masses of the solvent and intermolecular anharmonic oscillators and $U(Q,Q^{(i)})$, is a bath potential energy surface. For the classical dynamics of a solvent and the intermolecular motion the tunneling process takes place in the time-independent field. Thus, after integrating out with respect to the monentum of the classical modes, the action may be obtained from

$$S = \int\limits_{-\beta/2}^{\beta/2} dt \left[\frac{M\dot{q}^2}{2} + V(q) - F(q,Q_\alpha,Q^{(i)}) \right] + U(Q,Q^{(i)})\beta$$

(5)

The first term describes the reversible motion of the particle in the static field of the bath molecules.

In the instanton approximation, it is necessary to find the solution of the equation of motion as

$$\delta S/\delta q = 0$$

or

$$-M\ddot{q} + \partial V/\partial q - \partial F(q,Q,Q^{(i)})/\partial q = 0.$$

(6)

This equation may be easily solved with the help of the following integral of motion, the energy

$$E = -M\dot{q}^2/2 + V(q) - F(q,Q,Q^{(i)}).$$

(7)

This function is time-independent. Hence it is possible to find an exact solution of eq.(6)

$$t = \int_{q_A}^{q} dq \left[\frac{2}{M} \left(V(q) - F(q,Q,Q^{(i)}) - E \right) \right]^{-1/2} \tag{8a}$$

and

$$\beta = 2 \int_{q_A}^{q_B} dq \left[\frac{2}{M} \left(V(q) - F(q,Q,Q^{(i)}) - E \right) \right]^{-1/2} \tag{8b}$$

The latter equation is a periodicity condition for the bounce trajectory and determines the energy E. q_A and q_B are the turning points obtained from the equation

$$V(q) - F(q,Q,Q^{(i)}) - E = 0. \tag{9}$$

We then evaluate the instanton action S_B from

$$S_B = 2 \int_{q_A}^{q_B} dq \left[2M \left(V(q) - F(q,Q,Q^{(i)}) - E \right) \right]^{+1/2} + (E+U)\beta \tag{10}$$

The appropriate configuration of the solvent and intermolecular coordinates may be found from the minimization of the action with respect to Q and $Q^{(i)}$

$$\partial S_B / \partial Q \Big|_{Q_{min}} = 0; \qquad \partial S_B / \partial Q^{(i)} \Big|_{Q^{(i)}_{min}} = 0. \tag{11}$$

Equations (8a) - (11) allow us to solve exactly the tunneling problem for an arbitrary potential V(q), nonlinear coupling F, a potential energy surface of a solvent and intermolecular subsystem.

Now, we demonstrate this formalism for our particular cases of F and U. The coupling function F is chosen to be in a bilinear form

316

$$F = q \sum_\alpha C_\alpha Q_\alpha + q C^{(i)} Q^{(i)}, \qquad (12)$$

or

$$U = \sum_\alpha U(Q_\alpha) = \sum_\alpha \left[m_\alpha \omega_\alpha^2 Q_\alpha^2 / 2 + a_\alpha Q_\alpha^4 / 4 \right]. \qquad (13)$$

For simplicity we adopt the following potential $V(q)$ of two parabolic wells:

$$V(q) = \frac{M\omega_0^2}{2} q^2 \theta(q_0 - q) + \frac{M\omega_0^2}{2} (q - q_0 - q_1)^2 \theta(q - q_0), \quad (14)$$

where ω_0 is the frequency of the initial and final well parabola and θ is the well-known step function.

Now, let us consider the following two particular cases:

1. <u>The pure harmonic case</u>(a linear response theory) $a_\alpha = 0$.

For the potential (14) it is possible to calculate the semiclassical action, which has the following form

$$S_B(Q) = \frac{1}{2} M\omega_0 (q_1 + q_0)^2 [1 - a - x + (x + a) \ln(x + a)$$

$$- \frac{M\omega_0^2}{8} (q_1 + q_0)^2 x^2 \beta + \beta \sum_\alpha m_\alpha \omega_\alpha^2 Q_\alpha^2 / 2 \qquad (15)$$

where we have introduced the following definition

$$x = \frac{2 \sum_\alpha C_\alpha Q_\alpha}{M\omega_0^2 (q_0 + q_1)}, \qquad (16)$$

and the asymmetry parameter a

$$a = \frac{q_1 - q_0}{q_1 + q_0} \qquad (17)$$

The value of x is determined from the following equation

$$x = (1-B)\omega_0 \beta/2B = -ln(a+x) \qquad (18)$$

where

$$B = \frac{C_{(i)}^2/m_{(i)}\omega_{(i)}^2 + \sum_a C_a^2/\omega_a^2 m_a^2}{M\omega_0^2} \qquad (19)$$

For a polar solvent which may be described by the dielectric loss function $\varepsilon(\omega)$

$$B = \frac{E_{(i)} + E_c}{E_q} , \qquad (20)$$

where

$$E_c = \frac{1}{8\pi^2} \sum_k |\Delta D_k|^2 \int_{-\infty}^{\infty} \frac{d\omega}{\omega} \cdot \frac{Im\,\varepsilon(\hat{k},\omega)}{|\varepsilon(\hat{k},\omega)|^2} , \qquad (21a)$$

$$E_{(i)} = \frac{C_{(i)}^2 (q_1+q_0)^2}{2m_{(i)}\omega_{(i)}^2} , \qquad (21b)$$

$$E_q = M\omega_0^2(q_0+q_1)^2/2 \qquad (21c)$$

are the reorganization energies along the solvent, intermolecular and tunneling coordinates respectively. The expression for S_1/S_0 is

$$S_1/S_0 = \left[x(1+aK)+x^2K/2+alna\right]/(1-a+alna), \qquad (22)$$

where $S_B = S_0 - S_1$, S_0 being the action in the case where there is no coupling with the bath, and

$$K = \left(\frac{1-B}{2B}\right) \omega_0 \beta. \tag{23}$$

We have numerically calculated the value of S_1/S_0 as a function of K. From Fig.1, we see the strong dependence of the semiclassical action on the parameter K. For the strong polarity B there is a large enhancement of the rate constant. Let us note that the parameter K includes the simultaneous dependence on temperature as well as polarity.

2. The anharmonic potential energy of a solvent ($a_\alpha \neq 0$).

For this case, we do not consider the dependence of the action S_B on intermolecular vibration, focusing our attention on the solvent effect. Taking into account eq.(13), equation (18) may be written in the following form

$$\tilde{a} \, x^3 + x(1-B)B = - \frac{2B^2}{\omega_0 \beta} \, ln(a+x), \tag{24}$$

where

$$\tilde{a} = a(q_0 + q_1)^2 \, \frac{M\omega_0^2}{12M\omega^2} , \tag{25}$$

is a dimensionless anharmonicity parameter. Here we have assumed that the solvent consists of six molecules symmetrically located around the tunneling system, i.e. parameters a,Q,C,m and ω are independent of the molecule index α.

Fig.2 shows the dependence of the normalized action S_1/S_0 on the polarity B at the different values of the anharmonicity parameter $\tilde{a}$. The action S_B/S_0 decreases with the increase of coupling B. It means that, in the case of a strong interaction

319

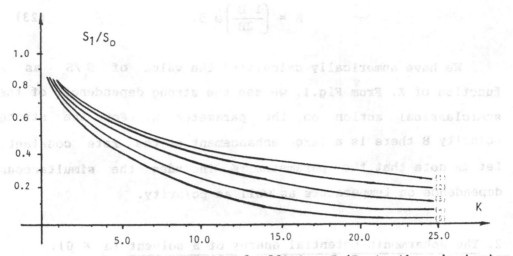

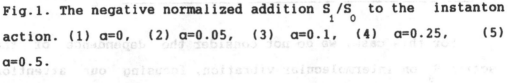

Fig.1. The negative normalized addition S_1/S_0 to the instanton action. (1) $\alpha=0$, (2) $\alpha=0.05$, (3) $\alpha=0.1$, (4) $\alpha=0.25$, (5) $\alpha=0.5$.

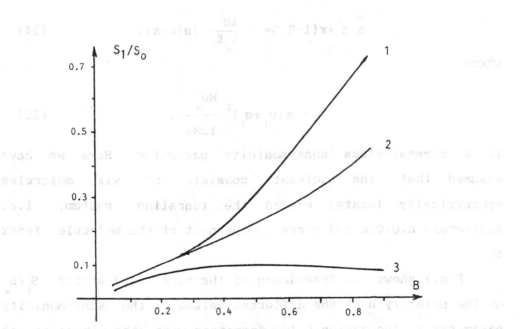

Fig.2. The dependence of S_1/S_0 on the parameter B at $\omega_0\beta=15$, $\alpha=0.1$ (1) $\tilde{a}=0.2$; (2) $\tilde{a}=0$; (3) $\tilde{a}=3$.

of a particle with a bath, the tunnel barrier decreases, when the transfer probability is increased. Meanwhile, the potential energy of the medium is enhanced, upon increasing the semiclassical action of the system. The optimal configuration of the medium coordinates is governed by eq.(11). The negative anharmonicity (a<0) effectively decreases the medium oscillator frequency increasing the value of polarity B. The positive anharmonicity enhances the solvent effective oscillator frequency, while at the same time decreasing the value of B.

Thus, in this paper we have considered the problem of quantum tunneling in the presence of an arbitrary interaction with the classical anharmonic modes of a condensed medium. As in [24-25], it has been possible to introduce the integral of motion, the energy , of a tunneling proton, and therefore to find the exact solution of the problem for an arbitrary potential energy surface of a solvent as well as an interaction of a particle with a bath. As an example, we have focused on the particular case of the interaction term (bilinear coupling). The high temperature bath has been considered as a set of nonlinear oscillators with a quartic anharmonic potential energy. For the case of a proton transfer the intermolecular mode is assumed to be classical.

In accordance with the review of Hynes [26], there is a vagueness about the role of the polar modes of a solvent on the tunneling rate. The theory proposed here gives an unambiguous answer to this question. The rate constant enhances when the influence of the solvent is taken into account, i.e. the stronger the polarity of a solvent the larger the tunneling rate. For the true evaluation of the rate constant it is necessary to take into account the anharmonicity of a solvent (or the nearest neighbors). This effect may increase or decrease the rate constant for many orders of magnitude (usually, $S_B/n \cong 2.8 \div 4.5$ and at $\tilde{a} = -0.2$, $S_1/h \cong 20 \div 30$ (see Fig.2)).

REFERENCES

[1] R.P.Bell, @The Tunnel Effect in Chemistry@, (Chapman and Hall, London, 1980)

[2] T.Werner, J.Chem. Phys., 83, 320 (1979)

[3] A.L. Huston, G.W. Scott and A. Gupta, J.Chem. Phys. 76 4978 (1982).

[4] G. Woessner et al. J.Chem. Phys. 89, 462 (1985).

[5] M. Lee, J.T. Yardley and R.M. Hochstrasser J.Chem. Phys. 91, 5802 (1983).

[6] Y.R. Kim, J.T. Yardley and R.M. Hochstrasser J.Chem. Phys. 36, 311 (1989).

[7] V.A Benderskii, U.I. Goldanskii and A.A. Ovchinnikov Chem. Phys. Lett. 73, 492 (1982).

[8] V.A.Benderskii, P.G. Phillipov, Yu.I. Dakhnovskii and A.A. Ovchinnikov Chem. Phys. 67, 301 (1982).

[9] L.I. Trakhtenberg, V.L. Klochikhin and S.Ya. Pshezhetsky, Chem. Phys. 69, 121 (1982).

[10] V.I. Goldanskii, V.N. Fleurov and L.I. Trakhtenberg, Sov. Sci. Rev. Chem. B.9, 59 (1987).

[11] A.A. Ovchinnikov and V.A. Benderskii, Journ, Electroanal. Chem. 100, 563 (1979).

[12] M.Ya. Ovchinnikova, Chem. Phys. 36, 85 (1979).

[13] M.Morillo and R.I. Cukier, J.Chem. Phys. 92, 4833 (1990).

[14] W. Siebrand, T.A. Wildman and M.Z. Zgierski, J.Am. Chem. Soc. 106, 4089 (1984).

[15] A.Warshel and Z.T. Chu, J.Chem. Phys. 93, 4003 (1990).

[16] A. Warshel, J.Phys. Chem. 86, 2218 (1982).

[17] D. Borgis, S. Lee and J.T. Hynes, Chem. Phys. Lett. 162, 19 (1988).

[18] R.P. Feynman and V.L. Vernon, Ann Phys. (N.Y.) 24, 118 (1963).

[19] J.S. Langer, Ann Phys. (N.Y.) 41, 108 (1967).

[20] A.I. Larkin and Yu.N. Ovchinnikov, Zh. Exp. Teor. Fiz. 86, 719 (1984) [Sov. Phys. JETP 59, 420 (1984)];Pis'ma Zh. Exp. Teor. Fiz. 37, 322 (1983) [Sov. JETP Lett. 37, 382 (1983)].

[21] A.O. Caldeira and A.J. Leggett, Phys. Rev. Lett. 46, 211 (1981); Ann. Phys. (N.Y.) 149, 374 (1983).

[22] Yu.I. Dakhnovskii, A.A. Ovchinnikov and M.B. Semenov, Zh.Exp.Teor. Fis. 92 955 (1987) [Sov. Phys. JETP 65, 541 (1987); Mol. Phys. 63, 497 (1988)].

[23] V.Weiss, H.Grabert, P.Hanggi and P. Riseborough, Phys. Rev. 35 B, 9535 (1987).

[24] Yu. I. Dakhnovskii and A.A. Ovchinnikov, Phys. Lett. 149 A, 39 (1990).

[25] Yu.I. Dakhnovskii, A.A. Ovchinnikov and Z.K. Smedarchina, Phys. Lett. 149 A, 43 (1990).

[26] J.T. Hynes, Ann. Rev. Chem 36 573 (1985).

[27] A. Schmid, J.Low Temp. Phys. 49, 609 (1982); A.J.Leggett et al. Rev. Mod. Phys. 59, 1 (1987).

A DENSITY-FUNCTIONAL STUDY OF HYDROGEN BONDS

Michael Springborg

Fakultät für Chemie, Universität
W 7750 Konstanz,
Federal Republic of Germany

ABSTRACT

The density-functional formalism of Hohenberg and Kohn offers a parameter-free scheme for calculating electronic ground-state properties of finite and infinite systems, but has only been little applied to hydrogen-bonded systems. We present here a method based on this formalism for studying polymeric systems and apply it specifically on hydrogen-bonded materials. A detailed study of hydrogen fluoride is presented for which the density-functional calculations mainly are used in providing informations about structural properties. Moreover, the results of the calculations form the basis for evaluating the quality of model potentials for investigating statistical and dynamical properties, for calculating vibrational properties, for studying interchain couplings, and for examining solitons. Finally, some results of a recent study of hydrogen cyanide are presented.

1. INTRODUCTION

Any quantitatively correct description of proton transfer in specific hydrogen-bonded systems requires detailed informations about the bonds between the atoms of the materials. But often advanced experiments on not completely defined samples are required in order to gain the required insight. One may alternatively study the systems of interest theoretically using accurate, parameter-free methods. This second approach, which is the one we will take, has the advantage that *in principle* one may study any existing or imaginary system thereby exploring any feature at desire, but *in practice* one is limited to studying idealized situations and/or systems. Therefore, the calculations may only be regarded a supplement — although important — to the experiments.

In parameter-free studies of the electronic properties determining the interatomic bonds of the hydrogen-bonded systems the common approach is to consider either finite (clusters of smaller) molecules or periodic, infinite polymers. The systems are often assumed isolated in space although electrostatic effects due to the surroundings have been included. Moreover, the Born-Oppenheimer approximation is invoked, such that quantum effects of the protons are excluded. Furthermore, only static properties at $T = 0$ are explored. The results — often as functions of structural parameters — may subsequently form a data base that can be used in estimating parameters of models that finally can be applied in studying dynamical or temperature-dependent effects as well as those of symmetry-breaking structural defects (e.g., solitons).

The Hartree-Fock approximation forms the basis for most studies, and only for finite systems have correlation effects been included on top of the Hartree-Fock calculations. An alternative approach, with which correlation effects relatively easily can be incorporated, is to make use of the Hohenberg-Kohn density-functional formalism.[1] During the last 2–3 decades methods based in this formalism have been used in a very large number of theoretical studies of the electronic properties of covalent and ionic crystals and of surfaces as well as of covalently bonded, finite or infinite molecules, but have only been little applied to hydrogen-bonded systems. It is the purpose of the present contribution to report results of a density-functional study of extended hydrogen-bonded systems.

We will (in Sec. 2) briefly outline the basic ideas behind the theoretical approach. A detailed discussion of a simple hydrogen-bonded system, hydrogen fluoride, forms the main body of this paper. For this system we discuss structural properties in Sec. 3, vibrations in Sec. 4, and solitons in Sec. 5. Another system, hydrogen cyanide, is the topic of Sec. 6, and we conclude in Sec. 7.

2. THE DENSITY-FUNCTIONAL METHOD

According to the theorem of Hohenberg and Kohn[1] one can write the total energy E of the ground state of the electrons of a system, for which the Born-Oppenheimer approximation is assumed valid, as a unique functional of the electron density ρ:

$$E = E[\rho]. \tag{1}$$

In contrast to its existence, the precise form of the functional is, however, not known, and an approximation is introduced. In order to do so, it is convenient to transform the problem of calculating the total energy into that of solving the so-called Kohn-Sham[2] single-particle equations. In Ry atomic units these are

$$[-\nabla^2 + V_N(\vec{r}) + V_C(\vec{r}) + V_{xc}(\vec{r})]\psi_i(\vec{r}) = \epsilon_i \cdot \psi_i(\vec{r}). \tag{2}$$

V_N, V_C, and V_{xc} are the electrostatic potential of the nuclei, that of the electrons, and the remaining so-called exchange-correlation potential, respectively. We see that Eq. (2) resembles the Hartree equation with V_{xc} entering as an external potential.

The fact that $E[\rho]$ of Eq. (1) is unknown manifests itself therein that V_{xc} is unknown. V_{xc} is therefore approximated; most often through a local approximation in which V_{xc} at the point $\vec{s}$ is taken as for a homogeneous electron gas of the constant density $\rho(\vec{s})$. We will here use the analytical form given by von Barth and Hedin.[3]

It is intuitively clear that as long as the electron density $\rho(\vec{r})$ is slowly varying or as exchange and correlation effects are small compared with the energies of interest, this approach is well justified. However, for the hydrogen-bonded systems for which the energies of interest are small and for which the electrons are localized, it is less obvious that such a (local-density) density-functional approach will work. The two systems we will discuss here represent both possibilities: we obtain accurate results for hydrogen fluoride but inaccurate for hydrogen cyanide.

It is finally realized that all potentials entering Eq. (2) depend solely on the density:

$$\rho(\vec{r}) = \sum_{i=1}^{\text{occ}} |\psi_i(\vec{r})|^2. \tag{3}$$

The density-functional equations (1)–(3) are computationally easier to solve accurately than the many-electron Schrödinger equation. On the other hand, one should keep in mind that the starting point is an approximated equation.

As is common practise the solutions $\psi_i(\vec{r})$ are expanded in a basis, and we use here the so-called linear muffin-tin orbitals (LMTO's). Inside non-overlapping, atom-centered (muffin-tin) spheres these are given numerically as the solutions to Eq. (2) with the total potential replaced by its spherically symmetric component and with ϵ_i replaced by a 'reasonable' value. Outside all spheres they are represented analytically as spherical Hankel functions. As discussed elsewhere,[4-6] the LMTO's offer an accurate basis of limited size. It should be added that the muffin-tin approximation is used *solely* in defining the basis functions.

In their present form we have designed the computer codes for calculations on polymers that can be considered isolated, infinite, periodic, and having a helical (including translational and zigzag) symmetry. Using the symmetry we form Bloch waves from the atom-centered LMTO's, and define a dimensionless k variable with $k = 0$ and $k = 1$ being the zone center and zone edge, respectively. Using a large lattice constant we may also study finite molecules. The method has been applied on a number of polymeric systems and is in detail discussed in Refs. 4–6.

As mentioned in the Introduction, density-functional methods have been applied during a number of years on many systems (see, e.g., Ref. 7). From a conceptual point of view the eigenvalues ϵ_i need not be related to electronic excitation energies and the eigenfunctions $\psi_i(\vec{r})$ not to electronic wavefunctions. However, experience has shown that it is a good approximation to ignore these inconcistencies. From our point of view a more severe problem is that the local approximation tends to result in bond strengths that are overestimated (see, e.g., Ref. 7). Typically, this overestimate is up to some few eV per bond, and its value depends mainly on the type of the bond. This has some important consequences for some of the systems of our interest as we shall see.

3. STRUCTURE OF HYDROGEN FLUORIDE

Hydrogen fluoride can not be considered the typical hydrogen-bonded system as it has one of the strongest hydrogen bonds known. However, this feature as well as its structural simplicity make it an excellent compound for theoretical studies. Crystalline hydrogen fluoride consists of weakly interacting chains of the form shown in Fig. 1.[8-10] The fluorine atoms occupy the apexes of planar zigzag chains and the hydrogen atoms are placed asymmetrically between the fluorine atoms forming alternating shorter, covalent bonds with the nearest neighbours and longer, hydrogen bonds with the more distant neighbours. The structure of a single chain can thus be described by the three parameters d_F, d_H, and α shown in Fig. 1.

For an isolated monomer ($d_F \to \infty$; α irrelevant) we found[11] the total energy as a function of d_H as shown in Fig. 2. The minimum occurs for $d_H = 1.70$ a.u. which agrees well with the experimental values of 1.71–1.73 a.u.[12,13] For the infinite

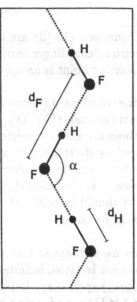

Fig. 1. The geometry of a single hydrogen fluoride chain. The solid and dashed lines represent covalent and hydrogen bonds, respectively.

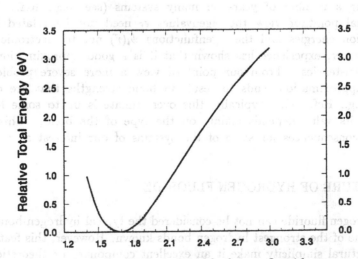

Fig. 2. Total energy of a single, isolated hydrogen fluoride monomer as a function of d_H.

polymer we found the lowest total energy for $\alpha = 125°$, $d_F = 4.72$ a.u., and $d_H = 1.85$ a.u. The experimental values[8-10] are here $\alpha = 116° - 120°$, $d_F = 4.71-4.72$ a.u., and $d_H = 1.80-1.83$ a.u. We observe that our density-functional method gives a good description of the structure of both the isolated HF monomer and of the $(HF)_x$ polymer. We finally notice the relatively large increase in d_H when passing from the monomer to the polymer. Referring to Fig. 2 we see that this increase costs about 0.2 eV for the isolated monomer, which obviously must be compensated by the creation of the hydrogen bonds for the polymer.

We performed a more detailed study of the polymer for two fixed values of α: $\alpha = 120°$ and $\alpha = 180°$. As functions of $d_F/2$ and d_H we found in these cases the total energies shown in Fig. 3. In Fig. 3(b) we observe two local total-energy minima, which, however, are separated by a barrier so low, that we consider its existence beyond the numerical accuracy. We moreover notice that the total energy in (a) is 0.61 eV and in (b) 0.31 eV per monomer below that of the isolated monomer. These values represent our calculated energy of the hydrogen bonds.

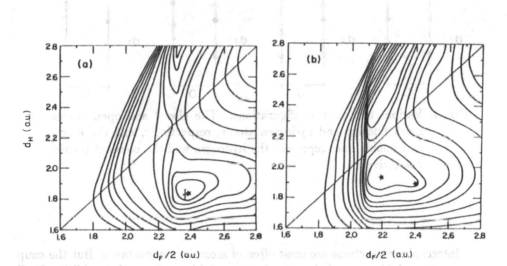

Fig. 3. Relative total energy per monomer for the $(HF)_x$ polymer of Fig. 1 for (a) $\alpha = 120°$ and (b) $\alpha = 180°$ as functions of $d_F/2$ and d_H. The asterisks mark local minima and the cross in (a) the experimental structure with error bars. The local energy zero is in (a) 0.30 eV below that in (b), which in turn is 0.31 eV below that of the isolated monomer. The contour values are 0.05, 0.1, 0.2, 0.3, 0.4, 0.5, 0.75, 1.0, 1.25, 1.5, and 2.0 eV.

In Fig. 3 we also notice that when decreasing d_F (which corresponds to applying pressure) the value of d_H for which the total energy is lowest increases until it equals $d_F/2$. This corresponds to the hydrogen atoms being placed symmetrically between the two neighbouring fluorine atoms. Such pressure-induced transitions have been observed experimentally for hydrogen fluoride.[14]

Liquid hydrogen fluoride consists of chains as that of Fig. 1 but with the atoms no longer placed in a common plane.[15,16] Also gaseous hydrogen fluoride contains $(HF)_n$ oligomers, in particular $n = 6$.[17] Theoretical studies of dynamical or statistical properties of gaseous and liquid hydrogen fluoride are most often carried through using models incorporating approximate potentials for the interactions between the monomers. These potentials are often derived from ab initio calculations on the $(HF)_2$ dimer. However, as is evident from the discussion above, the potentials should also be able to describe the properties of $(HF)_n$, $n > 2$, oligomers; including the total energies of Fig. 3 for the polymers. It turned out, however,[11] that the so far proposed model potentials could not reproduce the essential features of Fig. 3. Thus, there is a definite need for improved model potentials for the study of dynamical and statistical properties of hydrogen fluoride.

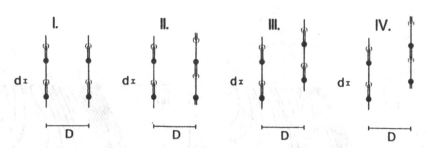

Fig. 4. Various two-chain configurations. The closed and open circles mark fluorine and hydrogen atoms, respectively, and the single and double lines represent the hydrogen and the covalent bonds, respectively.

Interchain interactions are most often of secondary importance. But the couplings between neighbouring chains may have crucial impacts on the stability of solitons. Consider, e.g., Fig. 4. We show here pairs of linear chains being either parallel (structure I and III) or antiparallel (II and IV). The fluorine atoms of the two chains are for structure I and II in-phase and for structure III and IV out-of-phase (i.e., displaced half a unit cell length on one chain with respect to the other). A soliton is a domain wall on a single chain between two parts of the chain only differing in an interchange of the covalent and the hydrogen bonds. Occurring in, e.g., one of the chains of structure I of Fig. 4 it will leave halfpart of the pair of chains with the antiparallel configuration and the other halfpart with the parallel configuration. In case of a total-energy difference between the structures I and II, the single soliton will move in that direction that leads to the largest segment of the structure with the lowest total energy.

Since the single chains with $\alpha = 120°$ and with $\alpha = 180°$ have comparable structural properties (see Fig. 3), we studied two linear chains as shown in Fig. 4. The unit-cell length was kept at 5.0 a.u., and the displacement d of the hydrogen atoms from the midpoint between neighbouring fluorine atoms was varied in unison on both chains. This resulted[18] in the total energies of Fig. 5(a)–5(d). For comparison we also show [Fig. 5(e)] the energies for the single, isolated chain as extracted from Fig. 3(b). Finally [Fig. 5(f)], we fixed $d = -0.6$ a.u. on one chain and varied it on the other such that $d = -0.6$ a.u. is the parallel and $d = +0.6$ a.u. is the antiparallel configuration.

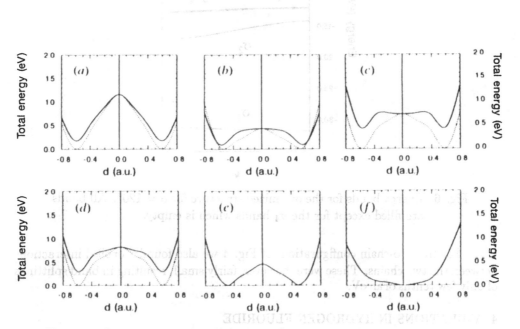

Fig. 5. Relative total energy per HF dimer [except for (d) for which it is per monomer] for (a),(b) $D = 5.0$ a.u. and (c),(d),(f) $D = 8.0$ a.u. for (a),(c) structure I and II and (b),(d) for structure III and IV of Fig. 4. The dashed and full curves corresponds to the antiparallel and parallel chains, respectively.

The results indicate a clear preference for the antiparallel arrangement over the parallel arrangement. Moreover, as not can be seen in Fig. 5, the total energy was lowest for the in-phase configuration and for large D. All features may be ascribed to electrostatic interactions between the dipoles of the monomers, and may as such be due to some modifications when having a true three-dimensional crystal of zigzag chains lying in different non-parallel planes. In total, interchain interactions are, however, observed to be present and one should address these in future theoretical studies, especially when looking at proton transfer via solitons.

In closing this section we show in Fig. 6 the band structures for the optimized structure for $\alpha = 120°$. The lowest (σ_1) valence band is mainly a F 2s band and the other three have dominating F 2p components, with, however, some admixing of H 1s contributions for the σ bands. Especially the σ_2 band shows some dispersion indicating that the inter-monomer bonds in $(HF)_x$ have a partially covalent character. Moreover, since there is some orbital interactions between the monomers, the band structures will change as a function of structure.

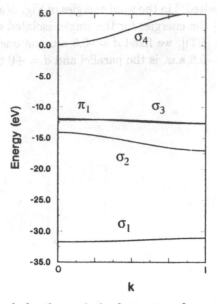

Fig. 6. Energy bands for the optimized structure for $\alpha = 120°$. All bands
are filled except for the σ_4 bands which is empty.

For the two-chain configurations of Fig. 4 we also found[18] orbital interactions
between the two chains. These were, however, fairly small, resulting in band splittings
of only few tenths of an eV.

4. VIBRATIONS IN HYDROGEN FLUORIDE

The total energies for the isolated HF monomer (Fig. 2) are well fit with a
Morse potential. This can be used in calculating frequencies of bond-stretch vibrations giving 3865, 7506, 11034, 14118, and 17088 cm^{-1} for the lowest five modes. The
experimental values[12] are 3961, 7750, 11372, 14831, and 18130 cm^{-1}, and the agreement between theory and experiment is accordingly seen to be good. The energies
are not evenly spaced as a consequence of the anharmonicity of the potential.

The total energies of Fig. 3(a) can be expanded around the minimum to second
order in the bond lengths. Using this we can calculate the frequency of the bond-
stretch, in-plane, Γ phonon resulting in a value of 3535 cm^{-1}. Compared to the
isolated monomer we observe a softening of almost 10%, which is due to the increased
covalent bond length and to the hydrogen bond. A similar reduction has been observed
experimentally by Lisy et al.[19] on $(HF)_n$, $n = 3$–6. The experimental frequencies for
the polymer are[20-22] 3360–3404 cm^{-1} in good agreement with our calculated value.

However, the potential experienced by this phonon is strongly anharmonic due
to the existence of two energetically degenerate structures. Moreover, the frequency
is comparable to the barrier height [see Fig. 3(a)] for a collective shift of all hydrogen atoms from one minimum to the other. We therefore replaced the harmonic
potential by an anharmonic one[11] which was designed (i) to have the same curvature
around the minima as the harmonic one, (ii) to have the correct barrier height for the
transition between the two energetically degenerate structures, and (iii) to give the
right distance between the two minima. With this potential we solved numerically
the one-dimensional Schrödinger equation for the hydrogen atoms by expanding the

solutions in a basis set consisting of two subsets each formed by the eigenfunctions to the harmonic potential at one of the minima.

Relative to the lowest state we found the first three modes at 470±100, 2050±45, and 3960±110 cm^{-1}, where the uncertainties originate from slightly different models. Within the simplest picture the basis functions will combine pairwise giving a low-energy 'bonding' combination and an 'antibonding' combination with a higher energy. Our calculated modes correspond accordingly essentially to the antisymmetric $n = 0$, the symmetric $n = 1$, and the antisymmetric $n = 1$ combination, where n labels the quantum number for the harmonic-oscillator basis function.

The lowest mode falls in a region where many modes are observed, and the highest corresponds fairly well — taking the limitations of the model into account — to the result within the harmonic approximation. The middle one occurs in a region where no mode so far has been observed, but we strongly recommend an experimental search for it. However, due to its form it may be difficult to observe.

5. SOLITONS IN HYDROGEN FLUORIDE

In the preceeding section we observed how the anharmonic potential experienced by the protons in the polymer led to unusual features in the phonon spectrum. Also the occurrence of topological solitons in the chains is related to the anharmonic potential. As already indicated, the two structures characterized by (α, d_F, d_H) and $(\alpha, d_F, d_F - d_H)$ are energetically degenerate (see Fig. 1), and a soliton, being a domain wall on a single chain between the two structures, may exist, be highly mobile, and be responsible for charge and/or proton transfer through the system.

We used the results of Fig. 3 in defining a model potential with which we subsequently studied the solitons.[11] It turned out that the lattice distortions of the solitons were very localized and that the creation energy of the neutral solitons was equal to the barrier height for a collective shift of all the protons from one structure to the other. In the model we included a tight-binding description of the electrons in order to see at which energies soliton-induced orbitals would occur. These were found just outside the band regions; up to 1 eV from the band edges. Their occurrence may provide an experimental tool for observing the existence of solitons, although also other effects like interchain couplings, finite lengths of the chains, excitons, defects, etc. may be held responsible for these features. Another outcome of the model calculations was that, whereas the soliton-induced lattice distortions were very localized, the soliton-induced electronic orbitals were spread out over more sites. This means that a continuum approximation may be justified for the electrons but not for the lattice.

Fig. 3 offers information about total-energy changes when changing the positions of *all* hydrogen atoms and/or *all* fluorine atoms simultaneously. As such it provides information about the extreme situation of one soliton per HF unit. However, the solitons are supposed to be much more dilute, and their presence in one unit cell may moreover influence the structure of the neighbouring unit cells. It has thus been proposed (see, e.g., Ref. 23) that not only the positions of the hydrogen atoms but also of the so-called heavier sublattice (here, the fluorine atoms) will be modified under the propagation of a soliton. In order to study this we have recently initiated a more detailed study of hydrogen fluoride and will here report some preliminary results.

We studied single, isolated chains of the type shown in Fig. 7; i.e., linear chains consisting of periodically repeated $(HF)_2$ dimers. The F_1 atoms were kept fixed with an interatomic distance of 9.6 a.u. The H_1 atoms were moved in unison from being closest to the F_1 atoms to being closest to the F_2 atoms. The positions of the F_2 and the H_2 atoms were allowed to relax. The results of the density-functional calculations are shown in Fig. 8. We show those positions of the various atoms, which result in the lowest total energy, as functions of the distance between the H_1 and F_1 atoms. Since the F_1 atoms are fixed and the H_1 atoms are those being moved, the main results concern the positions of the F_2 and H_2 atoms.

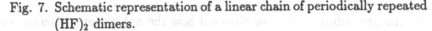

Fig. 7. Schematic representation of a linear chain of periodically repeated $(HF)_2$ dimers.

The results of Fig. 8 can be expressed as follows. At the initial stage when shifting the H_1 atoms, the F_2 atoms stay fixed and the H_2 atoms move more or less parallel to the H_1 atoms. This continues until when the H_1 atoms are roughly in the middle between the F_1 and F_2 atoms, at which point the F_2 atoms approach the F_1 atoms slightly. Referring to Fig. 3 we see that this leads to a lowering of the barrier for the H_1 atoms which then easier can approach the F_2 atoms. Simultaneous with the displacements of the F_2 atoms, the H_2 atoms 'jump' towards the F_1 atoms. As a consequence of this 'jump' we obtain a chain that to some extent consists of alternating F (i.e. F_2) and HFH (i.e. $H_2F_1H_1$) units. An intuitively more correct picture may be alternating F^{-q} and $(HFH)^{+q}$ ions. The Mulliken populations indicate that the H_1 atoms each carry about half an electron and that the charge q is about 0.2. Finally, the displacements of the F_2 atoms can be considered an indication of the above-mentioned coupling to the heavier sublattice. Moreover, compared with the collective shift of all hydrogen atoms simultaneously, the barrier height is roughly halved when moving the atoms as in Fig. 8.

Accompanying the transition to the F^{-q} and $(HFH)^{+q}$ ions the electrons become partially localized as is indicated in Fig. 9. We show here the band structures for various structures of the periodic $(HF)_2$-chain. The structure of Fig. 9(a) corresponds roughly to the minimum and is thus related to that of Fig. 6. In Fig. 9(b) we show the bands for the chains with all F–F and F–H distances equal to 4.80 and 2.40 a.u., respectively. This is the structure of relevance for the periodically repeated solitons that could be studied with the informations of Fig. 3. The bands resemble those of Fig. 9(a) except for a broadening of the $\sigma_3 + \sigma_4$ bands. However, when we pass to the bands of the F^{-q}–$(HFH)^{+q}$ chain [Fig. 9(c)] we notice that the bands split at the zone edge and all become narrow indicating the localization of the electrons. It moreover turned out that of each pair of bands the lower one has largest components on the (HFH)$^{+q}$ complexes and the upper one largest components on the F^{-q} ions.

We have constructed a model that reproduces the results of Fig. 8 as well as the related total energies. We assume the hydrogen atoms to move in a double well. The form of this double well depends on the positions of the two neighbouring hydrogen atoms such that the energy is lowest for the hydrogen atom of interest when it is in

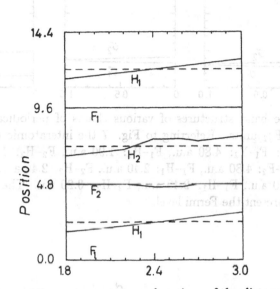

Fig. 8. Positions of the various atoms as functions of the distance between the F$_1$ and H$_1$ atoms (see Fig. 7). The dashed lines represent the positions of the fluorine atoms for the periodic, undistorted chain as well as the midpoints between those.

a minimum closest to a fluorine atom to which no other hydrogen atom is covalently bonded. The energy is higher when it is in a minimum close to a fluorine atom to which another hydrogen atom is covalently bonded. This part of the potential is modelled through a modified double-Morse potential. To this we finally add harmonic potentials between the fluorine atoms and between the hydrogen atoms, of which the former is to assure the correct lattice period.

Using the numbering and interatomic distances of Fig. 10, our model becomes

$$E = A \cdot \sum_n \left[\left(1 - e^{-\alpha(d_n - d_0)}\right)^2 \cdot e^{\beta(d_{n+1} + d_{n-1} - 2d_0)} \right.$$

$$\left. + \left(1 - e^{-\alpha(D_n - d_n - d_0)}\right)^2 \cdot e^{\beta(D_{n+1} - d_{n+1} + D_{n-1} - d_{n-1} - 2d_0)} \right]$$

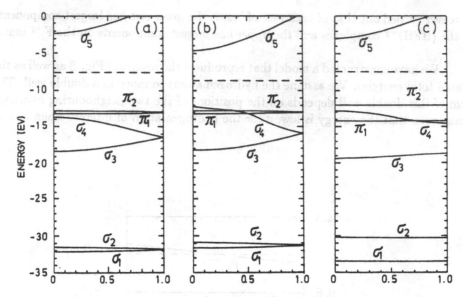

Fig. 9. The band structures of various chains of periodically repeated $(HF)_2$ units. Referring to Fig. 7 the interatomic distances are (a): F_1–F_2: 4.80 a.u., F_1–H_1: 1.90 a.u., F_2–H_2: 1.90 a.u.; (b): F_1–F_2: 4.80 a.u., F_1–H_1: 2.40 a.u., F_2–H_2: 2.40 a.u.; (c): F_1–F_2: 4.60 a.u., F_1–H_1: 2.40 a.u., F_2–H_2: 3.20 a.u. The dashed lines represent the Fermi level.

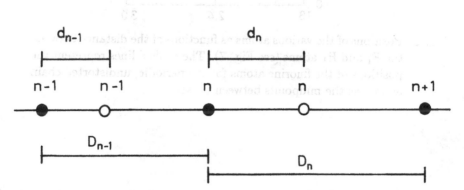

Fig. 10. Definition of the interatomic distances used in the model potential.

$$+k_F \sum_n (D_n - D_0)^2 + k_H \sum_n (D_n - d_n + d_{n+1} - d_H)^2. \qquad (4)$$

The parameter values were found to be (in Ry atomic units): $A = 0.20$, $\alpha = 1.50$, $d_0 = 1.72$, $\beta = 0.02$, $k_F = 0.06$, $k_H = 0.10$, $D_0 = 5.552$, and $d_H = 4.80$.

The potential (4) contains three-body interactions. Restricting to pair potentials (e.g., by omitting the exponentials multiplying the Morse potentials) did not lead to satisfactory potentials, but we can, however, not exclude the existence of a pair potential reproducing the density-functional results acceptably. Finally, the potential (4) will form the basis for studies of the properties of solitons in $(HF)_x$. The results of these studies as well as of related studies on zigzag chains will be published elsewhere.

6. HYDROGEN CYANIDE

Also hydrogen cyanide is structurally simple. In the solid phase it consists of parallel, linear chains of linear HCN monomers.[24] The monomers are bonded together via hydrogen bonds, and the interchain bonding is due to weak van-der-Waals forces. Compared with hydrogen fluoride, the hydrogen bonds are longer and weaker. Moreover, hydrogen cyanide is lacking the existence of two energetically degenerate structures; interchanging the covalent and hydrogen bonds of the hydrogen atoms results in a structure that need not be energetically degenerate with the initial one. Therefore, only under the influence of external forces like electromagnetic fields may solitons become (meta-)stable. As such, hydrogen cyanide may serve as a model compound for the many compounds with hydrogen atoms asymmetrically between two different nearest neighbours.

Our density-functional study of hydrogen cyanide[25] was motivated by one further aspect. Polycarbonitrile is a covalently bonded polymer with the same three atoms per unit cell as hydrogen cyanide. It is, however, a polymer closely related to the prototype of the conjugated polymers, trans polyacetylene, that show a very large doping-induced in electrical conductivity (see, e.g., Ref. 26). A comparison between hydrogen cyanide and polycarbonitrile could therefore directly illustrate the similarities and differences between hydrogen-bonded and conjugated polymers. Here we will, however, focus on the results for the hydrogen-bonded compound.

It turned out that our density-functional method could not account for the structural properties of hydrogen cyanide. For the isolated, linear HCN monomer we found[25] bond lengths of $d_{HC} = 1.93$ a.u. and $d_{CN} = 2.22$ a.u., which should be compared with the experimental values[27,28] of $d_{HC} = 2.01$ a.u. and $d_{CN} = 2.18$ a.u. For an isolated, linear CNH molecule we calculated $d_{CN} = 2.21$ a.u. and $d_{NH} = 1.75$ a.u., whereas the experimental values[29] are $d_{CN} = 2.22$ a.u. and $d_{NH} = 1.88$ a.u. Thus, the C–N bond lengths are well described but the lengths of the covalent bonds to the hydrogen atoms are underestimated.

More serious is it that the relative stability of the two molecules is not given correctly. We find the CNH molecule to be stabler by 0.1–0.2 eV, whereas the HCN molecule should have a total energy of about 0.6 eV below that of the CNH molecule (see, e.g., Ref. 30). We believe the observed discrepancy to be due to the local approximation. As discussed in Sec. 2, the strengths of the chemical bonds are overestimated by an amount that mainly depends on the type of the bond. Since the two molecules have different types of bonds, this overestimate will be different for the two monomers resulting in uncertainties of roughly ±1 eV in the relative stability of the two. We are therefore excluded from making any realistic statements about the relative stability of the HCN and CNH molecules.

For the linear $(HCN)_x$ polymer the lowest total energy was found for covalent bond lengths of $d_{HC} = 1.83$ a.u. and $d_{CN} = 2.28$ a.u. and a hydrogen-bond length of $d_{NH} = 3.51$ a.u. The total lattice period becomes thereby 7.62 a.u., which is considerably smaller than the experimental value[24,31] of 8.20 a.u. This underestimate is mainly due to too short H–C and N$\cdots$H bonds, and we will ascribe this to the tendencies of overestimating bond strengths. Related to this is the reduction in d_{HC} upon passing from the monomer to the polymer, which has not been observed experimentally and which may be an artifact of the calculations. We also believe the finding that the polymer is only metastable compared to dissociation into isolated monomers to be related to incorrect descriptions of the bonds of the hydrogen atoms.

For the $(CNH)_x$ polymer (with the hydrogen atoms closest to the nitrogen neighbours) the optimized bond lengths became $d_{CN} = 2.23$ a.u., $d_{NH} = 1.78$ a.u., and $d_{HC} = 3.67$ a.u. Also this is found to be only metastable but since the values of d_{CN} and d_{NH} are found to change less for this system upon polymerization the total energies per monomer of the isolated monomer and of the polymer are more close.

Finally, as for the isolated monomer we also for the polymers found the CNH-based system to be stabler than the HCN-based one.

The reasons for the failure of the density-functional method in describing the structural properties of hydrogen cyanide are more. First of all, the energies of the hydrogen bonds are small. Moreover, the relative stability of the CNH- and HCN-based systems is comparable with the uncertainties of the density-functional method when comparing systems with different types of bonds and as such one should not expect the currently applied local-density methods to be able to reproduce this. Furthermore, whereas the hydrogen bonds of hydrogen fluoride have a (although small) covalent component (cf. the band widths in Figs. 6 and 9), this is not the case for hydrogen cyanide which essentially consists of electrostatically interacting units. However, the local-density method will tend to overestimate any covalent component of a bond and accordingly also those of the hydrogen bonds in hydrogen cyanide.

The localization of the electrons in hydrogen cyanide is indicated in Fig. 11 that shows the band structures for the $(HCN)_x$ polymer, the $(CNH)_x$ polymer, and a 'transition state' for which the hydrogen atoms have been shifted to roughly the midpoints between the carbon and nitrogen atoms [this 'transition state' corresponds to that of Fig. 9(b) for hydrogen fluoride]. For all three structures the band widths are very small and not even the 'transition state' leads to delocalized electrons. The orbitals of the σ_1 (π_1) band correspond to the σ (π) bonds between the carbon and nitrogen atoms and vary therefore only little for the three structures of Fig. 11. On the other hand, the σ_2 band is formed by C–H orbitals for $(HCN)_x$, by N–H orbitals for $(CNH)_x$, and approaches thus a hydrogen $1s$ band for the 'transition state'. Finally, the σ_3 band is formed by lone-pair orbitals on nitrogen for $(HCN)_x$ and on carbon for $(CNH)_x$.

In closing this section we stress that although the calculated structural properties of hydrogen cyanide are related with some inaccuracies, others are more accurately described by the density-functional calculations. These include, e.g., the band structures as those shown in Fig. 11. We have used those and the accompanying wavefunctions in theoretically studying[25] experimentally accessible quantities for which we found only minor differences between the polymers and the isolated monomers. Significantly different properties, however, were observed for the conjugated polymer due to its delocalized electrons. For further details, the reader is referred to Ref. 25.

7. CONCLUSIONS

We have presented results of a theoretical, parameter-free study of the electronic properties of extended, hydrogen-bonded systems. The calculations were based on the exact density-functional formalism of Hohenberg and Kohn with a local approximation for exchange- and correlation-effects.

We examined two systems in detail: hydrogen fluoride and hydrogen cyanide, and found that structural properties were well described for the formed but not for the latter. It was argued that the local approximation was responsible for this difference, but since the present study represents one of the first and only on density-functional

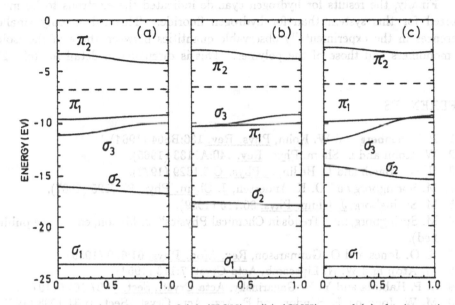

Fig. 11. Band structures of (a) $(HCN)_x$, (b) $(CNH)_x$, and (c) the 'transition state' between the other two. The dashed lines represent the Fermi level.

calculations on extended, hydrogen-bonded systems, we recognize here a definite need for further studies in order to more clearly understand the accuracies and limitations of such methods.

The most detailed investigation was undertaken for hydrogen fluoride. Using the results of the density-functional calculations we proposed that model potentials for describing dynamical and statistical properties of liquid and gaseous hydrogen fluoride should be improved, as they could not account for the properties of the polymers. The interactions between two parallel, linear hydrogen fluoride chains were found to be non-negligible. This may have crucial impacts on solitons: these may become non-existing, metastable, only existing in pairs, or not truly confined to single chains. However, other effects may reduce the importance of the interchain interactions. These include temperature fluctuations, 3D-effects, and effects due to the relative orientation of neighbouring planes of the zigzag chains. Anharmonic effects led us predict not yet observed features in the phonon spectra. These effects were also responsible for the possibility of creating solitons. For these we argued that the lattice distortions were localized, but that the accompanying electronic orbitals were more delocalized. These orbitals gave rise to extra peaks in the electronic density of states that might provide an experimental finger print for the existence of solitons, although also other features can be held responsible for their presence. Preliminary results of a recent study on linear chains of hydrogen fluoride with a doubled unit cell indicated that the propagation of solitons is assisted by a coupling to the heavier sublattice. We are presently extending this study to more realistic structures as well as to the calculation of experimentally accessible quantities in order to examine the possibilities of observing solitons experimentally.

Finally, the results for hydrogen cyanide indicated the electrons to be more localized for this system than for hydrogen fluoride. This led to only smaller differences in the experimentally observable quantities between those of the isolated monomers and those of the polymer. This is discussed in detail in Ref. 25.

REFERENCES

1. P. Hohenberg and W. Kohn, Phys. Rev. 136:B864 (1964).
2. W. Kohn and L. Sham, Phys. Rev. 140:A1133 (1965).
3. U. von Barth and L. Hedin, J. Phys. C 5:1629 (1972).
4. M. Springborg and O. K. Andersen, J. Chem. Phys. 87:7125 (1987).
5. M. Springborg, J. Chim. Phys. 86:715 (1989).
6. M. Springborg, in: "Trends in Chemical Physics", J. Menon, ed. (to be published).
7. R. O. Jones and O. Gunnarsson, Rev. Mod. Phys. 61:689 (1989).
8. M. Atoji and W. N. Lipscomb, Acta Cryst. 7:173 (1954).
9. S. P. Habuda and Y. V. Gagarinsky, Acta Cryst., Sect. B 27:1677 (1971).
10. M. W. Johnson, E. Sándor, and E. Arzi, Acta Cryst., Sect. B 31:1998 (1975).
11. M. Springborg, Phys. Rev. B 38:1483 (1988).
12. G. Di Lonardo and A. E. Douglas, Can. J. Phys. 51:434 (1973).
13. K.-P. Huber and G. Herzberg, "Constants of Diatomic Molecules", Van Nostrand Holland, New York (1979).
14. D. A. Pinnick, A. I. Katz, and R. C. Hanson, Phys. Rev. B 39:8677 (1989).
15. R. H. Maybury, S. Gordon, and J. J. Katz, J. Chem. Phys. 23:1277 (1955).
16. J. W. Ring and P. A. Egelstaff, J. Chem. Phys. 51:762 (1969).
17. D. F. Smith, J. Chem. Phys. 28:1040 (1958).
18. M. Springborg, Phys. Rev. B 40:5774 (1989).
19. J. M. Lisy, A. Tramer, M. F. Vernon, and Y. T. Lee, J. Chem. Phys. 75:4733 (1981).
20. J. S. Kittelberger and D. F. Hornig, J. Chem. Phys. 46:3099 (1967).
21. A. Anderson, B. H. Torrie, and W. S. Tse, Chem. Phys. Lett. 70:300 (1980).
22. B. Desbat and P. V. Huong, J. Chem. Phys. 78:6377 (1983).
23. E. W. Laedke, K. H. Spatschek, M. Wilkens Jr., and A. V. Zolotariuk, Phys. Rev. A 32:1161 (1985).
24. W. J. Dulmage and W. N. Lipscomb, Acta Cryst. 4:330 (1951).
25. M. Springborg, Ber. Bunsenges. Phys. Chem. (in press).

26. Proceedings of "International Conference on Science and Technology of Synthetic Metals, 1990, Tübingen", <u>Synth. Met.</u> 41-43 (1991).

27. I. Suzuki, M. A. Pariseau, and J. Overend, <u>J. Chem. Phys.</u> 44:3561 (1966).

28. G. Winnewisser, A. G. Maki, and D. R. Johnson, <u>J. Mol. Spectrosc.</u> 39:149 (1971).

29. R. A. Creswell and A. G. Robiette, <u>Mol. Phys.</u> 36:869 (1978).

30. T. J. Lee and A. P. Rendell, <u>Chem. Phys. Lett.</u> 177:491 (1991).

31. M. Maroncelli, G. A. Hopkins, J. W. Nibler, and T. R. Dyke, <u>J. Chem. Phys.</u> 83:2129 (1985).

THEORETICAL ANALYSIS OF ISOTOPIC SCRAMBLING IN

ION-MOLECULE REACTIONS INVOLVING PROTON TRANSFERS

E. M. Evleth and E. Kassab

Dynamique des Interactions Moléculaires
Université Pierre et Marie Curie
75240 Paris, France

INTRODUCTION

The structures of gas phase ion-molecule complexes are generally poorly experimentally characterized in spite of the fact that their energetics are well known.[1,2] In fact, one is often not sure what structural species one is dealing with. A large amount of information exists from mass spectrometry as to what kinds of reactions occur, mostly in the cases of highly vibrationally excited species, and some mechanisms can be established using isotopically labeled materials. Ironically, the use of isotopic labeling is not general in those studies in which the finer details of the energetics have been determined.[1,2] With regard to what is discussed here, theoretical treatment of small cluster dynamics has not proceeded very far and will certainly be an area of increased activity over the next 10 years. On the other hand, ab initio calculations are now capable of giving fairly accurate descriptions of the structures and energies of small ion systems.

An important example where proton transfer plays a key role in isotopic scrambling is found in ion-molecule gas phase reactions of water and protonated water. Understanding this particular system is important with regard to isotope fractionation in upper atmosphere ion chemistry. The experimental work has been done by Henchmann and coworkers[3,4] and has been recently analyzed by ourselves.[5] This reaction has been analysed under single collision conditions at room temperature where the estimated life of the collision complex is several nanoseconds. The initial proton transfer reaction is:

$$D_3^{16}O^+ + H_2^{18}O \rightarrow D_2^{16}O + H_2^{18}OD^+ \tag{1}$$

Additional scrambling reactions involve more than a simple proton transfer. In fact, the all products pairs, $H^{18}OD, D_2^{16}OH^+$, $H^{16}OD, D_2^{18}OH^+$, $D_2^{18}O, H_2^{16}OD^+$, etc., up to $D_3^{18}O^+, H_2^{16}O$ are formed but not in fully statistical amounts.

The best level of theoretical treatment finds no barrier for either the formation of H-bonded $H_2O_5^+$ complex or proton transfer[6] within the complex, the minimum on the hypersurface describes the bridging proton moving in a single potential well. The $H_2O_5^+$ structure is best described as $H_2O - H^+ -$

OH_2. The $D_3^{16}O^+ + H_2^{18}O$ system involves 5 individual complexes and transition states between each of these structures. Scrambling involves transition states having bifurcated structures of the type, $HOH_2 > OH_2$, in which the oxygen of OH_2 is bound to two hydrogens of the H_3O^+ unit. This is a proton site to another proton site transition state involving two hydrogen bonds. The energy surface[5] for the reaction of H_2O with D_3O^+ giving $DOH + D_2OH^+$ is shown in Fig 1.

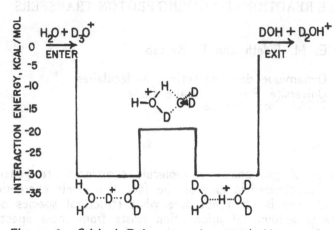

Figure 1. Critical Points on the $H_5O_2^+$ Hypersurface

In order for scrambling to occur, the initially formed $H_2O - D^+ - OD_2$ complex must rearrange to $HDO - H^+ - OD_2$. This can be viewed as occurring by the rotation of the $H_2O - D^+$ unit in the presence of a rigid OD_2 or the migration of the latter from the OD binding site to an OH binding site on $H_2O - D^+$. These motions replace the D in the hydrogen bond by an H. The decomposition of $HDO - H^+ - OD_2$ will give either $DOH + D_2OH^+$ or $H_2OD^+ + D_2O$. In addition, the initially formed H-bonded complex, $H_2O - D^+ - OD_2$, can either revert to the entrance channel reactants, $D_3O^+ + H_2O$, which is an experimentally undetectable event in this case, or also give $H_2OD^+ + D_2O$. Therefore, the appearance of $DOH + D_2OH^+$ establishes an event more complicated than proton transfer.

The double labeling experiments[3,4] provide more information than obtained using only deuterium labeled materials. In these experiments, in order for statistical scrambling to occur, all 5 complexes must be completely dynamically accessible within the lifetime of the collision complex. In the absence of collision, the complex has an internal energy equivalent to the complexation energy, 32 kcal/mol.[1,2] The bifurcated transition state for proton site exchange is only about 8-10 kcal/mol above the bottom of the well and qualitatively one might think complete equilibration would occur throughout all the wells in this system. In principle, a dynamical treatment of this system is possible using a multichannel RRKM method but there is doubt on its validity with regard to other methods used in the literature for much simpler systems (for example, see the work of Troe).[7] In any case, no dynamical treatment of this system has yet been performed which quantitatively rationalizes the experimental work. This is also the case for the isotopic scrambling occurring in the ND_4^+, NH_3 system.[3-5]

It should now be noted that there are few reliable theoretical estimates even for equilibrium constants for the formation of ion-molecule or more weakly bound neutral complexes. Although theoretical chemistry can now provide some reliable enthalpic estimates, the calculations of reliable entropies are not possible using harmonically derived frequencies for the loose vibrations. Loosely bound complexes will have up to 6 of these, some of which can be described as free or hindered rotor motions, others as occurring in broad very anharmonic single or double potential wells. For example, a recent experimental and theoretical study of the H_2O, OH^- system[8] demonstrated the impossibility of using standard quantum chemical codes to generate accurate entropic estimates for loose complexes. Generating an authentic estimate of both the vibrational manifolds and the partition functions necessary to estimate equilibrium constants will probably be developed using vibrational quantum Monte Carlo methods; a recent treatment of the HF dimer is a case in point.[9]

On the other hand, accurate structural determination of ion-molecule complexes, their enthalpies, and reaction mechanisms is an area where theoretical chemistry can make a significant contribution. Here, we will characterisize the gas phase reaction of protonated formaldehyde with water in order to show why oxygen scrambling does not occur. Earlier experimental work[10] misinterpreted the reason for non-scrambling and a detailed theoretical analysis will have general value in demonstrating what types of experiments are necessary to fully characterize the reaction mechanism of these types of proton transfer systems.

ANALYSIS OF THE $H_2COH^+ + H_2O$ INTERACTION

The interaction of $H_2COH^+ + H_2O$ can conceptually lead to the formation of several structures, (i) an ion-molecule H-bonded complex, $H_2COH^+ - OH_2$, (ii) its proton transfer equivalent, $H_2CO - HOH_2^+$, and (iii) a tetrahedral adduct, $HOCH_2OH_2^+$, protonated methanol. The formation of the latter structure would be implied if oxygen exchange occurred in the following reaction.

$$H_2C^{16}O + H_2^{18}O \rightarrow H_2C^{18}O + H_2^{16}O \tag{2}$$

Since oxygen exchange does not occur it was concluded[10] that protonated methandiol is not formed in this gas phase reaction. It should be noted that in bulk water, oxygen isotope exchange is rapid under acid catalyzed conditions with the formation of methandiol.[11]

Here we employ a standard ab initio protocol of selecting a particular basis set, performing a geometry optimization at the SCF level, computation of the SCF frequencies at this level, and finally computing the correlation energy at some level. Since the proton affinities are known for formaldehyde and water[2] one measure of the precision of the calculations reported here is their capacity for reproducing these values. The results reported in Table 1 demonstrate that a MP4/6-31+G**//6-31G* protocol is capable of yielding proton affinities within 1-2 kcal/mol of the experimental values. We note that further refinements (inclusion of basis set super position error corrections, use of scaled frequencies, larger basis sets) could produce poorer results. This basis set level is not intrinsically capable of giving 1-2 kcal/mol accuracy although it may appear at times to do so by a fortuitious cancelling of errors.

• Table 1. Proton Affinities of Formaldehyde and Water, Kcal/mol

Basis	6-31G*		6-31G**		6-31+G*		6-31+G**		Exptl.
	SCF	MP4	SCF	MP4	SCF	MP4	SCF	MP4	
H_2O	166	167	171	172	163	160	167	165	166.5
H_2CO	173	166	178	173	171	163	175	169	171.7
Diff.	+7	-1	+7	+1	+8	+3	+8	+4	+5

• Calculated from Electronic Energies corrected for thermal and ΔPV effects

The hydration energy of protonated formaldehyde has been redetermined most recently by Moet-Ner[12] and theoretically studied by Scheiner and Hillenbrand[13] at the 4-31G* SCF level. Experimentally, one does not know which species, the hydrate of protonated formaldehyde or protonated water complexed with formaldehyde, one is dealing with. In fact both might be present. The difference in the proton affinities is such that one would predict that the former complex is the more stable. Scheiner and Hillenbrand's calculations showed no minimum on the hypersurface for the $H_2CO - H_3O^+$ structure, there being no barrier for the proton transfer giving $H_2COH^+ - OH_2$. Our calculations of a larger portion of the surface are shown in Table 2.

• Table 2. Heats of Reaction of Various H_2COH^+ Hydrates Structures

Basis	6-31G*		6-31+G**		Exptl.
Method	SCF	MP4	SCF	MP4	
Structure					
$H_2COH^+ - OH_2$	-28.0	-32.7	-25.8	-29.0	-28.5
$^+H_2OCH_2OH$	-20.6	-23.1	-18.4	-20.5	
Transition State	+11.3	-2.9	+13.3	+0.8	

• Calculated from Electronic Energies corrected for thermal and ΔPV effects, Kcal/mol

As found in Table 1, the MP4/6-31+G**//6-31G* protocol gives a good estimate of the enthalpy of hydration of H_2COH^+. However, note that if one were not interested in other points on the hypersurface, an SCF/6-31G* enthalpy would produce accurate appearing results as is also found in the case of the proton affinities. In particular, the transition state for proton transfer is correlation sensitive. This is not normally the case for proton transfer transition states between oxygen centers. However, in this case the proton is being transferred in a 4-membered ring and this involves considerable ring strain as well as imposing non-linearity on the hydrogen bond.

Table 2 characterizes other structures in this hypersurface. First of all, a form of protonated methandiol, $^+H_2OCH_2OH$, does exist in addition to the intramolecular transition state for proton transfer. Only this latter structure is necessary for oxygen isotope scrambling: $^+H_2^{18}OCH_2^{16}OH => H^{18}OCH_2^{16}OH_2^+$. The key feature of the hypersurface is that the enthalpy of the transition state is nearly as that of the entry channel reactants. Dynamically, this is a 2-channel system with a loose transition state for the entry channel, $H_2COH^+ + OH_2$, and a tight one for proton transfer. Qualitatively, given the probably short lifetime of the collision complex (several nanoseconds) and the approximately equal energies of the two channels but differing entropies, one predicts little observable scrambling in this system. These theoretical calculations demonstrate that it is the high energy of the transition state for proton transfer which blocks scrambling not the lack of formation of a protonated methandiol intermediate. The latter structure is dynamically accessible in this system. This is shown schematically in Fig. 2.

An added complication to the above system is the transition state for the migration of water from the proton site on H_2COH^+ to the carbon site. The energy of this transition state is shown in Fig. 2 as only marginally above that of the protonated methandiol. In fact, this latter species is better described as an ion-molecule complex since it is not tetrahedral at the carbon site. Fig. 2 also shows the energy of a species which does have full tetrahedral symmetry. This structure is a transition state for internal rotation of the $-OH$ unit. Even so, all the structures in this part of the hypersurface have nearly the same energies and are dynamically accessible in the ion-molecule encounter complex in the absence of third body collisions. However, in the absence of isotopic scrambling there is no experimental proof that such structures are actually being dynamically sampled. Note, equally that the existence of minima in this structural region is not a dynamical necessity if scrambling actually experimentally or theoretically occurred. It is the location of the proton transfer transition state which represents the critical bottleneck for scrambling.

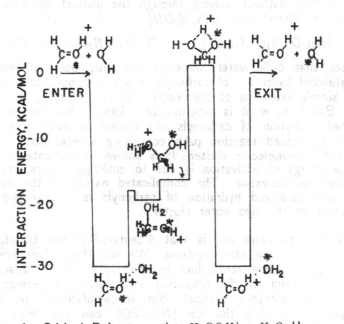

Figure 2. Critical Points on the $H_2COH^+ + H_2O$ Hypersurface

A broader view of the above system is to speculate on the possibility of scrambling as a function of cluster size n in the cluster:

$$(H_2C^{16}OH^+)(H_2^{18}O)_n \rightarrow (H_2C^{18}OH^+)(H_2^{16}O)_{n-1}H_2^{18}O \qquad (3)$$

This event is only measurable, using gas phase ion-molecule techniques, on the decomposition of the cluster giving $(H_2C^{18}OH^+)(H_2^{16}O)_{n-1} + H_2^{18}O$.

This kind of event is difficult to measure using gas phase ion-molecule techniques[13,14] for large clusters. In any case, other calculations we have performed[15] have shown that isotopic exchange should occur with relative ease for n = 1. Since no exchange occurs within the complex, $H_2C^{18}OH^+ - {}^{16}OH_2$, this is measurable if the following reaction occurs.

$$H_2^{16}O + (H_2C^{18}OH^+ - {}^{16}OH_2) \rightarrow H_2^{18}O + (H_2C^{16}OH^+ - {}^{16}OH_2) \qquad (4)$$

In the two water cluster we found that the most stable structure is H_3O^+ complexed by water and formaldehyde and that there is no minimum for H_2COH^+ dihydrated at the proton site. Therefore, even for the two water cluster case the basicity of formaldehyde is less than for water. There are other minima on this particular hypersurface, the overall system is much more complicated than in the one water case. In any case, we predict isotopic scrambling will occur for the two water cluster.

With regard to what occurs in bulk water under acidic conditions, the hydration of carbonyls is general acid catalyzed[17] and has been interpreted as a concerted event in which a proton is transferred from the solvent onto the carbonyl oxygen as water is added to the carbon atom and the proton is retransferred back into the solvent. Such a mechanism involves several water molecules[18] in the transition state in such a manner that the entropy of activation will play an important role in determining the free energy of activation. In a supercluster environment, this means that the following reaction occurs directly without passing through the distinct hydrated protonated intermediates, $H_2C^{18}OH^+$ and $HOCH_2OH_2^+$.

$$(H_2CO)(H_2O)_n(H_3O^+) \rightarrow (H_2C(OH)_2)(H_2O)_{n-1}(H_3O^+) \qquad (5)$$

One notes that bulk water modeling has only been attempted[19-21] on the base catalyzed hydration of carbonyls which is conceptually easier to describe as a simple extension of the reaction of OH^- adding to H_2CO giving $HOCH_2O^-$. Based on what is experimentally known, bulk water modeling of acid catalyzed hydration of carbonyls will require working from a quantum mechanically determined reaction path containing several water molecules in the protonated supermolecule cluster. It is known experimentally[18] that about half the free energy of activation is due to enthalpy effects, the other half due to entropy of activation. The complicated nature of the mechanism of bulk water acid catalyzed hydration of carbonyls is already implied by the results obtained on the two water cluster.

The final item to point out is that 4-centered proton transfer transition states are encountered in other systems. We will show elsewhere[15] that the observed[14] oxygen isotope scrambling in the $OH^- + H_2CO$ reaction is due to a 4-centered proton transfer transition state having an energy well below the energy of the entrance channel. Similar calculations[22] on the reaction of $PO_3^- + H_2O$ show that the ion $(HO)_2PO_2^-$ can not form in the ion-molecule reaction due to too high a barrier for the proton transfer transition state. Similar calculations have demonstrated that the slow rate of reaction[23] of the neutral system $H_2O + SO_3$ giving H_2SO_4 is due to the proximity of the energy of the 4-centered proton transfer transition state to that of the entrance channel.

Acknowledgements

The work reported here is part of a broader investigation of the role of proton transfer in the catalytic reactions occurring at the active sites in zeolites. This work is partially sponsored by the French Petroleum Institute, contract CNRS-IFP 510-135.

References

1. R. G., Keesee and A. W. Castleman, Jr., J. Phys. Chem. Ref. Data, 15:1011 (1986).
2. S. G. Lias, J. F. Liebman and R. D. and Levin R. D., J. Phys. Chem., Ref. Data, 13:695 (1984).
3. M. Henchman and J. E. Paulson, J. Chem. Soc., Faraday Trans., 2, 85:1673 (1989).
4. N. G. Adams, N. G. Smith and M. Henchman, J. Mass. Spectrom. Ion., Phys., 42:11 (1982).
5. E. M. Evleth, Z. D. Hamou-Tahra and E. Kassab, J. Phys. Chem., 95:1213 (1991).
6. J. E. Del Bene, J. Phys. Chem., 92:2874 (1988).
7. J. Troe, J. Mass. Spectrom. Ion. Phys., 80:17 (1987).
8. G. J. C. Paul and P. Kebarle, J. Phys. Chem., 94: 5184 (1990).
9. M. Quack and M. A. Suhm, J. Chem. Phys., 95:28 (1991).
10. J. K. Pau, J. K. Kim and M. C. Caserio, J. Am. Chem. Soc., 100: 3831 (1978).
11. P. Greenzaid, Z. Luz, and S. David, J. Am. Chem. Soc., 89:749, 756 (1967).
12. M. Moet-Ner, J. Am. Chem. Soc., 106:1265 (1984).
13. S. Scheiner and E. A. Hillenbrand, J. Phys. Chem., 89:3053 (1985).
14. H. Van der Wel and N.M.M. Nibberling, Recl. Trav. Chim. Pays-Bas, 107:479, 491 (1988).
15. E. M. Evleth and E. Kassab, in preparation.
16. N.M.M. Nibberling, Accts. Chem. Res., 23:279 (1990).
17. W. P. Jencks, Accts. Chem. Res., 12 425 (1976).
18. H.-J. Buschman, E. Dutkiewicz, W. Knoche, W. Ber. Bunsengs. Phys. Chem., 86:129 (1982).
19. J. D. Madura and W. L. Jorgensen, J. Am. Chem. Soc., 108:2517 (1986).
20. A. E. Howard and P. A. Kollman, J. Am. Chem. Soc., 110:7195 (1988).
21. H.-A. Yu and M. Karplus, J. Am. Chem. Soc., 112:5706 (1990).
22. Y. Akacem, O. Ouamerali, E. Kassab, E. M. Evleth, Y.-D. Wu and K. N. Houk, to be published.
23. X. Wang, Y. G. Jin, M. Suto, and L. C. Lee, J. Chem. Phys., 89: 4853 (1988).

TWO SUBLATTICE MODEL OF HYDROGEN BONDING

AT FINITE TEMPERATURES

O. Yanovitskii†‡*, N. Flytzanis† and G. Vlastou-Tsinganos†‡

† Physics Department, University of Crete, Heraklion

‡ Foundation of Research and Technology Hellas FORTH

* Perm. Ads: Institute for Theoretical Physics, Kiev, 252130, USSR

I. INTRODUCTION

The high anisotropic proton conductivity in many hydrogen–bonded molecules (e.g. ice), has attracted a lot of theoretical interest[1-5]. This unusualy high mobility of protons, also plays a role in interpreting certain biological processes[6,7]. The transfer of protons in a hydrogen bonded system is done by ionic (to be considered here) and rotational defects along Bernal-Fowler filaments[8]. The effective potential that a hydrogen-bonded proton feels between two heavy ions can be described by a double well potential, with two equivalent equilibrium positions separated by a barrier. This provides a mechanism for the transfer of protons along the chain. As the basic hydrogen–bonded chain one considers a sequence of alternating heavy ions (OH^- ions for ice) and protons but the heavy ion sublattice must not be thought of as a fixed substrate. The motion of the heavy ions along with their coupling to the protons is very important, as it is shown by experiments of the $O - H$ distance as a function of the $O - O$ distance in the $O - H - -O$ system[9]. For a realistic hydrogen bonding potential the relative displacement of neighbouring heavy ions changes the barrier height drastically.

A detailed analysis of the problem involves the consideration of a multidimensional energy surface. To make the problem tractable, one considers a phenomenological model with effective interactions, that captures the essential features of the potential surface. Due to the strong competition between the various degrees of freedom, variations of pressure or temperature can cause phase transitions in the ground state. To assure that the equilibrium structure corresponds to global minimum, in view also of the uncertainty in the values for the effective hamiltonian parameters, it is necessary to study the phase diagram in the space of the interaction parameters. For most realistic potentials one cannot resort to analytic solutions, not even in the continuum approximation[5], which may not be valid for weak inter-proton coupling. If, however, the bistable potential is approximated by a double quadratic potential, analytic solutions are possible[9]. In all the previous cases one treats the classical problem at $T = 0$. For a light proton however, zero-point energy can be significant. Its full quantum mechanical description is difficult, so we resort to approximations. The free energy of the system is evaluated using a semi–quantum mechanical approach, i.e. a variational formulation of the SCPA (Self Consistent Phonon Approximation), and thus thermodynamic quantities are evaluated along with the mobility determining the Peierls–Nabbarro (PN) barrier as functions of temperature. In section II, we present the model and the method to evaluate the free energy and the the PN barrier, while in section III we present the results.

II. MODEL AND METHODS

Our work is based on the model proposed by Antonchenko, Davydov and Zolotariuk[1], and its later modifications to include acoustic oxygen motion with movable $O - H$ interaction[5,11]. The model consists of two one-dimensional interacting sublattices of protons (denoted H) and heavy ions (denoted O), with masses m, M and coupling constants K_{HH} and K_{OO} correspondingly. In addition the oxygens are submitted to an on-site harmonic potential (with strength k), to simulate the effect of the surrounding material. The interaction between the two sublattices is made via a double well Morse potential, being the sum of the proton interaction, with its neighbouring heavy ions. It is determined by three parameters D, r_e, d which correspond to the depth of the wells, the position of minima and a constant related to the restoring force $(k_M = 2d^2 D)$ of the single $O - H$ interaction, respectively.

Therefore the Hamiltonian which describes the chain consists of three parts:

$$\mathcal{H} = \mathcal{H}_H + \mathcal{H}_O + \mathcal{H}_{OH}. \qquad (1)$$

where

$$\mathcal{H}_H = \sum_n \left[\frac{1}{2} m \dot{a}_n^2 + \frac{1}{2} k_{HH} (a_{n+1} - a_n)^2 \right], \qquad (2)$$

$$\mathcal{H}_O = \sum_n \left[\frac{1}{2} M \dot{\rho}_n^2 + \frac{1}{2} k \rho_n^2 + \frac{1}{2} k_{OO} (\rho_n - \rho_{n+1})^2 \right], \qquad (3)$$

and

$$\mathcal{H}_{OH} = \sum_n \left[D \left(1 - e^{-d(a_n - \rho_n - r_e)} \right)^2 + D \left(1 - e^{-d(R - a_n + \rho_{n+1} - r_e)} \right)^2 \right]. \qquad (4)$$

Here, the reference points from which all displacements are measured, are the equilibrium oxygen sublattice sites. Thus, a_n and ρ_n are the displacements of the nth proton and oxygen respectively.

We are interested in the behaviour of the above two–sublattice chain at finite temperatures. The quantum mechanical free energy of the system, can be approximated by the aid of the Gibbs–Bogoliubov inequality[12], which gives an upper bound to the exact free energy function $\mathcal{F}$:

$$\mathcal{F} \leq \mathcal{F}_0 + < \mathcal{H} - \mathcal{H}_0 >. \qquad (5)$$

$\mathcal{H}_0$ is a trial Hamiltonian (and its respective free energy $\mathcal{F}_0$), that must be chosen to be an approximation of $\mathcal{H}$. The procedure that must be followed, is to chose $\mathcal{H}_0$, such that its density matrix and its free energy can be calculated exactly, then evaluate the expectation value appearing in the above inequality using the density matrix corresponding to $\mathcal{H}_0$, and thus an upper bound to $\mathcal{F}$ is found. Our choice for $\mathcal{H}_0$ is a set of independent displaced harmonic oscillators, which is considered to map the effective potential each atom feels, characterised by the position of the displaced oscillator α_n and its frequency ω_n. Therefore, one can use them as variational parameters, in order to minimize the expression of Eq.(5) and thus determine the best approximation to $\mathcal{F}$.

An efficient method to optimize a function of many variables is the Simulated Annealing Monte Carlo method (SAMC)[12]. The idea is to cool the system very slowly, so that it escapes from local minima, in the search for the global one. The method consists of a basic step, which is repeated many thousand times in order to simulate a collection of atoms in equilibrium at a given temperature. In each step, an atom is given a small random displacement, and the resulting change in energy is computed. Depending on the sign of this energy change, the new configuration is readily accepted or rejected with some probability. Then by slowly lowering the temperature and repeating the same number of basic steps, one achieves to find the global minimum. Being in the global minimum well, one proceeds with the Steepest Descent method and locates the exact point of vanishing gradient.

III. RESULTS

One important point in the study of a hydrogen-bonded chain is the choice of the couplings. The exact values of these parameters are not known exactly experimentaly, and therefore one has to find indirect ways to chose them, by fitting experimental data. Here we adopt the values given in reference 10 i.e. $k_{HH} = 22N/m$, $k_{OO} = 15N/m$, $k = 29N/m$ and for the Morse potential $D = 0.4eV$, $d = 7.2/AA^{-1}$ (which gives a barrier $u_B = 0.3eV$ at the oxygen equilibrium separation. To check the dependence of the ground state on the interaction parameters, in Fig 1 we present the classical phase diagram at $T = 0°K$, in the space of k_{HH} and k, holding $k_{OO} = 15N/m$ fixed. The largest area in the phase diagram is occupied by the period 1 phase, in which all hydrogen atoms in the chain are located within the right (or left) well of its Double-Morse potential. This is the ground state phase in ice, and is stabilized with strong hydrogen-hydrogen interaction and a deep oxygen substrate. Structures with high periodicity appear in the phase diagram, for relatively small and comparable couplings. In that region the hydrogens interact rather weakly competing with the other ordering mechanisms, while the oxygens move within a shallow harmonic well and thus high periodicity formations are stabilized. When the oxygen substrate is deep, then the structures with small period and high symmetry are favored: 2×1 and 3×1. The above phase diagram was done classically, (i.e. comparing energies) because it was much more computer efficient than the quantum one. For this set of parameters, the ground state given above has been tested quantum-mechanically and was found to be of period 1 and thus one may procceed which the calculation of chain soliton exitations.

The conductivity in hydrogen bonded systems is connected with the static and dynamic properties of domain walls. In a hydrogen-bonded chain, there is the possibility of positive (I^+) and negative (I^-) ionic defects, whose importance in conductivity cannot be accurately determined, due to many experimental difficulties. For a static oxygen substrate, the two defects have the same energy and Peierls-Nabarro barrier, which will not be the case for a movable oxygen sublattice.

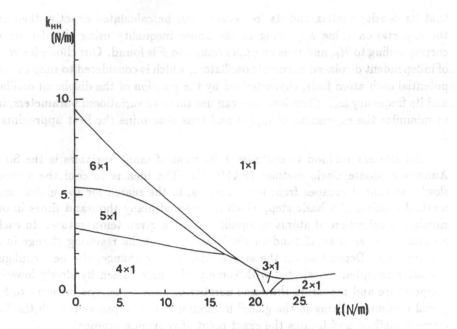

FIGURE 1. Classical phase diagram for the hydrogen bonded chain.

In Fig.2, we plot the PN barrier for the I^+ (solid circles) and I^- (open circles) as a function of temperature for the potential parameters given earlier. There is a strong temperature dependence for the I^- defects, while for the I^+ defects it is almost constant. In both cases however, it is much less than the barrier for a single proton jump. In the I^+ defect, the oxygen displacement is extremely small. The effect of temperature is to lower the effective barrier each proton sees. The PN barrier, i.e. the energy to move the defect, depends essentially on the centered configuration energy (E_C), where one proton sits halfway between two oxygens in a local minimum and the symmetric configuration energy (E_S), where two consecutive protons lie in opposite wells. For both configurations the lowering of the barrier decreases their respective energies by similar amounts at low temperatures giving the flat behavior.

For the I^- defect, there is a very strong oxygen displacement, which in the defect core can create a single minimum, lowering significantly the energy for the centered configuration (E_C^-). Increasing the temperature, decreases this energy (E_C^-) lowering. On the other hand the symmetric configuration decreases with T, so that the PN barrier decreases significantly. The minimum at $T = 400°K$ is due to the fact that the relative energy of the symmetric and centered configurations passes through unity. Thus a new minimum appears and the PN barrier (the difference between the minimum and maximum energy configurations) increases. The minimum is strongly dependent on the barrier height of the effective double well potential and weaker on other parameters. In Fig.2 at low T the I^- barrier is higher than the I^+, while above 260°K it is reversed. Due to the uncertainty in detemining some of the parameters, these results can only be considered qualitative, while the effect of the various parameters is being examined in detail.

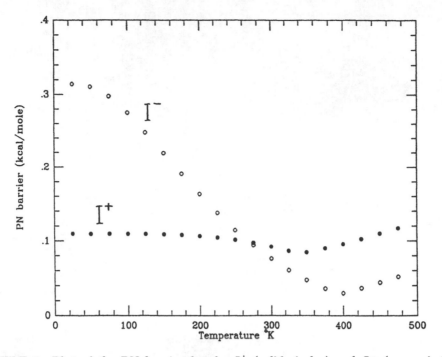

FIGURE 2. Plot of the PN barrier for the I^+ (solid circles) and I^- (open circles) defects, as a function of temperature for the parameters given in the text.

At $T = 0$ the ratio of the two is comparable to the value calculated by Savin *et al*, while their absolute values are significantly lower due to the zero-point energy, which can be very important for a light particle in a narrow potential well. Thus, at room temperature, for the parameters considered, the PN barrier is lower than $K_B T$, so that thermal fluctuations can easily give efficient charge transport under the application of a weak external field. The temperature T_0 at which $U_{PN} \sim k_B T_0$ depends strongly on the double well barrier U_B and its dependence on the other parameters is investigated. The values of the PN barrier are an important ingredient in any rate theory for kink mobility at a finite temperature.

Part of this work was supported by the EEC Science Project # SCI-0229C 89–1000 79/JU1. The authors benefited from discussions with E. Kryachko, A.V. Savin and A.V. Zolotariuk.

REFERENCES

1. V.Ya. Antonchenko, A.S. Davydov and A.V. Zolotariuk 1983 Phys. Stat. Sol.(b) *115* 631
2. E.Q. Laedke, K.H. Spatschek, M. Wilkens Jr. and A.V. Zolotariuk 1985 Phys. Rev. A *32* 1161
3. M. Peyrard, St. Pnevmatikos and N. Flytzanis 1987 Phys. Rev. A *36* 903
4. D. Hoshstrasser, H. Büttner, H. Desfontaines and M. Peyrard 1988 Phys. Rev. A *38* 5332
5. E.S. Kryachko 1990 Chem. Phys. *143* 359
6. J.F. Nagle, M. Mille and H.J. Morovits 1980 J. Chem. Phys. *72* 3959
7. H. Merz and G. Zundel 1981 Biochem. Biophys. Res. Comm. *101* 540
8. J.D. Bernal and R.H. Fowler 1933 J. Chem. Phys. *1* 515
9. M. Ichikawa 1978 Acta Cryst. B *34* 2074
10. P.T. Dinda and E. Coquet 1990 preprint
11. A.V. Savin and A.V. Zolotaryuk 1991 preprint
12. G. Vlastou-Tsinganos, N. Flytzanis and H. Büttner 1990 J. Phys. A *23* 4335
13. S. Kirkpatrick, C.D. Gelatt Jr and M.P. Vecchi 1983 Science *220* 671

CONTRIBUTORS

Dr. Luis G. Arnaut
Chemistry Department
University of Coimbra
3000 Coimbra
Portugal

Prof. Attila Askar
Dept. of Mathematics
Bosporos University
80815 Bebek-Istanbul
Turkey

Prof. Tassos Bountis
Dept. of Mathematics
University of Patras
Patras 26110
Greece

Prof. Giorgio Careri
Dept. of Physics
University of Rome
"La Sapienza", Piaz A. Moro
00185 Roma
Italy

Mr. Ioannis Chochliouros
Laboratoire de Modélisation en Mecanique
Université Pierre et Marie Curie
Tour 66
4 Place Jussieu
75252 Paris, Cedex 05
France

Professor P.L. Christiansen
Lab. of Appl. Math. Phys.
Building 306
Tech. U. of Denmark
DK - 2800 Lyngby
Denmark

Prof. Yuri I. Dakhnovskii
Institute of Chemical Physics
USSR Acad. Sci.
Kosigyn St. 4
117977 Moscow
USSR

Prof. David W. Deamer
Dept. of Zoology
University of California
Davis, Cal. 95616
USA

Prof. Norbert A. Dencher
Hahn-Meitner-Institut
Abt. Neutronenstreuung 1
Glienicker Str. 100
D-1000 Berlin
Germany

Prof. J. Paul Devlin
Dept. of Chemistry
Oklahoma State University
Stillwater, Okl. 74078
USA

Dr. Richard A. Dilley
Department of Biological Sciences
Purdue University
West Lafayette, In 47907
USA

Dr. Janez Dolinsek
Jozef Stefan Institute
University of Ljubljana
Jamova 39
61111 Ljubljana
Yugoslavia

Prof. S. Walter Englander
Dept. of Biochemistry and Biophysics
University of Pennsylvania
Philadelphia, Penn. 19014-6059
USA

Prof. E.M. Evleth
Dynamique des Interactions Moléculaires
ER 271
Université Paris VI
4 Place Jussieu
Paris 75230
France

Mr. Jan Florian
Inst. of Physics of Charles University
Ke Karlovu 5
121 16 Prague 2
Czechoslovakia

Prof. N. Flytzanis
Physics Dept.
University of Crete
Heraklion, Crete
Greece

Prof. John W. Glen
School of Physics and Space Research
University of Birmingham
Birmingham 15 2TT
UK

Dr. Thomas Goulet
Département de Médecine Nucléaire
et de Radiobiologie
Faculté de Médecine
Université de Sherbrooke
Sherbrooke, Quebec
Canada J1H 5N4

Dr. Joachim Heberle
Hahn-Meitner-Institut
Dept. for Neutron Diffraction, N1
Glienicker Str. 100
W-1000 Berlin 39
German

Prof. A. Karpfen
Inst. of Theor. Chemistry
University of Vienna
Währingerstrasse 17
A-1090 Vienna
Austria

Prof. Yuri Kivshar
Departamento de Fisica Teorica I
Facultad de Ciencias Fisicas
Universidad Complutense
28040 Madrid
Spain

Prof. E.S. Kryachko
Institute of Physical Chemistry
University of Munich
Therienstr. 41
D-8000 Munich 2
Germany

Mr. Nikos Lazarides
University of Crete,
Heraklion,
Crete, Greece

Prof. Y. Marechal
Dept. de Recherches Fondamentales
85X, CENG
F-38041 Grenoble CDX
France

Dr. Janez Mavri
Boris Kidric Inst. of Chemistry
P.O. Box 30
61115 Ljubljana
Yugoslavia

Prof. Mihaly Mezei
Department of Chemistry
Hunter College
The City University of New York
695 Park Avenue
New York, NY 10021
USA

Prof. Andrey Milchev
Inst. of Phys. Chem.
Bulgarian Academy of Sciences
Sofia
Bulgaria

Prof. John F. Nagle
Dept. of Physics and Biological Sciences
Carnegie-Mellon University
Pittsburg, Penn. 15213
USA

Dr. S. Tristram Nagle
Dept. of Physics and Biological Sciences
Carnegie-Mellon University
Pittsburg, Penn. 15213
USA

Prof. V.F. Petrenko
Thayer School of Engineering
Dartmouth College
Hanover, NH 03755

Prof. Michel Peyrard
Labo de Physique Nonlinéaire
Université de Bourgogne
6 Bd Gabriel
21000 Dijon
France

Prof. P. Pissis
Dept. of Physics
National Technical University of Athens
15773 Athens
Greece

Prof. C. Polymilis
Dept. of Physics
University of Athens
Panepistemiopolis
15783 Zografos
Athens
Greece

Dr. Alexander Savin
Institute for Physico-Technical Problems
119034 Moscow
USSR

Prof. Steve Scheiner
Dept. of Chemistry
Southern Illinois University
Carbondale, Ill. 62901
USA

Prof. N.D. Sokolov
Institute of Chemical Physics
Kosygin Street 4
Moscow 117977
USSR

Dr. Georg Souvignier
Max-Planck Inst. für
Ernährungsphysiologie
Rheinlanddamn 201
D-4600 Dortmund 1
FRG

Prof. Michael Springborg
Fak. für Chemie
Universität Konstanz
D 7750 Konstanz
FRG

Prof. Nikos Theodorakopoulos
Theor. and Phys. Chemistry Institute
The National Hellenic Research
Foundation
Athens 116 35
Greece

Mr. Yossi Tsfadia
Tel Aviv University
Faculty of Life Sciences
Department of Biochemistry
69978 Tel Aviv
Israel

Prof. Michalis Velgakis
Polytechnic School
University of Patras
Patras 26110
Greece

Prof. George E. Walrafen
Department of Chemistry
Howard University
500 College St. N.
Washington, DC 20016
USA

Prof. E. Whalley
Division of Chemistry
National Research Council of Canada
Ottawa, Ontario K1A OR9
Canada

Dr. O. Yanovitskii
Inst. for Theoretical Physics
Ukr.SSR Academy of Sciences
Kiev 130
USSR

Prof. Alexander V. Zolotaryuk
Inst. for Theor. Physics
Ukr.SSR Academy of Sciences
Kiev 130,
USSR

Prof. Georg Zundel
Inst. of Phys. Chemistry
University of Munich
Theresienstr. 41
D-8000 Munchen 2
FRG

Mr. Yannis Kourakis
Laboratoire D.S.C.
Université de Bourgogne
6 Bd. Gabriel
21000 Dijon
France

Dr. Alesandro Tani
Departamento de Chemica
Universita Pisa
Italy

Prof. V. Hugo Schmidt
Dept. of Physics
Montana State University
Bozeman, MT 59717
USA

INDEX

Ab initio studies, 19, 29, 37, 67, 73, 114, 159, 217, 220, 234, 276, 343
Acid, 4, 29, 155, 255, 287
  acetic, 217, 289
  catalysis, 282, 289, 348
  chloroacetic, 229
  solutions, 162, 285, 297
Acidity, 290
Adenine, 229, 233
Adiabatic transfer, 41, 49, 66, 314
ADZ model, 66, 122, 352
Anharmonicity, 53, 58, 139, 233, 313, 319, 332
Aqueous phase, 178, 191, 254
Aqueous solutions, 4, 162, 217, 233, 274, 287

Bacteriorhodopsin, 6, 25, 153, 162, 171, 175, 187
Base, 4, 29, 229, 235, 256, 287
  catalysis, 282, 289
Bethe ansatz, 131
Biological systems, 1, 17, 24, 39, 187, 189, 207, 229, 233, 351

Calorimetry, 209, 241
Charge
  carriers, 65, 72, 106
  partitioning, 17
  transport, 17, 162
Chemiosmotic, 24, 187
Collective modes, 261, 309
Collective variables, 105, 116
Conductivity, 7, 23, 79, 107, 167, 209, 255, 353
Connectivity, 167
Continuum limit, 69, 80, 85, 97, 109, 115, 126, 141, 333, 351

Crack, 139, 150
Crystallization, 252, 256, 259
Crystallography, 239, 282
Cyclic structure, 1, 7, 40

Defect
  activity, 249, 257
  Bjerrum, 17, 66, 79, 83, 106, 122, 240, 249, 254
  ionic, 17, 66, 79, 82, 106, 122, 240, 252, 351
  mobility, 249
Density functional, 325, 334
Depolarization, 208
Deuterium, 3, 174, 192, 249, 343
Diffusion, 194, 199
  coefficient, 202
  of protons, 199, 205
Dielectric constant, 19, 205
Dielectric relaxation, 207, 209, 241
Dipole moment, 18-21, 105, 110, 124
Dipole interactions, 114, 121, 208, 220
Disorder, 102, 240, 244
DNA, 2, 131, 233

Energy
  barrier, 9, 30, 35, 85, 108, 169, 282
  of activation, 208, 213, 240, 255
Entropy, 163, 208, 239, 241, 244, 346
Electron
  density, 38, 293
  distribution, 37
  systems, 153, 161, 326
Epitaxial layer, 252, 257

Fluorescence, 8, 200
Force constants 111, 306

Formaldehyde, 345
Free energy, 218, 315, 352
    of activation, 22, 220,
        282, 348
    of hydration, 225, 346
    of solvation, 273, 276
Frenkel-Kontorova model, 84,
        139, 142

Glass transition, 169, 207

Hamiltonian, 22, 53, 68, 85,
        97, 106, 123, 140,
        275, 352
Harmonic coupling, 52, 68,
        79, 110, 122, 317,
        332, 352
Harmonic model, 262, 268
Halobacterium halobium, 167,
        171
Heavy-ion coupling, 65, 80,
        95, 122, 125, 351
Hydrates clathrate, 249, 259
Hydrogen
    bond, 1, 19, 36, 49, 65,
        85, 95, 153, 231,
        270, 305, 325
    bonded, 19, 29, 65, 72,
        79, 105, 111, 154,
        217, 229, 262, 325,
        351
    chain, 25, 65, 68, 77,
        81, 95, 155, 174,
        187
    cyanide, 325, 337-340
    fluoride, 3, 90, 325,
        328-335
Hydrolysis, 11
Hydrophobic, 2, 163, 171,
        182, 188

Ice, 19, 105, 353
    amorphous, 249, 253
    conductivity, 255
    crystalline 21, 249, 254
    Ih, 239
    KOH doped, 240, 246, 254
Impurity, 8, 25, 96, 101,
        146, 240
Infrared spectra, 7, 153,
        187, 235, 250
Inhomogeneity, 95, 101, 146
Instability, 129
    modulational 73, 75
Isotopic exchange rates,
        249, 252
Isotopic scrambling, 249,
        343, 347

Kink, 69-71, 84, 86-89, 98,
        112, 125, 139, 142

Kink (continued)
    collisions, 146, 148
    stability, 98, 129, 145
Kinetic model, 281
Laser spectroscopy, 178
Lattice
    misfit, 139, 142
    solitons, 131
    Toda, 131, 140
Lipid
    bilayers, 26, 65, 171,
        194
    headgroups, 194
    vesicles, 173, 194
Lippincott-Schroeder poten-
        tial, 111
Low-frequency mode, 262, 299

Marcus equation, 34
Membrane
    protein, 24, 175, 187
    purple, 24, 167, 171,
        174, 188
Methylamine, 217, 274
Mobility, 22, 69, 88, 95,
        208, 258, 352
Molecular dynamics, 70, 145,
        179, 217, 261
Moller-Plesset theory, 33,
        276
Morse potential, 53, 81,
        113, 140, 332, 335,
        352

Neutron diffraction, 25, 73,
        172, 176, 243
NLS equation, 74, 75

O-H..O, 2, 49, 51, 108, 113,
        298, 351
OH-bond, 107, 111, 255, 352
Oxygen sublattice, 67-71,
        95, 124
Ordering transformation, 239

Peierls-Nabarro barrier,
        116, 352
Percolation, 7, 167
Phase transformation, 240
Ph-indicator, 173, 178, 187,
        189, 196
Pho-E channel, 199
Phonon, 54, 116, 132, 299,
        332, 352
Photosynthesis, 6, 187
Photocycle, 163, 175, 178,
        189, 192
Photolysis, 191, 256
Plant tissue, 207, 211
Polarizability, 299, 305
Polymers, 207, 211, 325, 337

Potential
  double-well, 8, 41, 44,
      49-52, 54, 59, 67,
      80, 122, 335, 345,
      351
  energy barrier, 51, 57,
      67, 70, 80, 97,
      123, 313
  energy surface, 29, 41,
      49, 313
  proton, 81, 97, 113, 160
Protein, 2, 25, 79, 169,
      171, 194
  hydrated, 23, 167, 207
  light-harvesting, 187
Proton
  collective motion, 153,
      158
  conductivity, 23, 79,
      105, 121, 162, 167, 351
  hydration, 297, 306
  mobility, 121, 255, 351
  polarizability, 153, 162
  pumping, 25, 171, 178
  transfer, 1, 5, 7, 29,
      39, 49, 79, 105,
      171, 189, 224, 233,
      273, 281, 313
  transport, 17, 24, 26,
      65, 121, 207
  tunneling, 55-57, 231,
      298
Protonation, 229, 233, 287,
      345

Quantum, 314, 326, 353
  calculations, 19, 32, 35,
      49, 52, 276, 345
  jumps, 49, 55
  tunneling, 5, 22, 30,
      167, 240, 298, 321

Radiation, 96, 99-101, 145,
      151
Raman spectra, 229, 297, 301
Reaction, 11, 159, 200, 220,
      281, 314, 345
  ion-molecule, 343
  rates, 286, 289

Resonance, 8, 284, 299

Schiff's base, 163, 172-182,
      188
Simulated annealing, 353
Solitary wave, 109, 121
Soliton, 6, 23, 29, 65, 105,
      131, 325, 330, 333,
      353
  modeling, 79
  scattering, 99, 101, 147
  statistics, 132
  trapping, 100
Solvation, 281, 293
Solvent, 281
  dynamics, 314
  polar, 284, 313
Sublattice, 67, 80, 121
  coupling, 72
  model, 351

Thermal excitation, 10, 54,
      65, 72, 76, 131,
      240, 355
Thermal solitons, 131
Thermodynamic integration,
      218, 274
Tunneling, 313, 316
  barrier, 9, 321
  probability, 56-60, 169

Vibration
  -assisted tunneling, 9,
      49
  intermolecular, 299, 309,
      314, 319
Vibrational levels, 49, 54
Vibrational spectra, 233,
      235

Water, 1-4, 12, 105, 163,
      174, 179, 209, 261,
      348
  of hydration, 208, 305
  wire, 7, 11

X-ray diffraction, 172, 176,
      207, 229.